AF678374

Aspergillus flavus and Aflatoxins (3rd Edition)

Aspergillus flavus and Aflatoxins (3rd Edition)

Guest Editors

Shihua Wang
Yang Liu

MDPI

Basel • Beijing • Wuhan • Barcelona • Belgrade • Novi Sad • Cluj • Manchester

Guest Editors

Shihua Wang
School of Life Sciences
Fujian Agriculture and Forestry University
Fuzhou
China

Yang Liu
School of Food Science and Engineering
Foshan University
Foshan
China

Editorial Office
MDPI AG
Grosspeteranlage 5
4052 Basel, Switzerland

This is a reprint of the Special Issue, published open access by the journal *Toxins* (ISSN 2072-6651), freely accessible at: https://www.mdpi.com/journal/toxins/special_issues/4DCZ19EGC1.

For citation purposes, cite each article independently as indicated on the article page online and as indicated below:

Lastname, A.A.; Lastname, B.B. Article Title. *Journal Name* **Year**, *Volume Number*, Page Range.

ISBN 978-3-7258-5097-6 (Hbk)
ISBN 978-3-7258-5098-3 (PDF)
https://doi.org/10.3390/books978-3-7258-5098-3

© 2025 by the authors. Articles in this book are Open Access and distributed under the Creative Commons Attribution (CC BY) license. The book as a whole is distributed by MDPI under the terms and conditions of the Creative Commons Attribution-NonCommercial-NoDerivs (CC BY-NC-ND) license (https://creativecommons.org/licenses/by-nc-nd/4.0/).

Contents

Preface

Fungi and their toxins not only cause serious economic losses, but more importantly, they can directly endanger human health. Among mycotoxins, aflatoxins are the most notorious and have carcinogenic, teratogenic, and mutagenic properties. Aflatoxin is mainly produced by *Aspergillus flavus*. In order to better prevent and control *A. flavus* and aflatoxins, the molecular mechanisms of aflatoxins production and the effect of aflatoxins on the host, detection, and prevention methods of aflatoxins have always been a hot research topic in this field. Therefore, this Special Issue focuses on studies related to *A. flavus* and aflatoxins.

Shihua Wang and Yang Liu
Guest Editors

toxins

MDPI

Editorial

Aspergillus flavus and Aflatoxins (3rd Edition)

Tanvir Ahmad [1], Shihua Wang [2,*] and Yang Liu [1,*]

[1] School of Food Science and Engineering, Foshan University/National Technical Center (Foshan) for Quality Control of Famous and Special Agricultural Products (CAQS-GAP-KZZX043)/Guangdong Key Laboratory of Food Intelligent Manufacturing, Foshan 528231, China; tanvir@fosu.edu.cn

[2] Key Laboratory of Pathogenic Fungi and Mycotoxins of Fujian Province, College of Life Sciences, Fujian Agriculture and Forestry University, Fuzhou 350002, China

* Correspondence: wshmail@m.fafu.edu.cn (S.W.); liuyang@fosu.edu.cn (Y.L.)

1. *Aspergillus flavus* and Aflatoxins

Aspergillus flavus is a saprophytic fungus commonly found in grain crops. It seriously threatens agriculture and food safety due to its ability to produce aflatoxins, potent secondary metabolites classified as Group I carcinogens by the International Agency for Research on Cancer [1]. *A. flavus* growth and aflatoxin contamination can occur both pre- and post-harvest, particularly under favorable environmental conditions, leading to substantial economic losses and health risks [2,3]. Beyond its toxigenic nature, *A. flavus* is also an opportunistic human pathogen capable of causing invasive aspergillosis and other serious infections [4].

Aflatoxins are among the most significant mycotoxins affecting food and feed safety. These naturally occurring toxic compounds are known for their hepatotoxic and immunosuppressive effects [5–9]. Aflatoxin contamination has emerged as a significant challenge in food safety and security in recent years [7,10]. While various mycotoxins (ochratoxin A, patulin, fumonisins, zearalenone, deoxynivalenol, and T-2 toxin) are of concern, aflatoxins remain the most prevalent and hazardous [11,12]. They frequently contaminate cereal grains and nuts at varying concentrations [13]. Prolonged exposure to aflatoxins can lead to acute and chronic health conditions in humans. Environmental, agronomic, and socioeconomic factors influence the risk of aflatoxin contamination.

Ensuring food safety requires thorough monitoring and effective processing to control mycotoxigenic fungi and mycotoxin levels in food products [14]. Various methods have been developed to detect and remove mycotoxins. However, their structural diversity, high stability, and low concentrations in food matrices demand robust, sensitive, and eco-friendly detection techniques [15]. Chromatographic methods such as HPLC, LC-MS/MS, and GC-MS/MS are widely employed, with LC-MS/MS favored for its multiplex detection capability [10]. Immunoassay-based tools like ELISA and biosensors also offer rapid and reliable screening [16,17]. In addition, novel detection techniques—such as electronic noses, proteomics, Raman spectroscopy, molecular diagnostics, and hyperspectral imaging—are emerging as promising alternatives [18–21].

Implementation of Hazard Analysis and Critical Control Points (HACCP)-based methods can help reduce aflatoxin contamination. While conventional methods can detoxify aflatoxins in food, increasing resistance in *A. flavus* and efficacy limitations highlight the need for new strategies that preserve food quality. Emerging approaches—such as polyphenols, enzymes, essential oils, plasma, and nanoparticles—offer enhanced detoxification potential, especially when combined. Microbial detoxification is also gaining attention as a safe and eco-friendly alternative [9,22–24]. Despite significant advances in detection

 https://doi.org/10.3390/toxins17070326

methods, the mechanisms responsible for the effective *A. flavus* inhibition and aflatoxin detoxification remain poorly understood. This highlights the urgent need for integrated, multidisciplinary research approaches to address these challenges. Hence, this editorial summarizes the key findings from the Special Issue titled "*Aspergillus flavus* and Aflatoxins (3rd Edition)" (https://www.mdpi.com/journal/toxins/special_issues/4DCZ19EGC1, accessed on 4 April 2024), which includes 13 contributions (10 research articles and 3 reviews) from different groups of scientists around the world. The advances are discussed in five thematic areas: (a) aflatoxin detection technologies, (b) detoxification and post-harvest control strategies, (c) AFB_1 toxicity and health effects, (d) fungal biology and aflatoxin biosynthesis regulation, and (e) host susceptibility and crop/public health risks. We believe this editorial will be a valuable resource for researchers working on mycotoxins and food safety.

2. Aflatoxins Detection Technologies

Recent innovations in aflatoxin detection technologies have paved the way for faster, more sensitive, and environmentally sustainable food safety solutions. Kourti et al. (contribution 12) introduced an immersible silicon photonic immunosensor for rapid, on-site detection of aflatoxin M_1 (AFM_1) in whole milk, offering a practical and high-performance tool for global food monitoring. The sensor employs dual U-shaped silicon nitride waveguides configured as Mach–Zehnder interferometers (MZIs), functionalized with AFM1-specific biomolecules. The system maintains high sensitivity by eliminating pumps and microfluidics while enhancing portability and ease of use. It achieves a detection limit of 20 pg/mL—surpassing European Commission thresholds for both infant (25 pg/mL) and adult (50 pg/mL) milk—and delivers accurate (86–112% recovery), reproducible (CV < 10%) results within 20 min, without the need for sample pretreatment.

Wu et al. (contribution 9) present an eco-friendly supramolecular platform for detecting aflatoxin B_1 (AFB_1), based on host–guest interactions using acyclic cucurbit [6] uril (acCB6). The study confirms a stable 1:1 complex between AFB_1 and acCB6 in water, supported by ^1H NMR, isothermal titration calorimetry, and molecular dynamics simulations. This host was integrated into a scaffold-assembled bioluminescent enzyme immunoassay (SA-BLEIA), enabling a solvent-free detection method with sensitivity comparable to standard assays. This approach addresses environmental concerns associated with conventional techniques and underscores the potential of supramolecular chemistry in advancing sustainable food safety diagnostics.

3. Detoxification and Post-Harvest Control Strategies

Cuccato et al. (contribution 8) evaluated the efficacy of two novel smectite-based mycotoxin detoxifying agents (SeOX and CHS) in mitigating the harmful effects of AFB_1 in broiler chickens exposed to dietary AFB_1 levels near EU regulatory limits (0.02 mg/kg). Both additives significantly reduced AFB_1 bio-accessibility by ~30%, as evidenced by higher AFB_1 excretion in treated groups compared to the AFB_1-exposed control. This reduction correlated with diminished oxidative stress, including improved antioxidant capacity and reduced hepatic lipid peroxidation. While AFB_1 alone upregulated CYP2A6 (a pro-oxidant enzyme) and suppressed Nrf2 (a regulator of antioxidant defenses), both MyDA treatments reversed these molecular changes, restoring redox balance. The additives showed no adverse effects on growth, liver histology, or other health parameters, underscoring their safety. The findings highlight SeOX and CHS as practical, sustainable solutions to counteract AFB_1 toxicity in poultry farming, aligning with food safety regulations while safeguarding animal health and productivity. This work advances strategies for mycotoxin risk management in agriculture through non-toxic, binder-based interventions.

Sinelnikov et al. (contribution 1) developed a sustainable enzymatic method to degrade AFB_1 in cereal grains using a recombinant oxidase (AFO) from *Armillaria tabescens*. Engineering *Pichia pastoris* with a chimeric signal peptide enhanced extracellular AFO production (0.3 g/L). AFO degraded 80% of AFB_1 in model solutions under optimal conditions (pH 6.0, 30 °C) within 72 h. In *A. flavus*-inoculated wheat and corn, AFO reduced AFB_1 by over twofold in moderately contaminated and ~40% in heavily contaminated samples. The study highlights AFO's promise as a scalable, eco-friendly bioremediation tool, especially for moderate contamination levels.

Javed et al. (contribution 11) reviewed innovative strategies to combat *Aspergillus* contamination and aflatoxin production in global grain and nut supplies. They highlighted atmospheric cold plasma (ACP), a non-thermal, residue-free technology that effectively inactivates *Aspergillus* spp. and degrades aflatoxins. The review also introduces a recombinant aflatoxin-degrading oxidase with strong detoxification potential in contaminated grains. These physical and biological interventions offer sustainable, scalable solutions for aflatoxin control, addressing food safety challenges and aligning with the demand for minimally processed, safe food.

Glesener et al. (contribution 5) investigated X-ray irradiation as a sterilization method to manage *A. flavus* and its associated AFB_1 in naturally contaminated maize. Irradiation up to 3.0 kGy effectively eliminated viable spores, as confirmed by culture-based methods. However, AFB_1 levels remained unchanged, showing that the toxin is not degraded while fungal viability is reduced. qPCR analysis confirmed persistent fungal DNA post-treatment, indicating that the method targets viability, not presence. This safe, reproducible approach is valuable for handling contaminated samples in research settings without altering toxin levels. Though unsuitable as a standalone food decontamination method, it supports integrated strategies for comprehensive mycotoxin control.

4. AFB_1 Toxicity and Health Effects

Chen et al. (contribution 6) presented a comprehensive multi-omics analysis of AFB_1-induced intestinal cytotoxicity using SW480 human colorectal cells. Exposure to 50 μM AFB_1 for 72 h caused significant cellular damage via oxidative stress, calcium overload, mitochondrial dysfunction and lipid metabolism disruption. Integrated transcriptomic, proteomic, and metabolomic data revealed activation of oxidative stress pathways and lipid dysregulation. Molecular dynamics simulations identified sterol carrier protein-2 (SCP2) as a key mediator, suggesting its role in amplifying AFB_1 toxicity through lipid transport. This study provides new insight into the gastrointestinal effects of AFB_1, expanding the focus beyond hepatic toxicity and highlighting the power of multi-omics in unraveling complex mycotoxin responses.

Yu et al. (contribution 3) examined AFB_1-induced hepatotoxicity in ducks, focusing on liver cholestasis and dietary mitigation strategies. Over four weeks, exposure to 90 μg/kg AFB1 caused significant oxidative stress. It disrupted bile acid metabolism, indicated by increased liver malondialdehyde (MDA) and total bile acid (TBA) levels and reduced antioxidant enzyme activity. AFB_1 upregulated cytochrome P450 family 7 subfamily A member 1 (CYP7A1) and cytochrome P450 family 8 subfamily B member 1 (CYP8B1), key enzymes in bile acid synthesis, while downregulating ATP-binding cassette subfamily B member 11 (BSEP), a critical bile acid transporter. Nutritional supplements, especially taurine and emodin, effectively countered these effects by restoring bile acid homeostasis and modulating gene expression. The study deepens the understanding of AFB_1-induced liver damage and highlights the potential of dietary interventions to protect poultry from contaminated feed.

Choi et al. (contribution 7) presented a comprehensive review of the adverse effects of AFB_1 on intestinal health and growth performance in monogastric animals, particularly poultry and swine. AFB_1-contaminated feed impairs gut microbiota balance, induces oxidative stress and immune activation, and compromises intestinal morphology and barrier function, reducing nutrient absorption and growth. A meta-analysis revealed a dose-dependent decline in average daily gain, with pigs showing greater sensitivity than chickens. The review highlights the small intestine as a primary site of AFB_1 damage. It emphasizes the efficacy of multi-component detoxifying agents, such as adsorbents and bio-transforming microbes like *Bacillus subtilis*, in mitigating these effects. By restoring gut integrity and promoting growth, these integrated dietary strategies offer practical and scalable solutions for reducing AFB_1's impact on livestock health and productivity. This approach has broad implications for feed safety and food security.

5. Fungal Biology and Aflatoxin Biosynthesis Regulation

Wang et al. (contribution 4) presented compelling evidence for the antifungal and anti-aflatoxigenic efficacy of rhein, a natural anthraquinone compound derived from *Rheum palmatum* L., against *A. flavus*. Rhein treatment at 50 μM inhibited mycelial growth and spore formation while reducing AFB_1 production by 87.2% in vitro. RNA-seq analysis revealed that rhein interferes with essential cellular processes, including energy metabolism, oxidative stress response, and cell wall integrity, leading to mitochondrial dysfunction, ATP depletion, and elevated ROS levels. These disruptions compromise fungal viability and toxin biosynthesis. Testing against peanut and maize seeds demonstrated strong in situ activity, with up to 90.4% reduction in spore growth and 99.43% reduction in AFB_1 levels. This dual-action mechanism underscores rhein's potential as a sustainable, natural strategy for controlling aflatoxigenic fungi and enhancing crop safety in post-harvest and storage systems.

Zhang et al. (contribution 2) explored the functional significance of the chitin deacetylase (CDA) gene family in *A. flavus*, revealing novel insights into fungal development and aflatoxin regulation. Although the six annotated CDA homologs (Cda1–Cda6) showed limited influence on chitin content or cell wall integrity, the targeted deletion of cda6 (*Δcda6*) produced a distinct phenotype, marked by enhanced conidiation, reduced mycelial growth, and significantly elevated aflatoxin production. Additionally, *Δcda6* colonization of host seeds was impaired. Subcellular localization studies confirmed that Cda6 functions within the cell wall and requires conserved CBD1 and CBD2 domains for its regulatory activity. Interestingly, the hyper-aflatoxigenic nature of *Δcda6* suggests that cda6 exerts a previously unrecognized role in modulating toxin biosynthesis, independent of classical chitin modification. These findings redefine the functional landscape of CDA proteins in fungal pathogens and highlight cda6 as a promising molecular target for mitigating aflatoxin contamination through innovative biocontrol strategies that avoid direct disruption of structural cell wall components.

6. Host Susceptibility and Crop/Public Health Risks

Elamin et al. (contribution 10) examined the susceptibility of *Ziziphus jujuba* var. *spinosa* (jujube) fruits and seeds to aflatoxin contamination by *A. flavus*, emphasizing the influence of fruit maturation and seed morphology. Mid-mature fruits were most prone to aflatoxin accumulation. Following fungal inoculation, internal seed tissues showed significant toxin levels even when physical barriers such as the exocarp, mesocarp, and endocarp remained intact. A key finding was the role of hilar region morphology in mediating fungal penetration and subsequent aflatoxin accumulation, accounting for contamination variability within seeds of the same developmental stage. The study reveals that *A. flavus* can breach jujube's structural defenses and colonize internal tissues, positioning hilar morphology as a critical susceptibility factor. These insights underscore the importance of integrating seed structural analysis and maturation-stage monitoring into food safety strategies for jujube-based medicinal products, aiding in preventing aflatoxin contamination in this high-value crop.

Chelenga et al. (contribution 13) presented a comprehensive review assessing the human health risks associated with aflatoxin residues in chicken products from birds consuming contaminated feed. Based on 33 global studies (1984–2023), this review shows that less than 1% of aflatoxins present in poultry feed are transferred to edible tissues. The average concentrations detected were 2.0 μg/kg in tissues and 0.3 μg/kg in eggs. Concentrations exceeding 20 μg/kg were detected in only 0.6% of cases and were limited to instances where feed contamination levels were above 300 μg/kg, the threshold set by the FDA for poultry feed. The findings support the concept of chickens acting as "biological filters", efficiently metabolizing or sequestering aflatoxins and limiting human dietary exposure. Despite prolonged feeding (>100 days), residue increases were modest and manageable. The study emphasizes that current feed regulations are primarily designed to protect poultry health rather than human consumers, for whom the risk remains minimal. The authors advocate for re-evaluating aflatoxin feed thresholds and propose exploring the filtering capacities of other poultry species, such as ducks and geese, to utilize contaminated grains safely. This review provides reassurance that chicken products pose a negligible aflatoxin exposure risk, even in regions burdened by feed contamination.

Author Contributions: T.A.: writing—original draft. S.W.: review and editing. Y.L.: writing and editing. All authors have read and agreed to the published version of the manuscript.

Funding: This research was funded by the National Key Research and Development Program of China 2022YFE0139500.

Acknowledgments: We express our sincere gratitude to all the authors who contributed their valuable work to this Special Issue.

Conflicts of Interest: The authors declare no conflicts of interest.

List of Contributions:

1. Sinelnikov, I.; Mikityuk, O.; Shcherbakova, L.; Nazarova, T.; Denisenko, Y.; Rozhkova, A.; Statsyuk, N.; Zorov, I. Recombinant Oxidase from *Armillaria tabescens* as a Potential Tool for Aflatoxin B1 Degradation in Contaminated Cereal Grain. *Toxins* **2023**, *15*, 678. https://doi.org/10.3390/toxins15120678.
2. Zhang, X.; Wen, M.; Li, G.; Wang, S. Chitin Deacetylase Homologous Gene *cda* Contributes to Development and Aflatoxin Synthesis in *Aspergillus flavus*. *Toxins* **2024**, *16*, 217. https://doi.org/10.3390/toxins16050217.
3. Yu, A.; Wang, H.; Cheng, Q.; Rajput, S.A.; Qi, D. The Effects of Aflatoxin B1 on Liver Cholestasis and Its Nutritional Regulation in Ducks. *Toxins* **2024**, *16*, 239. https://doi.org/10.3390/toxins16060239.
4. Wang, X.; Sahibzada, K.I.; Du, R.; Lei, Y.; Wei, S.; Li, N.; Hu, Y.; Lv, Y. Rhein Inhibits Cell Development and Aflatoxin Biosynthesis via Energy Supply Disruption and ROS Accumulation in *Aspergillus flavus*. *Toxins* **2024**, *16*, 285. https://doi.org/10.3390/toxins16070285.
5. Glesener, H.; Abdollahzadeh, D.; Muse, C.; Krajmalnik-Brown, R.; Weaver, M.A.; Voth-Gaeddert, L.E. X-ray Irradiation Reduces Live *Aspergillus flavus* Viability but Not Aflatoxin B1 in Naturally Contaminated Maize. *Toxins* **2024**, *16*, 329. https://doi.org/10.3390/toxins16080329.
6. Chen, M.; Wen, J.; Qiu, Y.; Gao, X.; Zhang, J.; Lin, Y.; Wu, Z.; Lin, X.; Zhu, A. Combining Multiple Omics with Molecular Dynamics Reveals SCP2-Mediated Cytotoxicity Effects of Aflatoxin B1 in SW480 Cells. *Toxins* **2024**, *16*, 375. https://doi.org/10.3390/toxins16090375.

7. Choi, H.; Garavito-Duarte, Y.; Gormley, A.R.; Kim, S.W. Aflatoxin B1: Challenges and Strategies for the Intestinal Microbiota and Intestinal Health of Monogastric Animals. *Toxins* **2025**, *17*, 43. https://doi.org/10.3390/toxins17010043.
8. Cuccato, M.; Amminikutty, N.; Spalenza, V.; Conte, V.; Bagatella, S.; Greco, D.; D'Ascanio, V.; Gai, F.; Schiavone, A.; Avantaggiato, G.; et al. Innovative Mycotoxin Detoxifying Agents Decrease the Absorption Rate of Aflatoxin B1 and Counteract the Oxidative Stress in Broiler Chickens Exposed to Low Dietary Levels of the Mycotoxin. *Toxins* 2025, *17*, 82. https://doi.org/10.3390/toxins17020082.
9. Wu, S.; Feng, K.; Niu, J.; Xu, J.; Mo, H.; She, X.; Yu, S.-B.; Li, Z.-T.; Yan, S. Acyclic Cucurbit[n]uril-Enabled Detection of Aflatoxin B1 via Host–Guest Chemistry and Bioluminescent Immunoassay. *Toxins* **2025**, *17*, 104. https://doi.org/10.3390/toxins17030104.
10. Elamin, A.; Sakuda, S. Susceptibility and Mechanism of Aflatoxin Contamination of *Ziziphus jujuba* var. *spinosa*. *Toxins* **2025**, *17*, 113. https://doi.org/10.3390/toxins17030113.
11. Javed, M.; Cao, W.; Tang, L.; Keener, K.M. A Review of Decontamination of *Aspergillus spp.* and Aflatoxin Control for Grains and Nuts with Atmospheric Cold Plasma. *Toxins* **2025**, *17*, 129. https://doi.org/10.3390/toxins17030129.
12. Kourti, D.; Angelopoulou, M.; Makarona, E.; Economou, A.; Petrou, P.; Misiakos, K.; Kakabakos, S. Aflatoxin M1 Determination in Whole Milk with Immersible Silicon Photonic Immunosensor. *Toxins* **2025**, *17*, 165. https://doi.org/10.3390/toxins17040165.
13. Chelenga, M.; Matumba, L.; Sitali, M.C.; Kachala, B.; Nambuzi, V.; Mwenifumbo, M.; Gama, A.P.; Mwanza, M.; Monjerezi, M.; Leslie, J.F. Are Aflatoxin Residues in Chicken Products a Real or Perceived Human Dietary Risk? *Toxins* 2025, *17*, 179. https://doi.org/10.3390/toxins17040179.

References

1. International Agency for Research on Cancer. *A Review of Human Carcinogens. F. Chemical Agents and Related Occupations: IARC Monographs on the Evaluation of Carcinogenic Risks to Humans*; WHO Press: Geneva, Switzerland, 2012.
2. Yan, L.; Song, W.; Chen, Y.; Kang, Y.; Lei, Y.; Huai, D.; Wang, Z.; Wang, X.; Liao, B. Effect of non-aflatoxigenic strains of Aspergillus flavus on aflatoxin contamination of pre-harvest peanuts in fields in China. *Oil Crop Sci.* **2021**, *6*, 81–86. [CrossRef]
3. Gachara, G.; Suleiman, R.; Kilima, B.; Taoussi, M.; El Kadili, S.; Fauconnier, M.-L.; Barka, E.; Vujanovic, V.; Lahlali, R. Pre-and post-harvest aflatoxin contamination and management strategies of Aspergillus spoilage in East African Community maize: Review of etiology and climatic susceptibility. *Mycotoxin Res.* **2024**, *40*, 495–517. [CrossRef]
4. Patterson, T.F.; Thompson, G.R., III; Denning, D.W.; Fishman, J.A.; Hadley, S.; Herbrecht, R.; Kontoyiannis, D.P.; Marr, K.A.; Morrison, V.A.; Nguyen, M.H. Practice guidelines for the diagnosis and management of aspergillosis: 2016 update by the Infectious Diseases Society of America. *Clin. Infect. Dis.* **2016**, *63*, e1–e60. [CrossRef] [PubMed]
5. Iqbal, S.Z. Mycotoxins in food, recent development in food analysis and future challenges; a review. *Curr. Opin. Food Sci.* **2021**, *42*, 237–247. [CrossRef]
6. Singh, S.; Kumar, V.; Dhanjal, D.S.; Dhaka, V.; Singh, J. Mycotoxin metabolites of fungi. In *New and Future Developments in Microbial Biotechnology and Bioengineering*; Elsevier: Amsterdam, The Netherlands, 2021; pp. 253–265.
7. Yang, Y.; Ren, M.-Y.; Xu, X.-G.; Han, Y.; Zhao, X.; Li, C.-H.; Zhao, Z.-L. Recent advances in simultaneous detection strategies for multi-mycotoxins in foods. *Crit. Rev. Food Sci. Nutr.* **2022**, *64*, 3932–3960. [CrossRef]
8. Tian, M.; Feng, Y.; He, X.; Zhang, D.; Wang, W.; Liu, D. Mycotoxins in livestock feed in China-Current status and future challenges. *Toxicon* **2022**, *214*, 112–120. [CrossRef]
9. Hamad, G.M.; Mehany, T.; Simal-Gandara, J.; Abou-Alella, S.; Esua, O.J.; Abdel-Wahhab, M.A.; Hafez, E.E. A review of recent innovative strategies for controlling mycotoxins in foods. *Food Control* **2023**, *144*, 109350. [CrossRef]
10. Alshannaq, A.; Yu, J.-H. Occurrence, toxicity, and analysis of major mycotoxins in food. *Int. J. Environ. Res. Public Health* **2017**, *14*, 632. [CrossRef]
11. Wokorach, G.; Landschoot, S.; Anena, J.; Audenaert, K.; Echodu, R.; Haesaert, G. Mycotoxin profile of staple grains in northern Uganda: Understanding the level of human exposure and potential risks. *Food Control* **2021**, *122*, 107813. [CrossRef]
12. Abrunhosa, L.; Morales, H.; Soares, C.; Calado, T.; Vila-Chã, A.S.; Pereira, M.; Venâncio, A. A review of mycotoxins in food and feed products in Portugal and estimation of probable daily intakes. *Crit. Rev. Food Sci. Nutr.* **2016**, *56*, 249–265. [CrossRef]
13. Gruber-Dorninger, C.; Jenkins, T.; Schatzmayr, G. Global mycotoxin occurrence in feed: A ten-year survey. *Toxins* **2019**, *11*, 375. [CrossRef]
14. Haque, M.A.; Wang, Y.; Shen, Z.; Li, X.; Saleemi, M.K.; He, C. Mycotoxin contamination and control strategy in human, domestic animal and poultry: A review. *Microb. Pathog.* **2020**, *142*, 104095. [CrossRef] [PubMed]

15. Janik, E.; Niemcewicz, M.; Podogrocki, M.; Ceremuga, M.; Gorniak, L.; Stela, M.; Bijak, M. The existing methods and novel approaches in mycotoxins' detection. *Molecules* **2021**, *26*, 3981. [CrossRef] [PubMed]
16. Tittlemier, S.; Cramer, B.; Dall'Asta, C.; Iha, M.; Lattanzio, V.; Malone, R.; Maragos, C.; Solfrizzo, M.; Stranska-Zachariasova, M.; Stroka, J. Developments in mycotoxin analysis: An update for 2017–2018. *World Mycotoxin J.* **2019**, *12*, 3–29. [CrossRef]
17. Tothill, I. Biosensors and nanomaterials and their application for mycotoxin determination. *World Mycotoxin J.* **2011**, *4*, 361–374. [CrossRef]
18. Femenias, A.; Gatius, F.; Ramos, A.J.; Sanchis, V.; Marín, S. Use of hyperspectral imaging as a tool for Fusarium and deoxynivalenol risk management in cereals: A review. *Food Control* **2020**, *108*, 106819. [CrossRef]
19. Zhai, W.; You, T.; Ouyang, X.; Wang, M. Recent progress in mycotoxins detection based on surface-enhanced Raman spectroscopy. *Compr. Rev. Food Sci. Food Saf.* **2021**, *20*, 1887–1909. [CrossRef]
20. Wu, Z.; Pu, H.; Sun, D.-W. Fingerprinting and tagging detection of mycotoxins in agri-food products by surface-enhanced Raman spectroscopy: Principles and recent applications. *Trends Food Sci. Technol.* **2021**, *110*, 393–404. [CrossRef]
21. Agriopoulou, S.; Stamatelopoulou, E.; Varzakas, T. Advances in analysis and detection of major mycotoxins in foods. *Foods* **2020**, *9*, 518. [CrossRef]
22. Kiš, M.; Milošević, S.; Vulić, A.; Herceg, Z.; Vukušić, T.; Pleadin, J. Efficacy of low pressure DBD plasma in the reduction of T-2 and HT-2 toxin in oat flour. *Food Chem.* **2020**, *316*, 126372. [CrossRef]
23. Ahmed, O.S.; Tardif, C.; Rouger, C.; Atanasova, V.; Richard-Forget, F.; Waffo-Téguo, P. Naturally occurring phenolic compounds as promising antimycotoxin agents: Where are we now? *Compr. Rev. Food Sci. Food Saf.* **2022**, *21*, 1161–1197. [CrossRef] [PubMed]
24. Abd-Elsalam, K.A.; Hashim, A.F.; Alghuthaymi, M.A.; Said-Galiev, E. Nanobiotechnological strategies for toxigenic fungi and mycotoxin control. In *Food Preservation*; Elsevier: Amsterdam, The Netherlands, 2017; pp. 337–364.

Disclaimer/Publisher's Note: The statements, opinions and data contained in all publications are solely those of the individual author(s) and contributor(s) and not of MDPI and/or the editor(s). MDPI and/or the editor(s) disclaim responsibility for any injury to people or property resulting from any ideas, methods, instructions or products referred to in the content.

toxins MDPI

Article

Recombinant Oxidase from *Armillaria tabescens* as a Potential Tool for Aflatoxin B1 Degradation in Contaminated Cereal Grain

Igor Sinelnikov [1,*], Oleg Mikityuk [2], Larisa Shcherbakova [2,*], Tatyana Nazarova [2], Yury Denisenko [1], Alexandra Rozhkova [1], Natalia Statsyuk [2] and Ivan Zorov [1]

[1] Federal Research Centre "Fundamentals of Biotechnology" of the Russian Academy of Sciences, 119071 Moscow, Russia; denisenkoyura@mail.ru (Y.D.); a.rojkova@fbras.ru (A.R.); inzorov@mail.ru (I.Z.)
[2] All-Russian Research Institute of Phytopathology, Bolshie Vyazemy, 143050 Moscow, Russia; mod-39@list.ru (O.M.); nataafg@gmail.com (N.S.)
* Correspondence: sinelnikov.i@list.ru (I.S.); larisaviniif@yahoo.com (L.S.)

Abstract: Forage grain contamination with aflatoxin B1 (AFB1) is a global problem, so its detoxification with the aim of providing feed safety and cost-efficiency is still a relevant issue. AFB1 degradation by microbial enzymes is considered to be a promising detoxification approach. In this study, we modified an previously developed *Pichia pastoris* GS115 expression system using a chimeric signal peptide to obtain a new recombinant producer of extracellular AFB1 oxidase (AFO) from *Armillaria tabescens* (the yield of 0.3 g/L), purified AFO, and selected optimal conditions for AFO-induced AFB1 removal from model solutions. After a 72 h exposure of the AFB1 solution to AFO at pH 6.0 and 30 °C, 80% of the AFB1 was degraded. Treatments with AFO also significantly reduced the AFB1 content in wheat and corn grain inoculated with *Aspergillus flavus*. In grain samples contaminated with several dozen micrograms of AFB1 per kg, a 48 h exposure to AFO resulted in at least double the reduction in grain contamination compared to the control, while the same treatment of more significantly (~mg/kg) AFB1-polluted samples reduced their contamination by ~40%. These findings prove the potential of the tested AFO for cereal grain decontamination and suggest that additional studies to stabilize AFO and improve its AFB1-degrading efficacy are required.

Keywords: aflatoxin B1; enzymatic toxin degradation; recombinant enzymes; grain decontamination

Key Contribution: A new chimeric signal peptide was used to improve the expression of recombinant aflatoxin-degrading oxidase in *Pichia pastoris*. A toxin-degrading effect of this recombinant enzyme was proven when treating aflatoxin B1 in wheat and corn grain inoculated with a toxigenic *Aspergillus flavus*.

1. Introduction

In spite of the different approaches developed for the prevention of and reduction in food and feed pollution with aflatoxin B1 (AFB1) [1–6], this widespread and very carcinogenic mycotoxin [7,8] continues to be a major concern as a contaminant of various feeds, including wheat, corn and other cereals [9,10]. Cereal grain and its primary processing products constitute the bulk of feed rations for many farm animals [11], especially monogastric ones, including poultry. Wheat and corn are the main components of the poultry feed diet in European countries and the United States, respectively [12,13]. At the same time, according to the Food and Agricultural Organization (FAO) and other sources, 25% of the total production of the world's food and feed crops, including wheat, corn and other cereals, are contaminated with mycotoxins [14,15]. In poultry farming, problems associated with the grain contamination with AFB1 are exacerbated by a high sensitivity of poultry, especially turkeys and ducks (LD_{50} 1.4–3.2 and 0.3–0.6 mg/kg of body weight, respectively) and, to a lesser extent, quails and chickens, to AFB1 [3,16–18], as well as by the fact that the feed cost can exceed 70% of the poultry meat cost [19]. Therefore, decontamination of cereal grain with the aim of providing the feed safety and cost efficiency still remains an urgent challenge.

In this regard, enzymatic degradation of AFB1 based on the ability of some fungi and bacteria [20–23] to synthesize various enzymes (including oxidoreductases) transforming this toxin into non- or less-toxic compounds is considered to be one of the most promising approaches [24–30]. The use of cell-free preparations containing such enzymes could make it possible to avoid the problems which may occur during the use of their producers (worsening of organoleptic properties of treated products, reduction in their nutritional value, etc.). Moreover, enzymatic preparations are more technologically suitable for the feed treatment [31] and, unlike similar preparations for the food industry, do not require a high level of purification of a target product.

Some xylotrophic basidiomycetes from the genera *Pleurotus*, *Phanerochaete*, *Armillaria*, *Peniophora* and *Trametes* [20,32–34] are well known as the sources of microbial oxidoreductases able to degrade or detoxify aflatoxins. For example, manganese-dependent peroxidase from *Phanerochaete sordida* [35], certain laccases from *Trametes versicolor*, *Pleurotus ostreatus*, and *Peniophora* sp., as well as recombinant laccase secreted by *A. niger* [20] are shown to actively degrade AFB1, AFB2, and G-type aflatoxins [27,36], and may be considered as possible tools for the decontamination of agroproducts. However, some limitations may hinder the development of decontaminating preparations based on these enzymes. For example, *P. sordida* was reported to be able to infect humans [37]. Laccases are well-known enzymes catabolizing many compounds including lignin, and are considered for use in wood processing and the textile industry [38] because of the broader specificity of this group of enzymes compared to other aflatoxin-degrading enzymes and their activity at extreme pH values and temperatures [39,40], which are not typical for feed processing. Moreover, in spite of the successful results demonstrated in some studies [27,36,40], the heterologous expression of laccases is rather difficult due to their complex copper-dependent active site [38,40] that results in a rapid proteolytic cleavage of these enzymes during their post-translational modification. Therefore, studies aimed at searching for new promising enzymes for feed decontamination still remain relevant.

Some studies suggest that one of such tools may be an intracellular enzyme possessing oxydase activity, which was isolated from the *Armillaria tabescens* mycelium as a component of a multi-enzyme complex of this edible fungus [34]; the authors of the study named it aflatoxin-detoxifizyme. This enzyme was later identified as a new AFB1 oxidase (AFO) able to catalyze the opening and the further hydrolysis of a bisfuran ring [41] (a structure directly associated with AFB1 toxicity [42,43]). An unsaturated C8–C9 bond was determined to be a putative reactive AFO site in the AFB1 molecule [43]. A further study showed that AFO represented a 76 kDa monomeric protein, which significantly differed from aforementioned oxidoreductases; the contact of this protein with the toxin caused a significant reduction in AFB1 toxicity and mutagenicity [34]. This enzyme possesses a low K_m value (0.334 μmol/L) for AFB1 that explains its high affinity to AFB1 [41,43]. In addition, the highly selective detoxifying properties of AFO were shown not only for AFB1, but also for sterigmatocystin and versicolorin A, the toxic AFB1 precursors synthesized by many *Aspergillus* species [42–44]. Moreover, the hydrogen peroxide production that occurs during the enzymatic reaction results in the detoxification of AFB1-epoxide, a highly toxic derivative formed at the first stage of AFB1 oxidation. These data evidence the good prospects for the development of detoxifying preparations based on AFB1 oxidase (AFO) from *A. tabescens*. However, their development requires an available technology of AFO production and secretion that can be achieved via the use of a system for a heterologous expression of a recombinant enzyme. A few earlier reports [45–47] as well as our preliminary study [48] on recombinant AFO production indicated the possibility of using transformed *Pichia pastoris* strains as the appropriate heterologous producers of this enzyme. The authors of these in vitro studies demonstrated the ability of the obtained recombinant AFOs to degrade AFB1 and determined some kinetic characteristics and the target activity level of the enzymes at different pH levels and temperatures [45–48]. To produce the recombinant

AFO in the *P. pastoris* GS115 strain in our earlier study [48], we optimized and used a previously developed expression system with a modified integration vector intended to increase the copy number variation of heterologous genes in yeast cell chromosomes. As a result, the most productive ADTZ-14 clone, which removed 50% of AFB1 after a 72 h incubation of the toxin in its culture liquid, was selected among 54 obtained transformants.

Continuing our investigation, in the current study, we modified the previously used *P. pastoris* GS115 expression system using a new chimeric signal peptide and generated a new highly productive recombinant strain to obtain extracellular heterologous AFO in an amount sufficient for forage grain decontamination. We also homogenously isolated the recombinant enzyme and selected optimal conditions by varying the pH and temperature of incubation medium to achieve the best target effect (AFB1 removal) in model solutions. In addition, we showed that the AFO treatment of wheat and corn grain, contaminated with AFB1 via inoculation with *A. flavus*, resulted under selected conditions in a significant reduction in the AFB1 content in grain.

2. Results

2.1. Recombinant Aflatoxin B1 Oxidase Obtained Using a Pichia Pastoris Expression System

The synthetic *afo* gene encoding aflatoxin-degrading oxidase from *Armillaria tabescens* (GenBank AY941095.1) was synthesized by Eurogen Ltd. (Moscow, Russia) according to the codon usage of *Pichia pastoris* (syn. *Komagataella phaffii*). The sequence was cloned by PCR using specially designed oligonucleotides; the size of the resulting PCR product was 2088 bp.

To construct a plasmid for the *afo* integration, a modified pPIG-1 vector was used. The pPIGa-1 vector contained an alcohol oxidase I (AOXI) promoter instead of a GAP promoter. In contrast to the previous study [48], the *afo* sequence was expanded with a chimeric signal peptide, which included a pro-region of the α-factor and a synthetic signal peptide (MKILSALLLLFTLAFA, https://pubmed.ncbi.nlm.nih.gov/25218497/, accessed on 22 November 2023) facilitating efficient co-translational transport into the endoplasmic reticulum. After transformation and selection (see Section 4.1), we obtained a recombinant *P. pastoris_afo* strain characterized by a 0.3 g/L yield of the target aflatoxin B1 oxidase (AFO) and reduced degradation of this enzyme.

The supernatant of the culture liquid (50 mL) was applied to a 15 mL Ni-NTA Sepharose column, resulting in the isolation of homogeneous AFO via chromatographic separation. The AFO fraction had a concentration of 1.3 mg/mL. The size and purity of the enzyme evaluated by means of denaturing gel electrophoresis (Figure 1) were 72 kDA and ~95%, respectively.

2.2. Enzymatic AFB1 Degradation in Model Solutions under Different Conditions

In order to confirm that the obtained recombinant AFO retained its target functionality and to select conditions promoting its toxin-degrading activity, the experiments on the AFB1 removal from the model solutions were arranged. In the course of these experiments, AFB1 degradation was examined through a 5-day toxin incubation in the enzyme-containing buffers at different pH and temperatures followed by a reverse-phase HPLC to determine a residual AFB1 content. The dynamics of the AFB1 removal from the model solutions significantly depended on both studied factors. The pH and temperature ranges used in these experiments were chosen on the basis of previous results obtained for AFO produced by *A. tabescens* [34,41] or transformed *P. pastoris* G115 clones [47,48], respectively.

In general, AFO showed the target activity at pH levels varying between 5.0 and 7.5, while more acidic (pH 4.0) or alkaline (pH 8.0 and 9.0) conditions resulted in an almost complete inactivation of the enzyme. The most effective AFB1 removal was observed in the case of a 3-day exposure of the model solution to the enzyme at pH 6.0 (Figure 2). In this case, less than half of the added toxin was detected in a 20 mM phosphate buffer after a 24 h exposure. After a 48 h co-incubation, the average AFB1 content reduced to 30% of the initial level; finally, after a 72 h exposure, the residual AFB1 content in the model solution was reduced to 20% of the initial content. No additional significant removal of AFB1 was observed in the course of the further incubation, and the toxin degradation level remained at 80–81% for the next two days until the completion of the experiment (Figure 2); even a 2.5-fold increase in the AFO concentration did not enhance or prolong the target effect. The same trend for the termination of the AFB1 removal from model solutions after 72 h of incubation was revealed for all other tested pH values (Figure 2). Based on these data, we further investigated the effect of three different temperatures (15, 30 and 50 °C) on the enzymatic toxin degradation at optimal pH (6.0) and recorded the 72 h toxin removal dynamics (Figure 3).

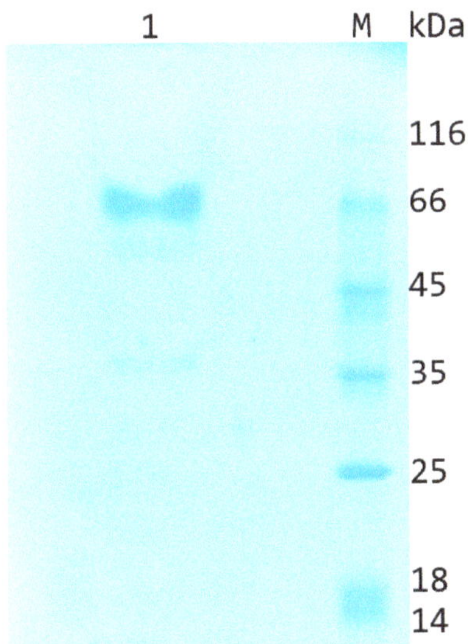

Figure 1. SDS-PAGE electrophoregram of a 72 kDa recombinant AFO (1) purified from the cultural liquid of *P. pastoris* GS115 by means of immobilized metal affinity chromatography. M, molecular weight marker.

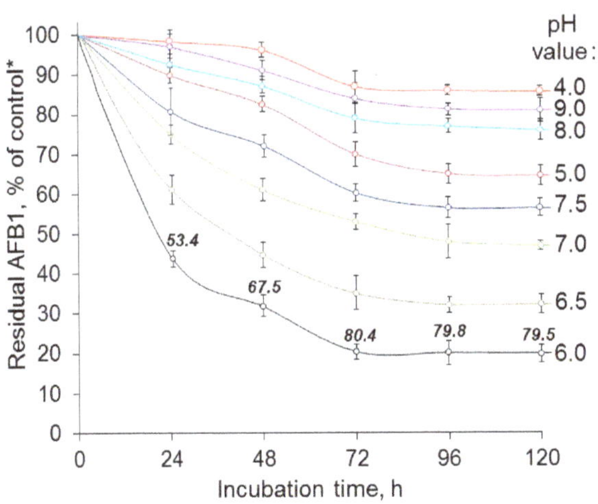

Figure 2. The effect of pH on the dynamics of AFB1 removal from model solutions exposed to recombinant aflatoxin oxidase (AFO) expressed in *Picha pastoris*. Numbers above the curve points indicate the average AFB1 reduction (%) in a buffer solution after exposure to AFO. All results were calculated as the mean of three experiments ± standard deviation (SD). * Here and in Figure 3: the AFB1 content detected by HPLC in the corresponding AFO-free buffer solution during a 120 h incubation was considered as a control.

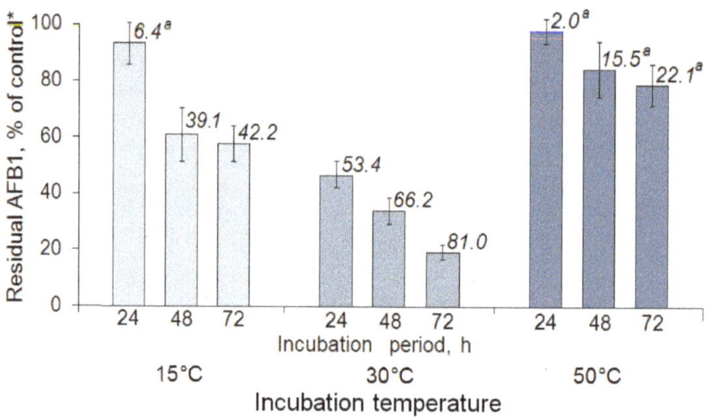

Figure 3. Effect of different temperatures on the rate of AFB1 removal from model solutions by recombinant aflatoxin oxidase (AFO). Results are presented as the mean of three experiments, each arranged into three replicates. Error bars indicate SD. Numbers above the columns show AFB1 degradation (%). Insignificant difference between the control and enzymatic degradation is indicated with "a". * See explanation of the control in Figure 2.

In our study, the most rapid decrease in the AFB1 content caused by the AFO treatment was observed at 30 °C. In this case, after 24, 48, and 72 h of incubation, ~50, 66 and 80% of the toxin was degraded, respectively. Incubation at 15 °C significantly slowed down the degradation process, resulting in a removal of only ~40% of AFB1 within a 48 h incubation. A temperature increase to 50 °C resulted in enzyme inactivation (Figure 3).

2.3. *Enzymatic Decontamination of Cereal Grain Inoculated with a Toxigenic Aspergillus flaus*

To examine if the obtained recombinant oxidase was able to reduce aflatoxin contamination of a polluted grain, a toxigenic *A. flavus* strain was grown on autoclaved wheat or corn grain for 3 or 7 days. Then, inoculated grain samples were treated with 20 mM phosphate buffer (control) or AFO, dissolved in the same buffer, for 24, 48 and 72 h under the aforementioned conditions with the further extraction and quantification of residual AFB1. The fungus actively developed on a grain; by the third day of cultivation, 47.0 ± 7.0 and 33.0 ± 5.8 ng of AFB1 per gram of grain was accumulated in the wheat and corn samples, respectively. By the end of cultivation, AFB1 concentrations in non-treated inoculated grain samples reached the microgram level (1.08 ± 0.20 and 0.99 ± 0.14 µg/g for wheat and corn grain, respectively). Thus, the average content of AFB1 in the analyzed samples calculated per kg of grain was ~30–50 and 990–1080 µg at the beginning and at the end of a 10-day experiment, respectively.

In contrast, a significantly lower amount of the toxin was determined in AFO-treated grain (Figure 4). For the majority of wheat and corn samples, which contained a few dozen micrograms of AFB1 per kg, a 48 h exposure to AFO reduced the grain contamination level to at least twice the extent compared to the control (Figure 4a), while the same treatment of more heavily contaminated samples reduced the AFB1 content by 40% on average (Figure 4b).

Regardless of the initial amount of AFB1, a general downward trend in the toxin content in contaminated wheat and corn grain was also observed after a 3-day treatment; in the case of a low-contamination corn grain, the achieved toxin degradation level was 61.2% (Figure 4a). However, the difference between the results of the 48 and 72 h treatments should be considered rather insignificant because of the high standard deviation values. A longer exposure of inoculated grain to AFO did not enhance the decontamination effect. Nevertheless, for some samples analyzed 3 days after inoculation, almost a 3-fold decontamination was reached in one of the performed experiments with a 72 h exposure to AFO. As a result, the content of residual AFB1 was reduced to 16.8 and 10.6 µg/kg in wheat and corn grain, respectively (Figure 4a). The obtained data indicate that the tested AFO expressed in yeast cells was able to affect the toxin not only in model solutions but also in a contaminated cereal grain.

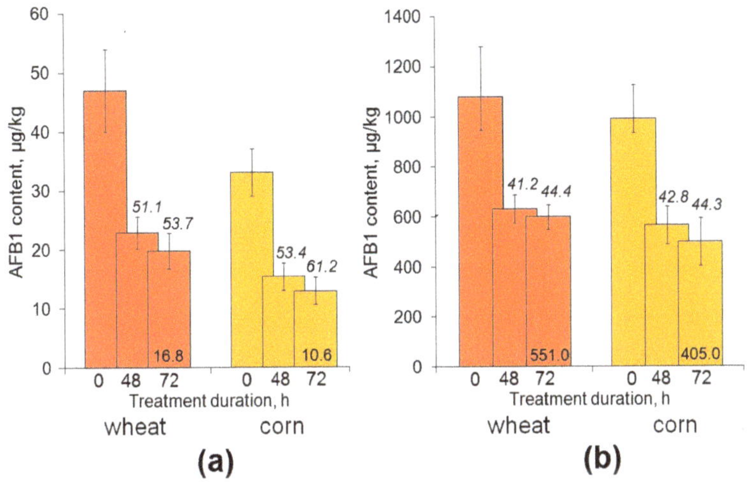

Figure 4. Enzymatic reduction in the AFB1 contamination level in wheat and corn grain artificially inoculated with a toxigenic *A. flavus* strain and incubated for 3 (**a**) and 7 (**b**) days. Numbers above the columns (in italic) indicate the average percent of the toxin degradation; numbers at the column bottoms show the minimum level of residual AFB1 (µg/kg) that was achieved after a 72 h treatment of grain with recombinant AFO. The difference between the control (non-exposed to the enzyme, 0 h) and the treated grain samples was significant at $p \leq 0.05$.

3. Discussion

To date, a wide range of recombinant enzymes able to attack mycotoxins have been produced using prokaryotic or eukaryotic expression systems [20,45,47,49–52]; for some of them, their decontaminating activity has been experimentally confirmed on grain and other food/feed products [45,49,51]. For example, a high decontamination potential was recently demonstrated for one recombinant AFO (Arm-ADTZ) in experiments on the enzymatic treatment of mold on corn (56.48% degradation within 24 h) and grain by-products obtained in the course of the distillation process in ethanol production [45].

AFO from *A. tabescens* seems to belong to the most studied AFB1-detoxifying enzymes including recombinant ones. Its crystal structure and the nucleotide sequence of a gene encoding a full-length protein have been determined [53], the safety of the resulting AFB1 degradation products has been confirmed [27], an additional dipeptidyl peptidase activity has been revealed [53], and at least three functional enzymes heterologously expressed in *P. pastoris* G115 have been reported [47,48]. Some kinetic characteristics, such as the K_m value, the specific activity, and the half-life period at 30–50 °C were described for one of these recombinant AFOs [45,47], as well as for an AFO isolated from *A. tabescens* [34,41]. In our study, the use of a new signaling peptide promoting effective AFO secretion allowed us to prevent a partial protein degradation observed in the previous experiments with the AFO-containing cultural liquid of transformed yeast cells [48] and to provide a sufficient production of the extracellular enzyme possessing the target activity.

In the performed model experiments, the target AFO activity was revealed within the range of acidic and alkaline pH values (5.0–7.5 with the optimum at pH 6.0) used in the feed industry. AFO exposure to different temperatures indicated that the target activity was the most stable at 30 °C (temperature favorable for the toxin production). Such characteristics of the pH dependence as well as the revealed optimum temperature coincide with the data obtained for the parental enzymes from *A. tabescence* [34] and *Trametes versicolor* [49], as well as with results reported for another recombinant aflatoxin oxidase expressed in *P. pastoris* G115 [45,47]. Under optimal conditions, AFO rapidly degraded 80% of AFB after 72 h of incubation, while only 50% of AFB1 was in vitro degraded after its 72 h co-incubation with the enzyme under the optimal conditions reported in our earlier study [48]. Since an increase in the enzyme concentration did not result in additional toxin removal during its co-incubation with AFB1 for 96 or 120 h, the lack of any additional decontamination effect after a 72 h exposure was related to the enzyme's destabilization rather than to the substrate substitution. This suggestion was confirmed by the significantly lower percentage of AFB1 degradation (max. 30.5%) in one of the 48 h incubation tests performed at the preliminary stage of our investigation; in this case, a prepared AFO solution was stored overnight prior to use. The addition of $BaCl_2$ to the reaction medium [34] used in our experiments did not improve the AFO stabilization and did not provide the prolongation of its degradation effect.

A worldwide monitoring of feed contamination with AFB1 suggests that cereal grain has been affected by toxigenic Aspergilli (especially during storage) much more often in recent years [54–56]. Currently, the AFB1 limit in feeds in the European Union has been set by the FAO at the level of 5 μg/kg; the United States' regulations on grain safety [10] and the Technical Regulations of the Customs Union (CU) restrict this level to 20 μg/kg (TR CU 015/2011). Because of the global warming-induced migration of *Aspergillus* fungi from tropical and subtropical regions, the incidence of finding cereal grain samples containing AFB1 and the level of the contamination tend to increase in zones with a temperate climate [56,57]; moreover, the further increase in the contamination risk is predicted [58]. Depending on the moisture level, *A. flavus* is able to grow on stored grain within the temperature range of 12–48 °C [59], with the most intense AFB1 production at 25–35 °C [60]. The AFB1 content in contaminated samples from various countries significantly varies. For example, the contamination of wheat and corn grain at the levels exceeding 30 μg/kg is often reported [10,61–63], and sometimes the maximum revealed AFB1 content can exceed 4 mg/kg [64]. Based on this information, we tested the obtained recombinant AFO on grain samples with the AFB1 content ranging from 33 to 1080 μg/kg and demonstrated that the enzyme can be effective even in the case of extremely high grain contamination. Despite the fact that the residual AFB1 content in wheat and grain samples, initially containing milligrams of AFB1 and exposed to AFO for 72 h, still significantly exceeded international and national maximum allowable limits, the similar treatment of less contaminated grain with the enzyme preparation resulted in a reduction in the residual AFB1 content below 20 μg/kg (the maximum allowable limit for CU countries and US). It is possible to suppose that a 2- or 3-day AFO application on non-heavily polluted samples may be more efficient and provide sufficient grain detoxification during this time period. Thus, we succeeded in obtaining a recombinant AFB1-degrading oxidase with promising properties of a potential grain decontamination agent. At the same time, the relatively low stability of the enzyme manifested via the loss of its target activity after 72- or 48 h incubation in model solutions or contaminated grain, respectively, constrains the prospects of its practical application at this stage and suggests that additional investigations are needed to improve the enzyme and to understand more about its detoxifying potential. In this regard, we will try to stabilize the obtained AFO via protein engineering methods in our further studies, and to test AFO activity on a contaminated grain of other cereals and crops characterized by a high risk of the AFB1 contamination.

According to one of the current concepts [27], a transition to a large-scale production of detoxifying enzymatic and microbial preparations is only a matter of time, so, along with other studies, our study can contribute to the advancement of this eco-friendly approach to improve feed safety.

4. Materials and Methods

4.1. Generation of a Yeast Strain Producing Recombinant AFO

A methylotrophic yeast strain *Pichia pastoris* GS115 (Thermo Fisher Scientific Inc., Waltman, MA, USA) was used to express the *afo* gene encoding aflatoxin-detoxiphysim (AFO).

Amplification of the *afo* gene was performed via routine PCR. The sequences for HindIII and NotI restriction endonucleases were added to the corresponding primers for the amplification of the *afo* gene. The corresponding primers were designed as follows:

Forward: 5′-gaagcttcttctATGGCTATGGCTACTACTACAACTG-3′ (Hind*III*);

Reverse: 5′-cgcggggccgcTTACAATCTTCTCTCTCTC-3′ (Not*I*).

The expression plasmid pPIGA-afo was developed using the earlier described pPIG-I vector [48].

Both the pPIG-I vector and the amplification product were double digested by HindIII and NotI restriction endonucleases (New England Biolabs, Ipswich, MA, USA) in accordance with the manufacturer's recommendations. The treated fragments were ligated with T4 DNA ligase (Eurogen Ltd., Moscow, Russia), and the ligated mixture (2 μL) was transformed into *E. coli* XL1-blue cells. Transformants were selected on the LB medium containing 2% agar supplemented with ampicillin (100 μg/mL, Sigma-Aldrich, St. Louis, MO, USA). To isolate plasmid DNA, *Escherichia coli* XL1-Blue strain (Agilent, Santa Clara, CA, USA) was grown at 37 °C on LB medium of the following composition (g/L): tryptone, 10; yeast extract, 5; NaCl, 5; pH 7.2–7.5). A pPIGA-afo plasmid was isolated from ampicillin-resistant transformants using a Plasmid Mini-prep kit (Eurogen Ltd., Moscow, Russia) and confirmed by means of PCR amplification, identification by double-enzyme digestion, and gene sequencing [65].

The pPIGA-afo plasmid (Figure S1) was linearized by ApaI restriction endonuclease (New England Biolabs, Ipswich, MA, USA) and integrated into *P. pastoris* GS115 by electroporation [66]. The selection of transformants was carried out on YPD medium with 1% agar supplemented by zeocin (200 μg/mL, Thermo Fisher Scientific Inc., Waltman, MA, USA). Total DNA samples were isolated from antibiotic-resistant colonies [67] and the presence of the *afo* insertion was verified by PCR followed by gene sequencing [65].

4.2. Recombinant AFO Production in Yeast Cells

The cultivation process included two stages: (1) biomass accumulation (24 h) and (2) protein synthesis induction by methanol.

Fermentation was carried out in a 1.5 L Sartorius A plus bioreactor (Sartorius, Gottingen, Germany) containing 1 L of fermentation medium of the following composition: glycerol, 88 g/L; KH_2PO_4, 18.8 g/L; $(NH4)_2SO_4$, 31.4 g/L; $MgSO_4 \times 7H_2O$, 9.2 g/L; $CaCl_2$, 0.7 g/L; biotin, 0.4 mg/L; Sofaxil defoamer, 1 g/L. After autoclaving, the medium was supplemented with 4 mL of PTM1 trace element solution [68]. During fermentation, the temperature of cultivation and a medium pH were maintained at the same level (30 °C and 5.5, respectively). The dissolved oxygen content (DO) was maintained within the range of 40–70% by adjusting the stirring rate within the range from 200 to 500 rpm.

A recombinant *P. pastoris* GS115-afo strain was grown in a 750 mL flask containing 100 mL of YPG medium. The strain was cultivated in an incubator shaker for 18–24 h at 30 °C and 250 rpm until reaching the optical density OD_{600} = 10; then, the whole volume of the culture was aseptically inoculated into a fermentation medium. After 16 h of growth on glycerol up to OD_{600} = 180–200, a periodic induction phase was started with the addition of 100% methanol with a feed rate of 1.5 mL/L/h. After 120 h of fermentation, cultural liquid was centrifuged for 20 min at 4000 rpm using an Avanti JXN-26 centrifuge (Beckman Coulter, Brea, CA, USA) and purified as described below.

4.3. AFO Purification Procedure

Immobilized metal affinity chromatography (IMAC) was used for AFO purification. The centrifuged cultural liquid was applied onto a 15 mL Ni-NTA Sepharose excel column (Cytiva, Little Chalfont, UK) with a flow rate of 1.5 mL/min. The column was equilibrated with the equilibration buffer (20 mM potassium phosphate buffer (pH 7.4) plus 500 mM NaCl). To remove nonspecific bound proteins, the column was washed with a washing buffer (20 mM potassium phosphate buffer (pH 7.4), 500 mM NaCl, 20 mM imidazole) with a flow rate of 2 mL/min. The one-step elution was carried out using the elution buffer (20 mM potassium phosphate buffer (pH 7.4), 500 mM NaCl, 400 mM imidazole) with a flow rate of 2 mL/min. The quality of the AFO separation was evaluated using SDS-PAGE (Bio-RAD, Hercules, CA, USA). Homogeneous enzyme samples were dialyzed against double distilled water at 4 °C, freeze-dried using a SP VirTis Freeze Dryer (SP Scientific, Warminster, PA, USA), and stored at −20 °C until use.

4.4. AFB1 Degradation by AFO in Buffer Solutions

To test the AFO's ability to remove AFB1 from model solutions, freeze-dried samples of purified AFO were dissolved in 20 mM Na-acetate buffer (pH 4.0 or 5.0), or 20 mM Na-phosphate buffer (pH varied within 6.0–8.0), or 20 mM Tris-HCl buffer (pH 9.0). These solutions were sterilized by means of filtration through a 0.22 μm Millipore filter and supplemented with AFB1 (Sigma-Aldrich, St. Louis, MO, USA) dissolved in a minimum volume of methanol up to a final concentration of 0.25 μg/mL. The corresponding buffers supplemented with the toxin at the same concentration were used as controls. The model solutions were incubated for 120 h at 27 °C. During incubation, 50 μL aliquots were sampled every 24 h, diluted with the mobile phase (see Section 4.8) up to a final volume of 1 mL, and analyzed via HPLC as described below. After the selection of the buffer and optimal pH for the target AFO activity, the effect of different incubation temperatures (15, 30, and 50 °C) on the enzymatic degradation process was evaluated after 24, 48, and 72 h of incubation under selected conditions. The experiments were arranged into at least three biological replications.

4.5. AFB1 Degradation by AFO in Artificially Inoculated Grain

4.5.1. Grain Inoculation with Aspergillus flavus

A toxigenic *A. flavus* strain AF24 earlier isolated from peanut and stored at the laboratory collection as a stock culture was recovered and grown on potato dextrose agar. To prepare the inoculum, AF24 was sub-cultured on the same medium in a 9 cm Petri plate at 25 °C. When the sporulating aerial mycelium covered the whole surface of the medium, spores were collected by flooding the plate with sterile distilled water (SDW) to obtain their suspension.

The experiment was arranged in three biological replications. Samples of wheat grain or cracked corn kernels (50 g) were placed in 250 mL Erlenmeyer flasks (9 flasks per each grain type per one experiment), wetted with distilled water (20 mL per flask) and autoclaved for 1 h at 120 °C. Afterward, 2 mL of a spore suspension containing 10^6 spores per mL of SDW was added into each flask. The flasks were shaken for a minute to thoroughly mix their content and evenly distribute the inoculum throughout the grain, placed into a thermostatic chamber, and incubated at 30 °C in the dark. Three days after inoculation, some of the flasks (9 flasks per each type of inoculated grain) were used for decontamination experiments, while the remaining flasks were left in the chamber for the next 4 days. Upon the end of the 3- or 7-day cultivation, the flasks were left overnight at a temperature below 12 °C to stop fungal growth and to prevent additional toxin production [59,69] during the subsequent treatment with enzyme at 30 °C.

Flasks containing wheat or corn grain without *A. flavus* spores were further used as non-inoculated (negative) controls.

4.5.2. Treatment of Inoculated Grain with AFO

A freeze-dried AFO preparation obtained by means of fermentation and purified as described above was used in grain decontamination experiments. On the day of treatment, a portion of the enzyme preparation was dissolved in 20 mM phosphate buffer (pH 6.0) to a final concentration of 50 mg/mL. Then, 100 mL of this solution was added into each flask and gently mixed with grain. In parallel, the same volume of the buffer was added into other flasks with the inoculated grain as inoculated (positive) controls. Thereafter, flasks were incubated at 30 °C with slow shaking for 48 or 72 h in the dark.

4.6. Isolation of Residual AFB1 from Cereal Grain after Enzymatic Decontamination

Upon completion of the incubation process, the flasks were supplemented with acetonitrile up to a final concentration of 80%. The grain was extracted for 1 h at room temperature and intensive shaking (100 rpm). Aliquots of the filtered acetonitrile extract equivalent to 5 g of grain were evaporated using a rotary evaporator up to the aqueous phase, which was further supplemented with a saturated NaCl solution up to the total volume of 20 mL. After the pH adjustment to 3.0, the solution was added to the equal volume of water-saturated hexane followed by the aqueous phase separation via extraction with dichloromethane. The final AFB1-containing extract was passed through a layer of anhydrous Na_2SO_4 and dried on a rotary evaporator. The residue was dissolved in a minimal volume of a mobile phase consisting of 0.3% H_3PO_4 and acetonitrile (1:1), filtered through a membrane with the pore size of 0.45 μm, and analyzed by means of reverse-phase HPLC. If necessary, the final extract was additionally purified prior to evaporation in a Silica Gel 60 column (0.063–0.2 mm, Merck, Darmstadt, Germany).

4.7. AFB1 Quantification in Buffer Solutions and AFO-Exposed Grain

The content of residual AFB1 in model buffer solutions and grain extracts was measured via HPLC using a Waters 1525 Breeze chromatograph (Waters Corp., Milford, MA, USA). An aliquot (10 μL) of each tested sample was applied in a Symmetry C18 temperature-controlled (27 °C) column (5 μm, 150 × 4.6 mm, Waters Corp., Milford, MA, USA). AFB1 was eluted with a 0.3% H_3PO_4:acetonitrile (1:1) mix and detected at 362 nm using a Waters 2487 UV detector (Waters Corp., Milford, MA, USA). The aforementioned commercial AFB1 was used as a reference sample (Figure S2, Supplementary Materials). The AFB1 content was calculated using a calibration curve plotted for the reference sample in the zone of a linear dependence of a peak area on the AFB1 amount. All samples were analyzed in triplicate. AFB1 reduction in contaminated samples was expressed as a percentage of the control (inoculated grain, which was not exposed to AFO). The toxin recovery level determined in experiments with the addition of reference AFB1 to the non-inoculated grain samples prior to their extraction reached 77 (wheat) and 75% (corn). The limit of detection for the grain extract was 0.005 μg/mL. All measurements were carried out in at least three analytical replicates.

4.8. Statistical Treatment

To confirm the significance of difference between the mean values of treated and control variants as well as between means of variants involving enzymatic treatments, the *t*-test for independent variables was used (STATISTICA v. 6.1 software package; StatSoft Inc., Tulsa, OK, USA). The difference was considered to be significant at $p \leq 0.05$.

Supplementary Materials: The following supporting information can be downloaded at: https://www.mdpi.com/article/10.3390/toxins15120678/s1, Figure S1: Map of the pPIGa-1-AFO plasmid; Figure S2: Chromatograms illustrating the reduction in the AFB1 content in extracts of wheat (a1,a2) and corn (b1,b2) grain inoculated with a toxigenic *A. flavus* (a,b) resulting from 24 h (a1,b1) and 48 h (a2,b2) grain exposure to recombinant AFO at pH 6.0 and 30 °C. Chromatograms (c,d) show the results of HPLC-analyses of extracts from non-inoculated wheat and corn grain, respectively, as well as commercial AFB1 standard (e) under the same conditions.

Author Contributions: Conceptualization, L.S., A.R. and I.Z.; methodology, I.S., Y.D., O.M. and L.S.; software, N.S.; investigation, I.S., O.M., Y.D. and T.N.; writing—original draft preparation, L.S., I.S. and Y.D.; writing—review and editing, L.S., A.R., N.S. and I.Z.; visualization, I.S. All authors have read and agreed to the published version of the manuscript.

Funding: This research was funded by the Russian Science Foundation, project number 22-16-00153.

Institutional Review Board Statement: Not applicable.

Informed Consent Statement: Not applicable.

Data Availability Statement: Data are contained within the article and supplementary materials.

Conflicts of Interest: The authors declare no conflict of interest.

References

1. Dzhavakhiya, V.G.; Voinova, T.M.; Popletaeva, S.B.; Statsyuk, N.V.; Limantseva, L.A.; Shcherbakova, L.A. Effect of various compounds blocking the colony pigmentation on the aflatoxin B1 production by *Aspergillus flavus*. *Toxins* **2016**, *8*, 313. [CrossRef] [PubMed]
2. Sipos, P.; Peles, F.; Brassó, D.L.; Béri, B.; Pusztahelyi, T.; Pócsi, I.; Győri, Z. Physical and chemical methods for reduction in aflatoxin content of feed and food. *Toxins* **2021**, *13*, 204. [CrossRef] [PubMed]
3. Umaya, S.R.; Vijayalakshmi, Y.C.; Sejian, V. Exploration of plant products and phytochemicals against aflatoxin toxicity in broiler chicken production: Present status. *Toxicon* **2021**, *200*, 55–68. [CrossRef] [PubMed]
4. Gómez-Salazar, J.A.; Ruiz-Hernández, K.; Martínez-Miranda, M.M.; Castro-Ríos, K. Postharvest strategies for decontamination of aflatoxins in cereals. *Food Rev. Int.* **2023**, *39*, 3635–3662. [CrossRef]
5. Marshall, H.; Meneely, J.P.; Quinn, B.; Zhao, Y.; Bourke, P.; Gilmore, B.F.; Zhang, G.; Elliott, C.T. Novel decontamination approaches and their potential application for post-harvest aflatoxin control. *Trends Food Sci. Technol.* **2020**, *106*, 489–496. [CrossRef]
6. Rajasekaran, K.; Sayler, R.J.; Majumdar, R.; Sickler, C.M.; Cary, J.W. Inhibition of *Aspergillus flavus* growth and aflatoxin production in transgenic maize expressing the α-amylase inhibitor from *Lablab purpureus* L. *J. Vis. Exp.* **2019**, *144*, e59169. [CrossRef]
7. Popescu, R.G.; Rădulescu, A.L.; Georgescu, S.E.; Dinischiotu, A. Aflatoxins in feed: Types, metabolism, health consequences in swine and mitigation strategies. *Toxins* **2022**, *14*, 853. [CrossRef] [PubMed]
8. Kumar, A.; Pathak, H.; Bhadauria, S.; Sudan, J. Aflatoxin contamination in food crops: Causes, detection, and management: A review. *Food Prod. Process. Nutr.* **2021**, *3*, 17. [CrossRef]
9. Mahato, D.K.; Lee, K.E.; Kamle, M.; Devi, S.; Dewangan, K.N.; Kumar, P.; Kang, S.G. Aflatoxins in food and feed: An overview on prevalence, detection and control strategies. *Front. Microbiol.* **2019**, *10*, 2266. [CrossRef]
10. Kumar, P.; Gupta, A.; Mahato, D.K.; Pandhi, S.; Pandey, A.K.; Kargwal, R.; Mishra, S.; Suhag, R.; Sharma, N.; Saurabh, V.; et al. Aflatoxins in cereals and cereal-based products: Occurrence, toxicity, impact on human health, and their detoxification and management strategies. *Toxins* **2022**, *14*, 687. [CrossRef]
11. Tadele, F.; Demissie, B.; Amsalu, A.; Demelash, H.; Mengist, Z.; Ambelu, A.; Yenew, C. Aflatoxin contamination of animal feeds and its predictors among dairy farms in northwest Ethiopia: One health approach implications. *Front. Vet. Sci.* **2023**, *10*, 1123573. [CrossRef] [PubMed]
12. Putman, B.; Thoma, G.; Burek, J.; Matlock, M. A retrospective analysis of the United States poultry industry: 1965 compared with 2010. *Agric. Syst.* **2017**, *157*, 107–117. [CrossRef]
13. Lee, J.; Nam, D.S.; Kong, C. Variability in nutrient composition of cereal grains from different origins. *SpringerPlus* **2016**, *5*, 419. [CrossRef]
14. Filazi, A.; Sireli, U.T.; Filazi, A.; Sireli, U.T. Occurrence of aflatoxins in food. In *Aflatoxins–Recent Advances and Future Prospects*; Razzaghi-Abyaneh, M., Ed.; IntechOpen: Rijeka, Croatia, 2013; pp. 143–171.
15. Eskola, M.; Kos, G.; Elliott, C.T.; Hajšlová, J.; Mayar, S.; Krska, R. Worldwide contamination of food-crops with mycotoxins: Validity of the widely cited "FAO estimate" of 25%. *Crit. Rev. Food Sci. Nutr.* **2020**, *60*, 2773–2789. [CrossRef] [PubMed]
16. Monson, M.S.; Coulombe, R.A.; Reed, K.M. Aflatoxicosis: Lessons from toxicity and responses to aflatoxin B1 in poultry. *Agriculture* **2015**, *5*, 742–777. [CrossRef]
17. Rawal, S.; Kim, J.E.; Coulombe, R. Aflatoxin B1 in poultry: Toxicology, metabolism and prevention. *Res. Vet. Sci.* **2010**, *89*, 325–331. [CrossRef] [PubMed]
18. Loi, M.; Paciolla, C.; Logrieco, A.F.; Mulè, G. Plant bioactive compounds in pre- and postharvest management for aflatoxins reduction. *Front. Microbiol.* **2020**, *11*, 243. [CrossRef]
19. Wahyono, N.D.; Utami, M.D. A review of the poultry meat production industry for food safety in Indonesia. *J. Phys. Conf. Ser.* **2018**, *953*, 012125. [CrossRef]
20. Alberts, J.F.; Gelderblom, W.C.A.; Botha, A.; van Zyl, W.H. Degradation of aflatoxin B(1) by fungal laccase enzymes. *Int. J. Food Microbiol.* **2009**, *135*, 47–52. [CrossRef] [PubMed]
21. Taylor, M.C.; Jackson, C.J.; Tattersall, D.B.; French, N.; Peat, T.S.; Newman, J.; Briggs, L.J.; Lapalikar, G.V.; Campbell, P.M.; Scott, C.; et al. Identification and characterization of two families of F420 H2-dependent reductases from mycobacteria that catalyse aflatoxin degradation. *Mol. Microbiol.* **2010**, *78*, 561–575. [CrossRef]
22. Zhao, L.H.; Guan, S.; Gao, X.; Ma, Q.G.; Lei, Y.P.; Bai, X.M.; Ji, C. Preparation, purification and characteristics of an aflatoxin degradation enzyme from *Myxococcus fulvus ANSM068*. *J. Appl. Microbiol.* **2011**, *110*, 147–155. [CrossRef] [PubMed]
23. Guo, Y.; Qin, X.; Tang, Y.; Ma, Q.; Zhang, J.; Zhao, L. CotA laccase, a novel aflatoxin oxidase from *Bacillus licheniformis*, transforms aflatoxin B1 to aflatoxin Q1 and epi-aflatoxin Q1. *Food Chem.* **2020**, *325*, 126877. [CrossRef] [PubMed]
24. Ji, C.; Fan, Y.; Zhao, L. Review on biological degradation of mycotoxins. *Anim. Nutr.* **2016**, *2*, 127–133. [CrossRef] [PubMed]
25. Adebo, O.A.; Njobeh, P.B.; Gbashi, S.; Nwinyi, O.C.; Mavumengwana, V. Review on microbial degradation of aflatoxins. *Crit. Rev. Food Sci. Nutr.* **2017**, *57*, 3208–3217. [CrossRef] [PubMed]
26. Lyagin, I.; Efremenko, E. Enzymes for detoxification of various mycotoxins: Origins and mechanisms of catalytic action. *Molecules* **2019**, *24*, 2362. [CrossRef]
27. Guan, Y.; Chen, J.; Nepovimova, E.; Long, M.; Wu, W.; Kuca, K. Aflatoxin detoxification using microorganisms and enzymes. *Toxins* **2021**, *13*, 46. [CrossRef] [PubMed]

28. Verheecke, C.; Liboz, T.; Mathieu, F. Microbial degradation of aflatoxin B1: Current status and future advances. *Int. J. Food Microbiol.* **2016**, *237*, 1–9. [CrossRef] [PubMed]
29. Xu, H.; Wang, L.; Sun, J.; Wang, L.; Guo, H.; Ye, Y.; Sun, X. Microbial detoxification of mycotoxins in food and feed. *Crit. Rev. Food Sci. Nutr.* **2022**, *62*, 4951–4969. [CrossRef]
30. Smaoui, S.; D'Amore, T.; Tarapoulouzi, M.; Agriopoulou, S.; Varzakas, T. Aflatoxins contamination in feed commodities: From occurrence and toxicity to recent advances in analytical methods and detoxification. *Microorganisms* **2023**, *11*, 2614. [CrossRef]
31. Kolosova, A.; Stroka, J. Substances for reduction of the contamination of feed by mycotoxins: A review. *World Mycotoxin J.* **2011**, *4*, 225–256. [CrossRef]
32. Loi, M.; Fanelli, F.; Zucca, P.; Liuzzi, V.C.; Quintieri, L.; Cimmarusti, M.T.; Monaci, L.; Haidukowski, M.; Logrieco, A.F.; Sanjust, E.; et al. Aflatoxin B1 and M1 degradation by Lac2 from *Pleurotus pulmonarius* and redox mediators. *Toxins* **2016**, *8*, 245. [CrossRef]
33. Branà, M.T.; Sergio, L.; Haidukowski, M.; Logrieco, A.F.; Altomare, C. Degradation of Aflatoxin B1 by a sustainable enzymatic extract from spent mushroom substrate of *Pleurotus eryngii*. *Toxins* **2020**, *12*, 49. [CrossRef]
34. Liu, D.L.; Yao, D.S.; Liang, Y.Q.; Zhou, T.H.; Song, Y.P.; Zhao, L.; Ma, L. Production, purification, and characterization of an intracellular aflatoxin-detoxifizyme from *Armillariella tabescens* (E-20). *Food Chem. Toxicol.* **2001**, *39*, 461–466. [CrossRef] [PubMed]
35. Wang, J.; Ogata, M.; Hirai, H.; Kawagishi, H. Detoxification of aflatoxin B1 by manganese peroxidase from the white-rot fungus *Phanerochaete sordida* YK-624. *FEMS Microbiol. Lett.* **2011**, *314*, 164–169. [CrossRef] [PubMed]
36. Liu, Y.; Mao, H.; Hu, C.; Tron, T.; Lin, J.; Wang, J.; Sun, B. Molecular docking studies and in vitro degradation of four aflatoxins (AFB1, AFB2, AFG1, and AFG2) by a recombinant laccase from *Saccharomyces cerevisiae*. *J. Food Sci.* **2020**, *85*, 1353–1360. [CrossRef] [PubMed]
37. Watanabe, N.; Ohkusu, K.; Okuda, M.; Imataki, O.; Ishii, T.; Negayama, K.; Tadokoro, A.; Kita, N.; Takagi, T.; Kanaji, N.; et al. *Phanerochaete sordida* as a cause of pulmonary nodule in an immunocompromised patient: A case report. *BMC Infect. Dis.* **2017**, *17*, 135. [CrossRef] [PubMed]
38. Brugnari, T.; Braga, D.M.; dos Santos, C.S.A.; Torres, B.H.C.; Modkovski, T.A.; Haminiuk, C.W.I.; Maciel, G.M. Laccases as green and versatile biocatalysts: From lab to enzyme market—An overview. *Bioresour. Bioprocess.* **2021**, *8*, 131. [CrossRef]
39. Rodríguez-Escribano, D.; Pliego-Magán, R.; de Salas, F.; Aza, P.; Gentili, P.; Ihalainen, P.; Levée, T.; Meyer, V.; Petit-Conil, M.; Tapin-Lingua, S.; et al. Tailor-made alkaliphilic and thermostable fungal laccases for industrial wood processing. *Biotechnol. Biofuels* **2022**, *15*, 149. [CrossRef]
40. Bao, C.; Liu, Y.; Li, F.; Cao, H.; Dong, B.; Cao, Y. Expression and characterization of laccase Lac1 from *Coriolopsis trogii* strain Mafic-2001 in *Pichia pastoris* and its degradation of lignin. *Appl. Biochem. Biotechnol.* **2023**, *195*, 6150–6167. [CrossRef]
41. Cao, H.; Liu, D.; Mo, X.; Xie, C.; Yao, D. A fungal enzyme with the ability of aflatoxin B1 conversion: Purification and ESI-MS/MS identification. *Microbiol. Res.* **2011**, *166*, 475–483. [CrossRef]
42. Schuda, P.F. Aflatoxin chemistry and syntheses. In *Syntheses of Natural Products. Topics in Current Chemistry*; Springer: Berlin/Heidelberg, Germany, 1980; Volume 91, pp. 75–111. [CrossRef]
43. Wu, Y.Z.; Lu, F.P.; Jiang, H.L.; Tan, C.P.; Yao, D.S.; Xie, C.F.; Liu, D.L. The furofuran-ring selectivity, hydrogen peroxide-production and low km value are the three elements for highly effective detoxification of aflatoxin oxidase. *Food Chem. Toxicol.* **2015**, *76*, 125–131. [CrossRef] [PubMed]
44. Schroeder, H.W.; Kelton, W.H. Production of sterigmatocystin by some species of the genus *Aspergillus* and its toxicity to chicken embryos. *Appl. Microbiol.* **1975**, *30*, 589–591. [CrossRef] [PubMed]
45. Li, L.; Mei, M.; Wang, J.; Huang, J.; Zong, X.; Wang, X. Expression and Application of Aflatoxin Degrading Enzyme Gene in *Pichia pastoris*. *Biotechnol. J.* **2023**, 2300167. [CrossRef]
46. Chen, J.; Liu, D.; Li, S.; Yao, D. Development of an amperometric enzyme electrode biosensor for sterigmatocystin detection. *Enz. Microb. Technol.* **2010**, *47*, 119–126. [CrossRef]
47. Wen, S.; Guan, M.; Zhou, T.; Cao, H.; Xie, C.; Liu, D.; Yao, D. Cloning, expression, purification and characterization of an aflatoxin-converting enzyme from *Armillaria tabescens*. *Wei Sheng Wu Xue Bao* **2011**, *51*, 1212–1221. [PubMed]
48. Sinelnikov, I.G.; Zorov, I.N.; Denisenko, Y.A.; Mikityuk, O.O.; Sinitsyn, A.P.; Shcherbakova, L.A. A new producer of a recombinant aflatoxin-degrading enzyme obtained via heterologous expression in *Pichia pastoris*. *Agric. Biol.* **2022**, *57*, 1166–1177. (In Russian) [CrossRef]
49. Yang, P.; Xiao, W.; Lu, S.; Jiang, S.; Zheng, Z.; Zhang, D.; Zhang, M.; Jiang, S.; Jiang, S. Recombinant expression of *Trametes versicolor* aflatoxin B1-degrading enzyme (TV-AFB1D) in engineering *Pichia pastoris* GS115 and application in AFB1 degradation in AFB1-contaminated peanuts. *Toxins* **2021**, *13*, 349. [CrossRef]
50. Li, C.H.; Li, W.Y.; Hsu, I.N.; Liao, Y.Y.; Yang, C.Y.; Taylor, M.C.; Liu, Y.F.; Huang, W.H.; Chang, H.H.; Huang, H.L.; et al. Recombinant aflatoxin-degrading F420H2-dependent reductase from *Mycobacterium smegmatis* protects mammalian cells from aflatoxin toxicity. *Toxins* **2019**, *11*, 259. [CrossRef]
51. Shcherbakova, L.; Rozhkova, A.; Osipov, D.; Zorov, I.; Mikityuk, O.; Statsyuk, N.; Sinitsyna, O.; Dzhavakhiya, V.; Sinitsyn, A. Effective zearalenone degradation in model solutions and infected wheat grain using a novel heterologous lactonohydrolase secreted by recombinant *Penicillium canescens*. *Toxins* **2020**, *12*, 475. [CrossRef]
52. Michlmayr, H.; Malachová, A.; Varga, E.; Kleinová, J.; Lemmens, M.; Newmister, S.; Rayment, I.; Berthiller, F.; Adam, G. Biochemical characterization of a recombinant UDP-glucosyltransferase from rice and enzymatic production of deoxynivalenol-3-O-β-D-glucoside. *Toxins* **2015**, *7*, 2685–2700. [CrossRef]

53. Xu, T.; Xie, C.; Yao, D.; Zhou, C.Z.; Liu, J. Crystal structures of aflatoxin-oxidase from *Armillariella tabescens* reveal a dual activity enzyme. *Biochem. Biophys. Res. Commun.* **2017**, *494*, 621–625. [CrossRef] [PubMed]
54. Smirnova, I.R.; Mikhalev, A.V.; Satyukova, L.P.; Borisova, V.S. Current state of feed quality and safety in Russia. *Veterinariya* **2009**, *2*, 3–7. (In Russian)
55. Borisova, T. Screening analysis of mycotoxins in grain and food products. *Analytics* **2017**, *2*, 98–103. (In Russian) [CrossRef]
56. Loi, M.; Logrieco, A.F.; Pusztahelyi, T.; Leiter, É.; Hornok, L.; Pócsi, I. Advanced mycotoxin control and decontamination techniques in view of an increased aflatoxin risk in Europe due to climate change. *Front. Microbiol.* **2023**, *13*, 1085891. [CrossRef] [PubMed]
57. Kovalenko, A.; Soldatenko, N.; Fetisov, L.; Sukhikh, E. Mycotoxicological monitoring of feed in the North Caucasus region. *Kombikorma* **2011**, *3*, 98–99. (In Russian)
58. Moretti, A.; Pascale, M.; Logrieco, A.F. Mycotoxin risks under a climate change scenario in Europe. *Trends Food Sci. Technol.* **2019**, *84*, 38–40. [CrossRef]
59. Hawkins, L.K.; Windham, G.L.; Williams, W.P. Effect of different postharvest drying temperatures on *Aspergillus flavus* survival and aflatoxin content in five maize hybrids. *J. Food Prot.* **2005**, *68*, 1521–1524. [CrossRef]
60. Siciliano, I.; Berta, F.; Bosio, P.; Gullino, M.L.; Garibaldi, A. Effect of different temperatures and CO_2 levels on *Alternaria* toxins produced on cultivated rocket, cabbage and cauliflower. *World Mycotoxin J.* **2017**, *10*, 63–71. [CrossRef]
61. Rushing, B.R.; Selim, M.I. Aflatoxin B1: A review on metabolism, toxicity, occurrence in food, occupational exposure, and detoxification methods. *Food Chem. Toxicol.* **2019**, *124*, 81–100. [CrossRef]
62. Toteja, G.S.; Mukherjee, A.; Diwakar, S.; Singh, P.; Saxena, B.N.; Sinha, K.K.; Sinha, A.K.; Kumar, N.; Nagaraja, K.V.; Bai, G.; et al. Aflatoxin B1 contamination in wheat grain samples collected from different geographical regions of India: A multicenter study. *J. Food Prot.* **2006**, *69*, 1463–1467. [CrossRef]
63. Summary of Food and Agricultural Statistics 2004. Available online: https://www.fao.org/3/ae881e/ae881e00.htm (accessed on 29 October 2023).
64. Dobolyi, C.; Sebők, F.; Varga, J.; Kocsubé, S.; Szigeti, G.; Baranyi, N.; Szécsi, Á.; Tóth, B.; Varga, M.; Kriszt, B.; et al. Occurrence of aflatoxin producing *Aspergillus flavus* isolates in maize kernel in Hungary. *Acta Aliment.* **2013**, *42*, 451–459. [CrossRef]
65. Sanger, F.; Nicklen, S.; Coulson, A.R. DNA sequencing with chain-terminating inhibitors. *Proc. Natl. Acad. Sci. USA* **1977**, *74*, 5463–5467. [CrossRef] [PubMed]
66. Kumar, R. Simplified protocol for faster transformation of (a large number of) *Pichia pastoris* strains. *Yeast* **2019**, *36*, 399–410. [CrossRef] [PubMed]
67. Lõoke, M.; Kristjuhan, K.; Kristjuhan, A. Extraction of genomic DNA from yeasts for PCR-based applications. *BioTechniques* **2011**, *50*, 325–328. [CrossRef]
68. Ghosalkar, A.; Sahai, V.; Srivastava, A. Optimization of chemically defined medium for recombinant *Pichia pastoris* for biomass production. *Bioresour. Technol.* **2008**, *99*, 7906–7910. [CrossRef]
69. Bernáldez, V.; Córdoba, J.J.; Magan, N.; Peromingo, B.; Rodríguez, A. The influence of ecophysiological factors on growth, *aflR* gene expression and aflatoxin B1 production by a type strain of *Aspergillus flavus*. *LWT-Food Sci. Technol.* **2017**, *83*, 283–291. [CrossRef]

Disclaimer/Publisher's Note: The statements, opinions and data contained in all publications are solely those of the individual author(s) and contributor(s) and not of MDPI and/or the editor(s). MDPI and/or the editor(s) disclaim responsibility for any injury to people or property resulting from any ideas, methods, instructions or products referred to in the content.

toxins

MDPI

Article

Chitin Deacetylase Homologous Gene *cda* Contributes to Development and Aflatoxin Synthesis in *Aspergillus flavus*

Xin Zhang †, Meifang Wen †, Guoqi Li † and Shihua Wang *

State Key Laboratory of Ecological Pest Control for Fujian and Taiwan Crops, Key Laboratory of Pathogenic, Fungi and Mycotoxins of Fujian Province, School of Life Sciences, Fujian Agriculture and Forestry University, Fuzhou 350002, China; zhangxin5319@gmail.com (X.Z.); meifangwen@hotmail.com (M.W.); liguoqi@fafu.edu.cn (G.L.)

* Correspondence: wshmail@m.fafu.edu.cn; Tel./Fax: +86-591-87984471

† These authors contributed equally to this work.

Abstract: The fungal cell wall serves as the primary interface between fungi and their external environment, providing protection and facilitating interactions with the surroundings. Chitin is a vital structural element in fungal cell wall. Chitin deacetylase (CDA) can transform chitin into chitosan through deacetylation, providing various biological functions across fungal species. Although this modification is widespread in fungi, the biological functions of CDA enzymes in *Aspergillus flavus* remain largely unexplored. In this study, we aimed to investigate the biofunctions of the CDA family in *A. flavus*. The *A. flavus* genome contains six annotated putative chitin deacetylases. We constructed knockout strains targeting each member of the CDA family, including Δ*cda1*, Δ*cda2*, Δ*cda3*, Δ*cda4*, Δ*cda5*, and Δ*cda6*. Functional analyses revealed that the deletion of CDA family members neither significantly affects the chitin content nor exhibits the expected chitin deacetylation function in *A. flavus*. However, the Δ*cda6* strain displayed distinct phenotypic characteristics compared to the wild-type (WT), including an increased conidia count, decreased mycelium production, heightened aflatoxin production, and impaired seed colonization. Subcellular localization experiments indicated the cellular localization of CDA6 protein within the cell wall of *A. flavus* filaments. Moreover, our findings highlight the significance of the CBD1 and CBD2 structural domains in mediating the functional role of the CDA6 protein. Overall, we analyzed the gene functions of CDA family in *A. flavus*, which contribute to a deeper understanding of the mechanisms underlying aflatoxin contamination and lay the groundwork for potential biocontrol strategies targeting *A. flavus*.

Keywords: *Aspergillus flavus*; aflatoxin; chitin deacetylase; *cda6*

Key Contribution: In this study, we aimed to analyze the functions of six *cda* genes in *A. flavus*. We found that the absence of CDA family member did not affect the chitin content and cell wall integrity. However, our results showed that the *cda6* gene plays a crucial role in regulating the development, aflatoxin synthesis, and pathogenicity of *A. flavus*.

1. Introduction

Aspergillus flavus is a fungus of the genus *Aspergillus* that is widely distributed and highly adaptable to its environment [1]. However, under improper storage conditions, *A. flavus* can proliferate in various crop seeds and produce aflatoxins [2]. On one hand, *A. flavus* can cause poison in humans and animals, leading to tumors, cancer, and even death [3,4]. On the other hand, *A. flavus* spores or mycelium can infect host tissues and cause invasive pulmonary aspergillosis and infectious aspergillosis in animals [5]. Moreover, *A. flavus* is the primary contaminant and major producer of AFB_1, which is one of the most hazardous carcinogenic toxins for humans and animals [6]. In summary, *A. flavus* poses a significant threat to food security, human, and animal health.

The fungal cell wall, comprising β-1, 6-glucan, β-1, 3-glucan, and chitin, represents a pivotal structure at the interface between fungi and their external surroundings [7]. Its composition dynamically changes during various growth and differentiation stages, exerting crucial roles in fungal survival, growth, and interaction with the environment [8]. Notably, the fungal cell wall safeguards cellular integrity, facilitates intercellular adhesion, and mediates cell–substrate attachment. Moreover, it significantly influences the pathogenicity and virulence of invasive fungi [9]. Due to the importance of the cell wall to fungi, it serves as a therapeutic target in clinical research and development of antifungal drugs [10,11].

Chitin is a crucial component of the cell wall for fungi, serving to maintain its stability. Chitin deacetylase (CDA) is prevalent across fungal species, and catalyzes the conversion of chitin to chitosan, modulating cell wall formation, integrity, and pathogenicity [12,13]. Structural biology studies have shown that CDAs share a similar catalytic mechanism with other members of the carbohydrate esterase 4 (CE4) family. Previous investigations conducted in model fungi like *Saccharomyces cerevisiae* and *Cryptococcus neoformans* have illuminated the essentiality of CDAs. In *S. cerevisiae*, the structure of chitosan is essential for spores to maintain their structural stiffness and resist various environmental pressures [13]. Two genes, *cda1* and *cda2*, play vital roles in the formation of correct ascospore walls in *S. cerevisiae* [14]. In *C. neoformans*, which harbors four *cda* genes, namely *cda1*, *cda2*, *cda3*, and *fpd1*, deletion of *cda* genes affects the chitosan content of the cell wall [15]. Deacetylated chitosan serves as a crucial structural component of the trophic mycelium cell wall of *C. neoformans*, contributing to the maintenance of structural integrity, with the enzyme activity of Cda1 being critical to the pathogenesis of this species [16]. In *Pyricularia oryzae*, the chitin deacetylase PoCda7 is essential for the pathogenicity of the rice blast fungus [17]. There are seven *cda* genes in the *A. fumigatus* genome, of which Cod4 and Cod7 play significant roles in polarity abnormality and conidiation [18]. While the seven candidate CDA proteins in *A. fumigatus* exhibit minor contributions to fungal cell wall synthesis and virulence, Cda2 is an exception, involved in conidiation [19]. The above results showed that CDAs play distinct roles in conidiation, cell wall formation, and pathogenicity across various fungal species.

All members of the CDA family proteins harbor conserved domains, including the NodB homology domain, pivotal for chitin deacetylation. The NodB domain is the landmark domain of carbohydrate esterase 4 (CE4) family, which can convert chitin into chitosan in various fungi. In *Rhizobium*, the NodB protein is a chitosan deacetylase capable of deacetylating the non-reducing N-acetylglucosamine residues of chitosan, but not the monosaccharide N-acetylglucosamine [20]. Furthermore, additional domains like chitin-binding domains (CBDs) and glycosylphosphatidylinositol (GPI) anchoring motifs enhance CDA functionality by facilitating cell surface interactions and membrane anchorage [21]. The GPI is involved in attaching protein to the fungal cell wall and cell membrane [22,23]. Notably, the CDA homologous gene of the *CBP1* gene in *Magnaporthe grisea* contains a chitin-binding domain (CBD) and the NodB domain, which is critical in response to external hydrophobic signals [24].

Based on studies in phytopathogenic filamentous fungi, we hypothesized that CDAs might play important roles in conidiation, aflatoxin biosynthesis, and host–pathogen interactions in *A. flavus*. In this study, we aim to comprehensively investigate the impact of the chitin deacetylase family on various physiological and pathological aspects of *A. flavus*. Another critical aspect of this research is to assess the impact of CDAs on the synthesis of aflatoxins in *A. flavus*. This study also investigated how the CDA family influences pathogenicity in *A. flavus*. In conclusion, this study aims to establish a theoretical framework for the prevention and management of *A. flavus*.

2. Results

2.1. Bioinformatics Analysis of CDA Family in A. flavus

Through homology comparison, we identified and named six chitin deacetylase genes in *A. flavus* as *cda1*, *cda2*, *cda3*, *cda4*, *cda5*, and *cda6*, respectively (Table S1). Protein sequences of CDA family in *A. flavus* were compared with the homologous sequences from *A. nidulans*, *A. fumigatus*, *A. niger*, *A. parasiticus*, and *A. oryzae*. Phylogenetic analysis revealed the presence of homologous proteins of CDA members across various *Aspergillus* genera. Notably, the highest percent identity (97.42%) was observed between *A. flavus* CDA6 (XP 041144085.1) and *A. oryzae* (EIT74861.1), underscoring the high conservation of the CDA family within *Aspergillus* species (Figure 1A). All members of the CDA family contain a conserved NodB homology domain, characteristic of the carbohydrate esterase 4 (CE4) family. This domain catalyzes the conversion of chitin to chitosan in diverse fungi. Additionally, the CDA6 protein contains chitin-binding domains (CBDs), a signal peptide (SP), and a glycosylphosphatidylinositol (GPI)-anchoring domain (Figure 1B).

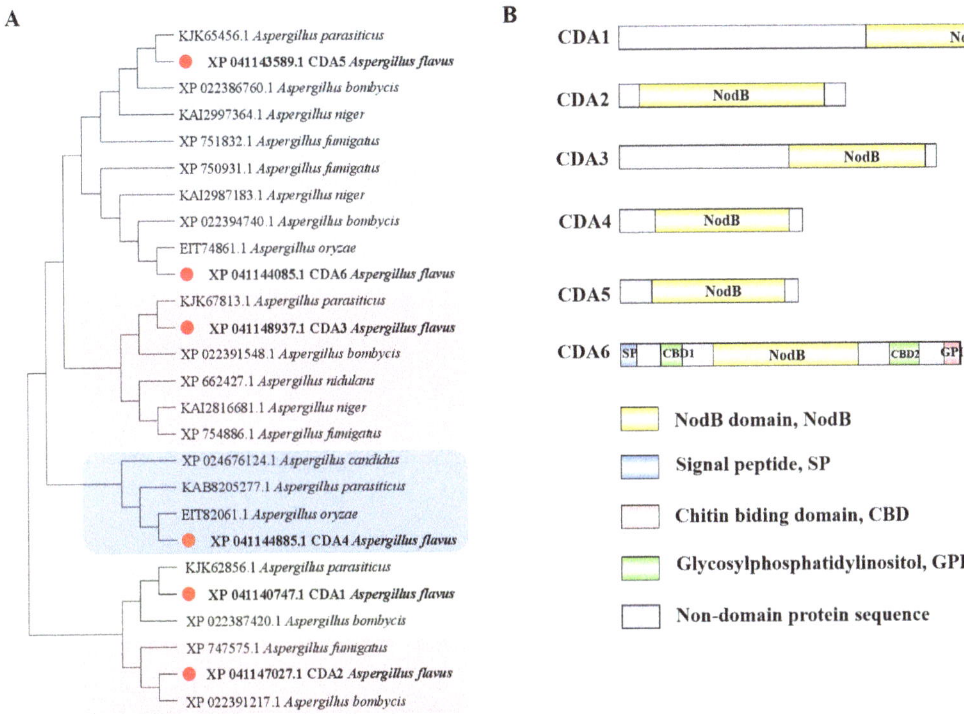

Figure 1. Bioinformatics analysis of CDA family in *A. flavus*. (**A**) Phylogenetic tree construction of CDA homologous proteins between different species. The red bullet represents *A. flavus* CDAs. (**B**) Domain analysis of CDA family in *A. flavus*.

2.2. Quantitative Analysis of Chitin in A. flavus

To investigate the biological functions of CDA members in *A. flavus*, gene deletion strains for *cda1*, *cda2*, *cda3*, *cda4*, *cda5*, and *cda6* were constructed using homologous recombination. After PCR verification, these strains were subjected to further testing (Figure S1). Given the high homology of CDA family members with those of other fungi, it was hypothesized that *A. flavus* CDA proteins may have chitin deacetylation activity. Quantitative analysis of chitin using Calcofluor White (CFW) staining revealed blue fluorescence on all mycelial cell walls. However, the fluorescence intensity of single *cda* knockout strains did not significantly differ from that of the WT strain (Figure 2A). The absorbance values of the developed color with p-dimethylaminobenzaldehyde were measured, and the data were analyzed for the relative content of cell wall chitin in *A. flavus*. The results showed no significant difference in chitin content between Δ*cda1*, Δ*cda2*, Δ*cda3*, Δ*cda4*, Δ*cda5*, and Δ*cda6* strains compared to WT strain (Figure 2B), indicating that deletion of *cda* genes did not affect chitin content. In summary, single *cda* gene deletion within the CDA family did not influence the chitin content of *A. flavus* cell wall.

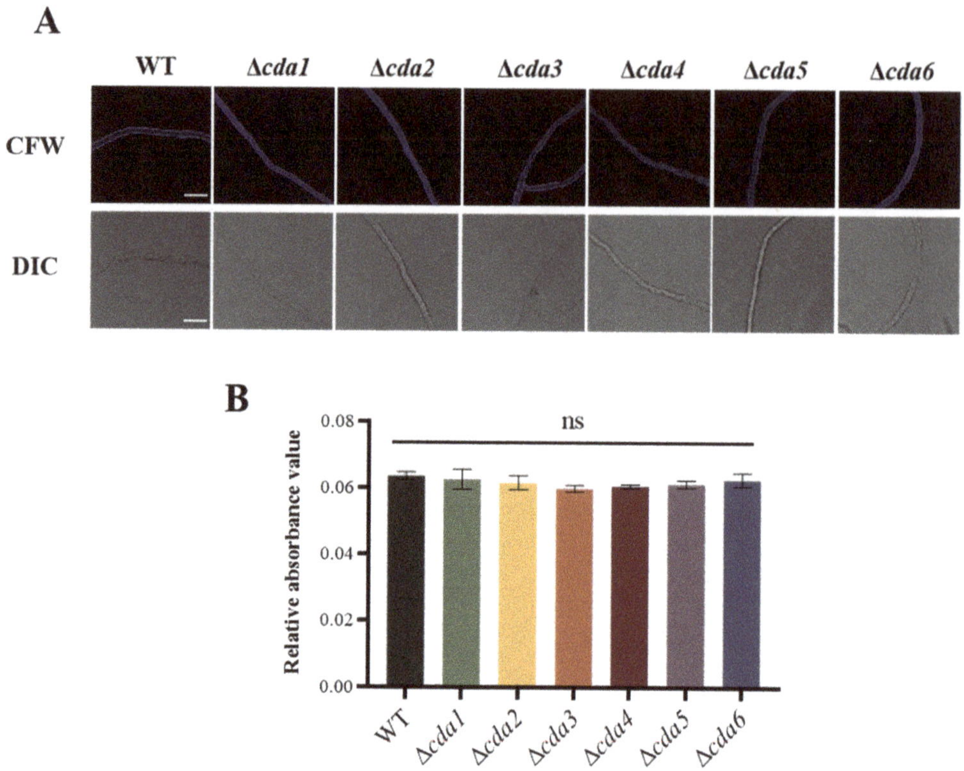

Figure 2. Quantitative analysis of chitin in *cda* single deletion strain of *A. flavus*. (**A**) Confocal microscope results of hyphae from Δ*cda1*, Δ*cda2*, Δ*cda3*, Δ*cda4*, Δ*cda5*, and Δ*cda6* dyed by CFW (CFW, Fluorescent Brightener 28, can bind to chitin and fluoresces blue. Bar = 25 μm). (**B**) Quantitative analysis of chitin in *cda* single knockout strain of *A. flavus* by chemical extraction. ns indicates no significance.

2.3. *Involvement of CDA Members in Regulating Conidial Formation in A. flavus*

To test the role of CDA family members in conidial formation in *A. flavus*, spore suspensions of WT, Δ*cda1*, Δ*cda2*, Δ*cda3*, Δ*cda4*, Δ*cda5*, and Δ*cda6* strains were inoculated on GMM basal medium and YGT nutrient-rich medium. The colony diameters of CDA member knockout strains in both GMM and YGT media were not significantly different compared to WT strain (Figure 3A), indicating that CDA family members do not regulate radial growth of *A. flavus*. However, the results revealed a significant increase in spore production in the Δ*cda6* strain compared to WT (Figure 3B,C), indicating involvement of the *cda6* gene in regulating conidial formation.

To further investigate the biofunctions of *cda6* gene in *A. flavus*, we also constructed a complement strain of this gene ($cda6^C$). Subsequently, WT, Δ*cda6*, and $cda6^C$ strains were further incubated in YGT media at 37 °C for 5 days. The result in Figure 3D,E indicated that the *cda6* gene negatively regulates the spore formation in *A. flavus*, and the complemented strain $cda6^C$ fully restored its function. To investigate the signaling pathway of the *cda6* gene regulating conidial formation in *A. flavus*, the transcript levels of regulatory genes for spore formation, *brlA* and *abaA*, were examined by qPCR. The results showed that the relative expression of *brlA* and *abaA* genes was significantly up-regulated in the Δ*cda6* strain compared to the WT strain (Figure 3F), indicating that the *cda6* gene regulated spore formation by affecting the transcript levels of the *brlA* and *abaA* genes.

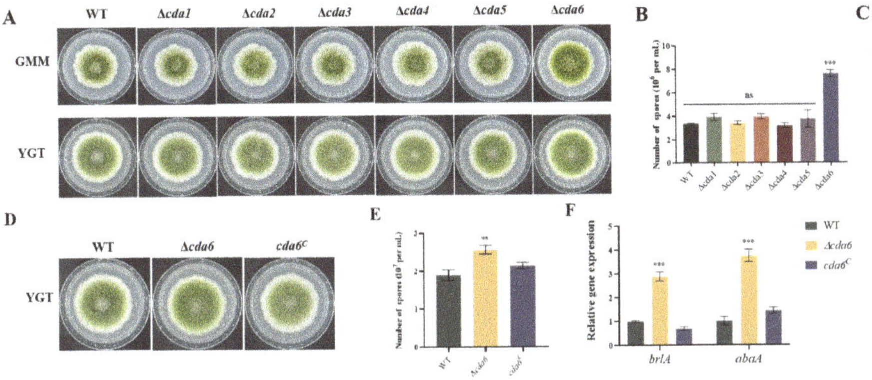

Figure 3. Role of CDA family members in radial growth and sporulation of *A. flavus*. (**A**) WT, Δ*cda1*, Δ*cda2*, Δ*cda3*, Δ*cda4*, Δ*cda5*, and Δ*cda6* strains were inoculated in GMM and YGT medium respectively to observe the growth status. (**B**) Statistics of conidial number of the above strains in GMM medium. (**C**) Statistics of conidial number of the above strains in YGT medium. (**D**) Growth status of WT, Δ*cda6*, and *cda6*^{C} strains on YGT medium. (**E**) Statistics of conidial number of WT, Δ*cda6*, and *cda6*^{C} in YGT medium. (**F**) Relative expression levels of the conidial formation regulatory genes *brlA* and *abaA* in the above strains. ns indicates not significant, ** indicates that the significance level was $p < 0.01$, and *** indicates that the significance level was $p < 0.001$ ($n = 3$).

2.4. Fluorescence Localization of CDA6 Protein in *A. flavus*

To elucidate the localization of CDA6 protein in *A. flavus* mycelia, we constructed a CDA6-mcherry strain using homologous recombination. Under a laser confocal microscope, the position of red fluorescent in the mycelia of the Δ*cda6* strain indicated the cellular localization of the protein. CFW is a non-specific fluorescent dye that binds cellulose and chitin in the cell wall. The results showed that the red fluorescence of CDA6-mcherry mycelia basically coincided with the blue fluorescence of Calcofluor White (Figure 4), indicating that CDA6 protein was mainly distributed in the cell wall of *A. flavus* mycelia.

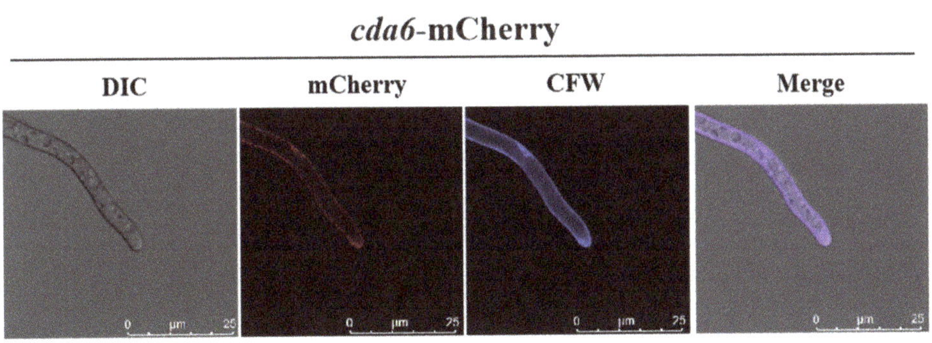

Figure 4. Subcellular localization of CDA6 protein in *A. flavus*. CDA6 protein was localized in the cell wall of *A. flavus* hypha. The hypha was stained with CFW solution, and the localization of CDA6 protein was observed under laser confocal microscope.

2.5. *cda6* Affects the Sclerotia Formation in *A. flavus*

To elucidate the role of the *cda6* gene in sclerotia formation in *A. flavus*, the spore suspensions of WT, Δ*cda6*, and *cda6*^{C} strains were inoculated with equal concentrations on

CM and GMM medium. Statistical analysis revealed that the Δ*cda6* strains did not produce any sclerotia on CM and GMM medium compared to WT (Figure 5A–D), indicating that the *cda6* gene was essential for sclerotia formation in *A. flavus*. To further investigate the mechanism of the *cda6* gene in regulation of sclerotia formation in *A. flavus*, the transcript levels of the *nsdC* and *nsdD* genes related to sclerotia formation were examined by qPCR. The result showed that the relative expression of the *nsdC* and *nsdD* genes in Δ*cda6* strain was significantly lower than that in WT and *cda6*^{C} strains (Figure 5E), suggesting that the *cda6* gene regulates sclerotia formation in *A. flavus* by affecting the transcript levels of *nsdC* and *nsdD* genes.

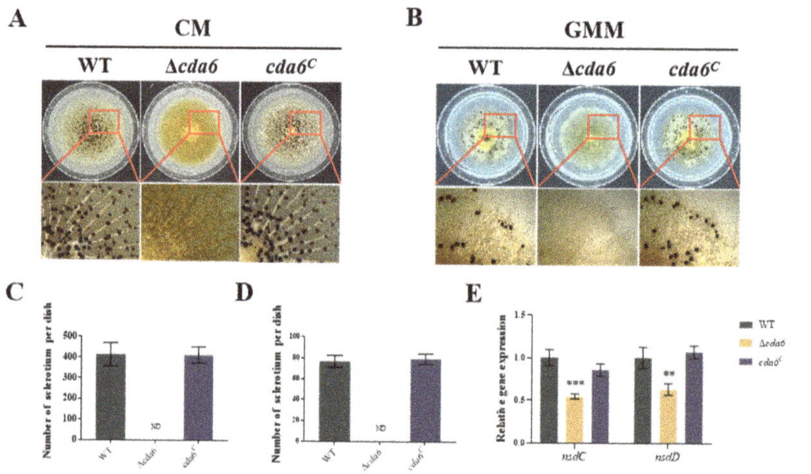

Figure 5. Effect of *cda6* gene on sclerotium formation of *A. flavus*. (**A**) Sclerotium state of WT, Δ*cda6*, and *cda6*^{C} strains on CM solid medium, the magnification is 75 times. (**B**) Sclerotium state of WT, Δ*cda6*, and *cda6*^{C} strains on GMM solid medium, the magnification is 75 times. (**C**) Sclerotium number of WT, Δ*cda6*, and *cda6*^{C} strains on CM solid medium. (**D**) Sclerotium number of WT, Δ*cda6*, and *cda6*^{C} strains on GMM solid medium. (**E**) Relative expression levels of *nsdC* and *nsdD* genes in WT, Δ*cda6*, and *cda6*^{C} strains. ND indicates no detection, ** indicates that the significance level was $p < 0.01$, and *** indicates that the significance level was $p < 0.001$ ($n = 3$).

2.6. Effect of *cda6* Gene on Aflatoxin Biosynthesis

To explore whether the *cda6* gene is involved in regulating the biosynthesis of AFB_1, equal amounts of spore suspensions of WT, Δ*cda6*, and *cda6*^{C} strains were inoculated on YES liquid medium, and cultured in the dark at 29 °C for 6 days. Dichloromethane was added to extract the aflatoxin, and thin-layer chromatography (TLC) was used for detection. The results showed that the aflatoxin production of the Δ*cda6* strain significantly increased compared to that of WT and *cda6*^{C} strains (Figure 6A,B), indicating that the *cda6* gene negatively regulated aflatoxin synthesis. In order to further explore how the *cda6* gene regulates aflatoxin synthesis, the transcription levels of aflatoxin synthesis related genes *aflQ*, *aflR*, and *aflS* were detected by qPCR. The qPCR data showed that the relative expression levels of *aflQ*, *aflR*, and *aflS* in the Δ*cda6* strain were significantly higher than those in WT and *cda6*^{C} strains (Figure 6C), which further proved that the *cda6* gene may negatively regulate aflatoxin biosynthesis by affecting the transcript levels of genes related to aflatoxin synthesis.

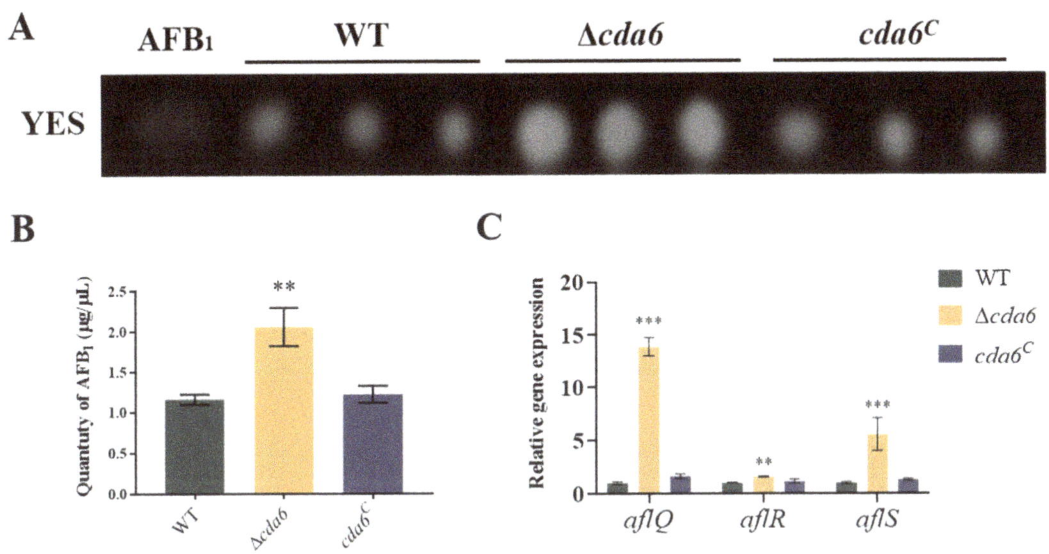

Figure 6. Effect of *cda6* gene on aflatoxin synthesis. (**A**) WT, Δ*cda6*, and *cda6*^{C} strains were cultured on YES liquid medium at 29 °C for 6 days, and aflatoxin AFB_1 was extracted and detected by thin-layer chromatography (TLC). (**B**) Semi-quantitative analysis of aflatoxin in the above strains. (**C**) Relative expression levels of aflatoxin synthesis-related genes *aflQ*, *aflR*, and *aflS* in the above strains. ** indicates that the significance level was $p < 0.01$, and *** indicates that the significance level was $p < 0.001$ ($n = 3$).

2.7. cda6 Gene Deletion Affects A. flavus Pathogenicity

To investigate the role played by the *cda6* gene in infesting peanut and maize seeds, we selected the same batch of peanuts and corn, then infected and cultured them at 29 °C for 7 days to quantitatively analyze conidia and aflatoxin. The data showed that the Δ*cda6*-contaminated peanut and maize produced significantly fewer spores compared to WT and *cda6*^{C} strains (Figure 7A,B,E,G). Aflatoxin was extracted and detected by TLC after infestation of peanut and maize, and the results of the semi-quantitative analysis showed that the aflatoxin production from Δ*cda6*-infected seeds was significantly higher than those from WT and *cda6*^{C} strains (Figure 7C,D,F,H), indicating that the *cda6* gene affects the pathogenicity of *A. flavus* to crop seeds peanut and maize.

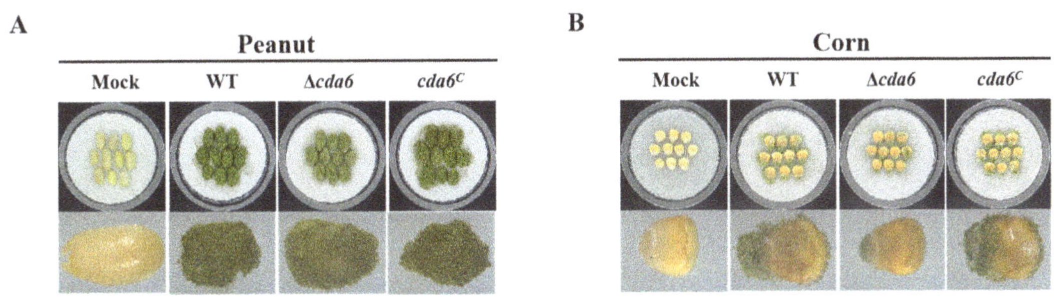

Figure 7. *Cont.*

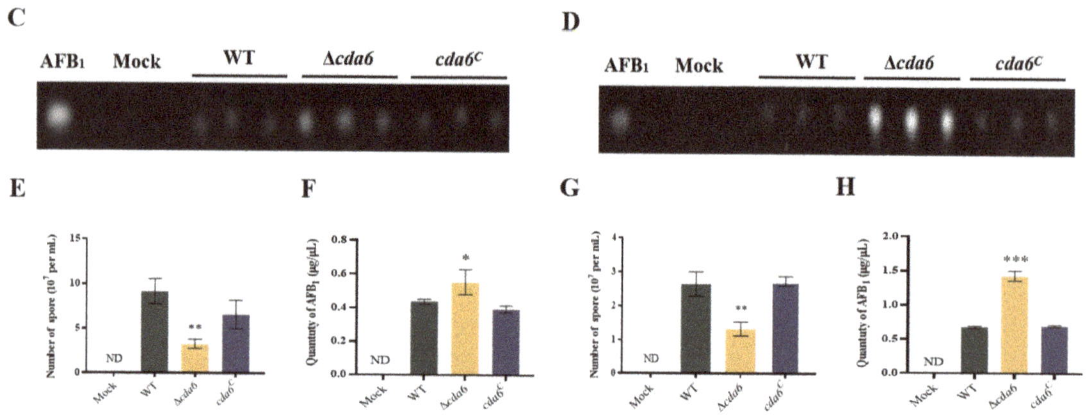

Figure 7. Effect of *cda6* gene on pathogenicity of *A. flavus* to peanut and corn. (**A**) Growth state of peanut seeds infected with WT, Δ*cda6*, and *cda6*^{C} strains cultured in the dark at 29 °C for 7 days. (**B**) Growth state of WT, Δ*cda6*, and *cda6*^{C} strains on corn kernels cultured in the dark at 29 °C for 7 days. (**C**) AFB_1 was extracted from peanut seeds and detected by TLC. (**D**) AFB_1 was extracted from corn and detected by TLC. (**E**) Quantification of conidia from the infected peanut seeds. (**F**) Semi-quantitative analysis of AFB_1 from peanut seeds. (**G**) Quantification of conidia from the infected corn seeds. (**H**) Semi-quantitative analysis of AFB_1 from corn. ND indicates not detection, * indicates that the difference was significant $p < 0.05$, ** indicates that the difference was significant $p < 0.01$, and *** indicates that the difference was significant $p < 0.001$ ($n = 3$).

2.8. The Role of cda6 Gene in the Stress Response

To investigate the potential involvement of the *cda6* gene in the stress response of *A. flavus*, various stress agents were introduced to the culture medium, including antifungal drugs, SDS, EtOH, and H_2O_2. It was observed that the deletion of the *cda6* gene did not result in significant differences in the inhibition rates compared to both the WT and *cda6*^{C} strains. This result suggested that the *cda6* gene does not play a discernible role in responding to cell membrane and cell wall stress (Figure S2A–F). Similarly, oxidative stress experimental results revealed no significant variation in the inhibition rate among the WT, Δ*cda6*, and *cda6*^{C} strains (Figure S2G,H), implying that the deletion of the *cda6* gene does not impact oxidative stress response in *A. flavus*.

2.9. Functional Studies of Structural Domains and Signal Peptide of CDA6 in A. flavus

The structural analysis of the CDA6 protein in related fungi revealed a conserved structure across different species (Figure 8A). To investigate the role of structural domains and signal peptide of CDA6 in *A. flavus*, domain-deficient strains were generated, including *cda6*$^{\Delta SP}$, *cda6*$^{\Delta NodB}$, *cda6*$^{\Delta GPI}$, cda6$^{\Delta CBD1}$, and *cda6*$^{\Delta CBD2}$. Growth observation revealed no difference in colony diameter between *cda6* knockout and domain deletion strains compared to WT (Figure 8B), indicating that the structural domains and signal peptides were not involved in regulating the radial growth of *A. flavus*. However, quantitative analysis revealed a significant increase in conidial production in the *cda6*$^{\Delta CBD1}$ and *cda6*$^{\Delta CBD2}$ strains compared to that in WT strain, while no significant difference was observed in conidial production in the *cda6*$^{\Delta SP}$, *cda6*$^{\Delta NodB}$, and *cda6*$^{\Delta GPI}$ strains (Figure 8D). These findings suggested that the structural domains CBD1 and CBD2 played a major role in the regulation of conidial formation in *A. flavus*. Previous experiments have shown that the deletion of the *cda6* gene causes *A. flavus* to be unable to form sclerotium normally. To investigate the specific effects of various domains and signal peptides of the CDA6 protein on sclerotium formation, WT and all domain deletion mutants were cultured on GMM medium. Compared to the WT, a significant decrease in sclerotial production was observed in the *cda6*$^{\Delta CBD1}$ and *cda6*$^{\Delta CBD2}$ strains (Figure 8C,E), while no significant changes were noted in other deletion mutants relative to the WT. These results suggested that the CBD1 and CBD2 domains of CDA6 play pivotal roles in regulating sclerotium formation in *A. flavus*.

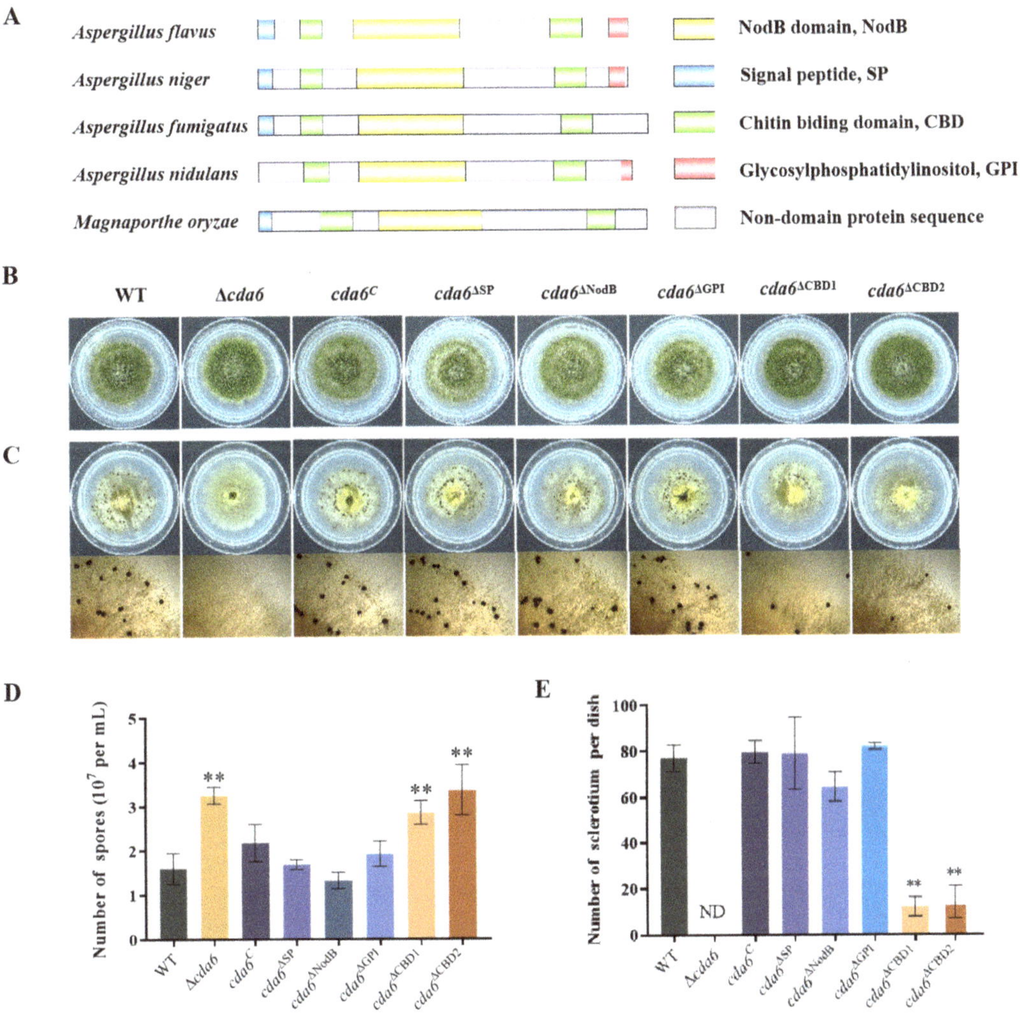

Figure 8. Effects of CDA6 domains and signal peptides on conidial and sclerotium formation in *A. flavus*. (**A**) Domain analysis of CDA homologous proteins in different species. (**B**) WT, Δ*cda6*, *cda6*^{C}, *cda6*$^{\Delta SP}$, *cda6*$^{\Delta NodB}$, *cda6*$^{\Delta GPI}$, *cda6*$^{\Delta CBD1}$, and *cda6*$^{\Delta CBD2}$ strains were inoculated on GMM medium to observe the growth status. (**C**) The sclerotial state of WT, Δ*cda6*, *cda6*^{C}, *cda6*$^{\Delta SP}$, *cda6*$^{\Delta NodB}$, *cda6*$^{\Delta GPI}$, *cda6*$^{\Delta CBD1}$, and *cda6*$^{\Delta CBD2}$ strains after being inoculated on GMM medium and cultured at 37 °C for 10 days. (**D**) The statistics of conidial number from above strains in GMM solid medium (**E**) Statistics of the sclerotium number in GMM solid medium of above strains. ND indicates not detection, and ** indicates that the significance level was $p < 0.01$ ($n = 3$).

3. Discussion

The fungal cell wall plays a crucial role in the protection, support, and maintenance of cellular morphology, with chitin being a key constituent. The cell wall is a dynamic cell structure, and chitin can be deacetylated to form chitosan, which is catalyzed by chitin deacetylase [25]. In various fungal species, the deacetylation modification mediated by chitin deacetylases is essential for maintaining cell wall morphology, structure, and integrity, particularly in pathogenic fungi [25]. Despite the significance of chitin deacetylation in fungal biology, the specific biological function of chitin deacetylase in *A. flavus* remains largely unexplored.

Previous studies have identified seven chitin deacetylase genes in *A. fumigatus* [19]. Homologs of the *A. fumigatus cda* genes were identified from the *A. flavus* genome and named as *cda1*, *cda2*, *cda3*, *cda4*, *cda5*, and *cda6*, respectively. According to the CAZy (Carbohydrate Active Enzymes database, http://www.cazy.org/, accessed on 15 November 2022), the CDA family is classified within the carbohydrate esterase 4 (CE4) family branch. Through the construction of the evolutionary tree, CDA family members were found to be highly conserved within the genus *Aspergillus*. Additionally, the domains of the CDA family exhibit a high degree of conservation across evolution. Notably, the NodB domain has been identified as functionally related to deacetylation activity [20]. This conservation suggested that the CDA family likely plays a fundamental role in fungal physiology and may have conserved functions across different species within the genus *Aspergillus*.

In this study, we aimed to investigate the function of the chitin deacetylase (CDA) family members in *A. flavus*. We constructed knockout strains for each member of the CDA family, including Δ*cda1*, Δ*cda2*, Δ*cda3*, Δ*cda4*, Δ*cda5*, and Δ*cda6*. We utilized the CFW staining method [26] and a chemical extraction method [27] to quantify the chitin content in the cell wall. Our findings revealed that deletion of individual *cda* genes did not significantly alter the cell wall chitin content or integrity in *A. flavus* (Figure 2), suggesting potential functional redundancy or the involvement of alternative chitin-modifying enzymes. Similar results were reported in *A. fumigatus*, where the deletion of a single *cda* gene or even all seven *cda* genes did not affect fungal cell wall synthesis or virulence. It was revealed that the deacetylation of cell wall chitin in *A. fumigatus* was not catalyzed by members of the CDA family, but probably mediated by other *A. fumigatus* CE4 family deacetylases [19]. The deletion of CDA family members in this study did not affect cell wall integrity in *A. flavus*, and similarly, deletion of the *A. fumigatus* chitin deacetylase *Afcod4* gene did not significantly affect *A. fumigatus* cell wall integrity [18]. In *S. cerevisiae*, two genes (*cda1* and *cda2*) are required for proper ascospore wall formation [14]. In contrast, the results in this study suggested that members of the CDA family may exhibit functional redundancy or may not be directly involved in the deacetylation of chitin; it is plausible that other CE4 family deacetylases could mediate chitin deacetylation in *A. flavus*.

A. flavus predominantly disseminates within the ecosystem through the generation of asexual spores, known as conidia. The process of conidiophore production is intricately governed by a network of regulators, including the central regulatory cascade consisting of *brlA-abaA-wetA* [28]. Our results revealed that the *cda6* gene exerts a negative regulatory influence on conidial formation in *A. flavus*. We observed that the *cda6* gene impacts the transcript levels of key spore formation regulatory genes such as *brlA* and *abaA* during spore development, thereby modulating conidial formation. However, the *cda6* gene did not affect the radial growth of *A. flavus*. However, the *cda6* gene did not affect the radial growth of *A. flavus* (Figure 3). In *A. fumigatus*, the mutant deficient for *cda7* can grow in normal medium, but exhibits reduced conidial production when grown on a medium with GlcNAc as the sole carbon source, indicating a crucial role of *cda7* in GlcNAc metabolism [19]. In the corn fungus *Ustilago maydis*, which contains seven *cda* genes, these genes play critical roles in various aspects of fungal biology, including virulence, adhesion, and plant defense activation [25]. In fission yeast *Schizosaccharomyces pombe*, *cda1* is required for spore formation, and knockdown of this gene results in the formation of a small number of abnormal spores [29]. Additionally, *A. flavus* forms sclerotia, dark brown and hard spherical bodies considered self-protective dormant structures to resist external environments [30]. Regulatory genes *nsdC* and *nsdD* are required for the production of sclerotia in *A. flavus* [31]. In this study, the *cda6* gene is required for the sclerotium formation in *A. flavus* (Figure 5). Meanwhile, the relative expression of *nsdC* and *nsdD* significantly decreased in the Δ*cda6* strain compared to the WT strain. Taken together, these findings suggested that the *cda6* gene plays a multifaceted role in the conidial and sclerotium formation of *A. flavus*.

Previous research has indicated that the distribution of CDA can vary among fungal species, suggesting species-specific localization patterns that may correlate with their respective functions. For instance, in *Absidia coerulea* and *Mucor rouxii*, CDA is distributed in the periplasmic space between the cell wall and the cell membrane [32,33]. Conversely, in *A. nidulans* and *Colletotrichum lindemuthianum*, CDA is secreted into extracellular matrix [34,35]. We hypothesize that the localization of CDA may be related to its specific function within each species. To investigate the localization of CDA6 protein in *A. flavus*, we constructed a cda6-mCherry strain. Our results revealed that CDA6 protein is localized in the mycelial cell wall and has the ability to bind to chitin (Figure 4). Interestingly, despite its presence in the cell wall, CDA6 protein did not exhibit any discernible effects on membrane or cell wall stress response. CDA6 protein in *A. flavus* is highly homologous to CBP1 (chitin-binding protein), a homologous to chitin deacetylase in *Magnaporthe grisea* [24]; we speculated that the function of CDA6 protein in *A. flavus* might be similar to that of CBP1. Considering its localization in the cell wall and its ability to bind to chitin, we hypothesized that CDA6 likely acts as a chitin-binding protein, potentially involved in cell wall integrity.

Aflatoxin, the secondary metabolite of *A. flavus*, includes stable chemical variants such as AFB_1, AFB_2, AFG_1, and AFG_2, with AFB_1 as one of the most toxic natural carcinogens [36]. Our findings revealed that *cda6* knockout increased AFB_1 production. Additionally, we examined the relative expression levels of two aflatoxin production regulatory genes, *aflR* and *aflS*, along with one structural gene, *aflQ*. We observed a significant increase in the expression levels of these genes (Figure 6), suggesting that the *cda6* gene may negatively regulate aflatoxin biosynthesis by affecting the transcript levels of genes related to AFB_1 synthesis. *A. flavus* is a pathogenic fungus capable of infesting a variety of oilseed crops, such as peanuts and maize, causing significant economic losses [37,38]. Our results showed that the knockout of the *cda6* gene resulted in reduced colonization ability of *A. flavus*. However, the AFB_1 produced by *A. flavus* in these two oil seed crop species was not reduced (Figure 7). This discrepancy may be attributed to the negative regulatory role of the *cda6* gene in aflatoxin synthesis. Despite the reduced colonization ability of the *cda6* knockout strain, the upregulation of aflatoxin production genes may compensate for the decreased fungal biomass, leading to similar levels of AFB_1 production in the oil seed crops.

When the external environments changed, *A. flavus* can constantly regulate its own growth and development mechanism to maintain survival. Different antifungal drugs have different mechanisms of action, such as Caspofungin which inhibits the cell wall β-1,3-glycoside synthase, which plays a role in breaking cell wall synthesis [39]. It was reported that the cell wall chitin content largely influenced the sensitivity of the strain to Caspofungin, with higher chitin content decreasing the sensitivity of the strain to Caspofungin [40,41]. Voriconazole inhibits ergosterol synthesis in fungal cell membranes and is used to treat severe fungal infections [42]. Amphotericin B binds to sterols on the fungal cell membrane, thus impairing the permeability of the cell membrane [39]. Our experiments showed that the deletion of *cda6* gene does not affect the sensitivity of *A. flavus* to Caspofungin, Voriconazole, or Amphotericin B, which is different from a previous report (Figure S2A,C). Additionally, we assessed the response of *A. flavus* to cell membrane stress agents such as SDS and EtOH. Our results showed no significant difference in the inhibition rate of the Δ*cda6* strain compared to the WT and $cda6^C$ strains (Figure S2E). Similarly, oxidative stress experimental results showed that there was no significant difference in the inhibition rate between WT, Δ*cda6*, and $cda6^C$ strains (Figure S2G). All above results showed that the *cda6* gene is not involved in the response to cell wall stress, cell membrane stress, or oxidative stress in *A. flavus*.

The signal peptide (SP) and structural domains NodB, GPI, CBD1, and CBD2 are conserved motifs in the CDA6 homologous protein. Previous experiments have demonstrated that the CDA6 protein is located in the cell wall. Initially, we presumed that the localization of this protein might be related to the function of the signal peptide, which guides the transfer and localization of newly synthesized proteins across the membrane. However, our experiments showed that the SP, GPI, and NodB domains are not involved in the regulation of growth and development, AFB_1 synthesis and pathogenicity of *A. flavus*. While the GPI-anchored structural domain is involved in cell wall integrity and intercellular interactions, and mislocalization of GPI-anchored proteins has a dramatic effect on cell wall composition and fungal virulence in *Candida albicans* [43], our findings suggested that the GPI domain does not play a role in regulating these aspects in *A. flavus*. On the other hand, the chitin-binding domain (CBD) of CDA6 is homologous to the chitin-binding domain of plant chitinase and plant lectins, such as wheat germ lectins (WGA) [44]. In this study, we found that structural domains CBD1 and CBD2 of CDA6 protein are primarily responsible for regulating *A. flavus* development (Figure 8). This suggested that two CBD domains of CDA6 protein play crucial roles in executing its function.

4. Conclusions

Our study provides comprehensive insights into the roles of the chitin deacetylase (CDA) family in *A. flavus*. Knockout of individual CDA family members did not impact the chitin content or cell wall integrity of *A. flavus*, suggesting potential functional redundancy within the CDA family or the involvement of alternative CE4 family deacetylases in chitin modification pathways. Moreover, our study identified CDA6 as a key protein localized within the mycelial cell wall. Deletion of the *cda6* gene led to a notable inhibition of conidia formation. Additionally, we observed a critical role for the *cda6* gene in sclerotium formation and its negative regulation of aflatoxin biosynthesis. Moreover, our findings indicated the involvement of the *cda6* gene in the pathogenicity of *A. flavus* towards oil seed crops such as maize and peanut. The structural domains CBD1 and CBD2 emerged as crucial components for the function of CDA6 in *A. flavus* development. In summary, our study showed the diverse functions of the CDA family within *A. flavus*, particularly highlighting the pivotal role of CDA6 protein in various aspects of development, aflatoxin synthesis, and pathogenicity. These findings significantly contributed to our understanding of the biological significance of CDA enzymes in filamentous fungi and offer valuable potential targets for future research and applications in agriculture and biotechnology.

5. Materials and Methods

5.1. Strains and Culture Conditions

To observe growth and conidial morphology, all strains were cultivated on solid agar media compositions including yeast extract–glucose–trace element (YGT), yeast extract–peptone–sucrose (CM), and glucose minimal medium (GMM) at 37 °C. The GMM medium composition comprises glucose, ammonium tartrate, potassium chloride (KCl), magnesium sulfate heptahydrate ($MgSO_4{\cdot}7H_2O$), and trace elements [45]. Colony diameters and conidial yields were quantified on the third day post-inoculation. Each experiment was repeated three times.

5.2. Sequence Analysis

BLASTP specifically compares a protein query sequence against a protein sequence database to find similar sequences. A BLASTP search was conducted through the National Center for Biotechnology Information (NCBI database, https://www.ncbi.nlm.nih.gov/, accessed on 15 November 2022), and the CDA homologous genes in *A. flavus* were found and named *cda1*, *cda2*, *cda3*, *cda4*, *cda5*, and *cda6*, respectively. Subsequently, BLASTP searches on the NCBI platform were performed to identify homologous protein sequences from diverse species using the conserved domain sequences of the CDA protein family from *A. flavus*. The downloaded protein sequence files were retrieved and imported into MEGA7.0 software for subsequent phylogenetic tree construction, employing the Neighbor-joining method to elucidate the evolutionary relationships among the identified sequences. The Smart (http://smart.embl-heidelberg.de/, accessed on 15 November 2022) and InterPro protein analysis web site (http://www.ebi.ac.uk/interpro/, accessed on 15 November 2022) were employed for comprehensive domain analysis, while DOG 2.0 was utilized for domain mapping. SignalP 6.0 (https://services.healthtech.dtu.dk/service.php?SignalP-6.0, accessed on 15 November 2022) was conducted to predict the presence of signal peptides within proteins. PredGPI (http://gpcr2.biocomp.unibo.it/predgpi/pred.htm, accessed on 15 November 2022) and TMHMM (https://services.healthtech.dtu.dk/service.php?TMHMM-2.0, accessed on 15 November 2022) were used to perform predictive analysis for the presence or absence of GPI-anchored proteins in proteins.

5.3. Construction of Mutant Strains

In this study, the knockout strains of CDA family members were constructed based on the principle of homologous recombination, and experiments were carried out according to a previously described method [46]. Screening marker (*pyrG*) and upstream and downstream homologous fragments were fused and then transferred into the CA14 protoplast of *A. flavus* [47,48]. CA14 was a *pyrG*-deficient strain, which could not grow normally in the medium without exogenous uridine and uracil. To construct the complemented strain, recombinant fragments of the *cda6* gene were introduced into the protoplasts of the *cda6* knockout mutant. Mutant strains with specific domain deletions (such as $cda6^{\Delta SP}$, $cda6^{\Delta NodB}$, $cda6^{\Delta GPI}$, $cda6^{\Delta CBD1}$, and $cda6^{\Delta CBD2}$) were also constructed using similar homologous recombination techniques as described above. By fusing mCherry red fluorescent protein as a tag with the target protein, the mCherry fluorescent label strain was obtained, allowing the localization of the protein molecule to be detected. All mutant strains were confirmed by PCR and sequencing.

5.4. Calcofluor White (CFW) Staining

An appropriate volume of *A. flavus* spore suspension was inoculated onto sterilized YGT solid medium. Following overnight incubation at 37 °C, mycelium was harvested, washed with phosphate-buffered saline (PBS) solution, and subsequently deprived of culture medium. Fluorescent whitening agent 28 (CFW) was introduced and allowed to incubate under light-protected conditions for 10 min [26]. After incubation, the mycelium was rinsed twice with PBS solution. Subsequently, mycelial samples were visualized under 405 nm excitation light using laser confocal microscopy. Each experiment was repeated three times.

5.5. Chitin Quantification

The method of chitin quantification has been reported in previous articles [27]. A standardized volume of *A. flavus* spore suspension was inoculated into YGT liquid medium and allowed to cultivate overnight. Following cultivation, the resultant mycelium was subjected to triple washing with sterile water to eliminate residual culture medium. Subsequently, the mycelium was desiccated, pulverized under liquid nitrogen, and precisely weighed (0.1–0.2 mg). The ground mycelium was then transferred into centrifuge tubes, to which 1 mL of cell wall lysate (comprising 50 mM Tris-HCl, 2% SDS, 0.3 M β-mercaptoethanol, and 1 mM EDTA) was added, followed by incubation at 100 °C for 15 min. To facilitate chitin quantification, an equivalent volume of concentrated hydrochloric acid (HCl) was introduced to acidify the lysate, which was subsequently subjected to drying at 100 °C for 17 h. The application of concentrated HCl serves to hydrolyze chitin and deacetylate the resulting para-acetaminosaccharide to aminoglucan. Chitin was analyzed by measuring the glucosamine released due to acidification using p-dimethylamino benzaldehyde as a chromogen. The filtered supernatant was treated with 4% acetyl acetone reagent, followed by heating at 100 °C for 10 min. Subsequently, 700 μL of 96% ethanol was added to the mixture, which was then supplemented with p-dimethylaminobenzaldehyde reagent and allowed to incubate at room temperature for 1 h. The absorbance of each solution at 520 nm was measured, and analytical data were recorded. Each experiment was repeated three times.

5.6. Extraction and Determination of Aflatoxin

A spore suspension with a concentration of 10^7 spores/mL was inoculated into yeast extract–sucrose agar medium (YES) and incubated at 29 °C for 6 days within a dark environment. Subsequently, the culture solution was subjected to equal-volume extraction with dichloromethane to isolate AFB_1 via shock extraction. An equivalent volume of dichloromethane was decanted from the lower layer, followed by evaporation to dryness. Dichloromethane was reintroduced to solubilize the toxin, and equal amounts of toxin samples were chromatographed on dry silica gel plates. Thin-layer chromatography was employed for the detection of the AFB_1, following the procedure detailed in a prior publication [49]. Each experiment was repeated three times.

5.7. qPCR Analysis

The experimental procedure for real-time fluorescent quantitative PCR (qPCR) analysis adhered to previously established protocols [50]. RNA extraction was performed using the Total RNA Extraction Kit (TIANMOBIO, Kannapolis, NC, USA), followed by reverse transcription into cDNA. Subsequently, qRT-PCR was conducted using Tip Green qPCR Super Mix (Transgen Biotechnology, Beijing, China) and a PikoReal 96 Real-Time Fluorescence Quantitative PCR System (Thermo Fisher Scientific, Waltham, MA, USA) in accordance with manufacturer's instructions. Primers targeting the gene of interest were employed for amplification, as delineated in Table S3.

5.8. Seed Infections Assay

Seed pathogenicity was assessed following established procedures [50]. Peanut and corn seeds exhibiting uniform morphology were selected for the assay. A blank control (Mock) devoid of infection was served as the control. Seeds underwent sequential washing with sterile water, 0.05% sodium hypochlorite solution, and 75% ethanol. Subsequently, seeds were inoculated with an equivalent concentration of *A. flavus* spores and incubated at 29 °C for 5 days under moist conditions. Following incubation, spores adhering to the seeds were enumerated and processed. Toxin extraction was achieved through thin-layer chromatography. Each experiment was repeated three times.

5.9. Stress Response Analysis

Stress response analysis was assessed following established procedures [51]. An appropriate volume of *A. flavus* spore was inoculated in YGT medium plates with different stress agents (0.1 ng/mL Caspofungin, 0.5 ng/mL Voriconazole, 0.1 ng/mL Amphotericin B, 300 μg/mL SDS, 2% EtOH, or 5 mM and 10 mM H_2O_2). All the plates were incubated at 37 °C for 5 days. The inhibition rate was calculated as follows: the diameter of the control group minus the diameter of the inhibition group, expressed as a percentage of the diameter of the control group. The stress response experiments were repeated three times.

5.10. Statistical Analysis

The data were presented as means (average values) ± SD (standard deviation) of at least three independent biological replicates. GraphPad Prism 8.0 was utilized for data analysis. Statistical significance was evaluated using a one-way analysis of variance (ANOVA). The least significant difference (LSD) test was used as a post hoc test following ANOVA. p-values less than 0.05 were considered statistically significant.

Supplementary Materials: The following supporting information can be downloaded at: https://www.mdpi.com/article/10.3390/toxins16050217/s1, Table S1: Information of *cda* homologous gene in *A. flavus*; Table S2: The strains used in this study; Table S3: Primers used for qPCR in this study. Figure S1: Construction of CDA mutants using homologous recombination; Figure S2: Role of *cda6* gene in response of *A. flavus* to stress. Ref. [52] is cited in supplementary materials

Author Contributions: Conceptualization, X.Z. and S.W.; methodology, X.Z.; software, M.W.; validation, X.Z., M.W. and G.L.; formal analysis, M.W.; investigation, M.W.; resources, X.Z.; data curation, X.Z.; writing—original draft preparation, S.W.; writing—review and editing, M.W.; visualization, G.L.; supervision, S.W.; project administration, S.W.; funding acquisition, S.W. All authors have read and agreed to the published version of the manuscript.

Funding: This research was funded by the National Natural Science Foundation of China (Grant No. 31972214).

Institutional Review Board Statement: Not applicable.

Informed Consent Statement: Not applicable.

Data Availability Statement: Data are contained within the article.

Acknowledgments: The authors thank Zhenhong Zhuang, Jun Yuan, Yu Wang, Kunzhi Jia, Xiuna Wang, and Xinyi Nie for their help in the experiments.

Conflicts of Interest: The authors declare no conflicts of interest.

References

1. Klich, M.A. *Aspergillus flavus*: The major producer of aflatoxin. *Mol. Plant Pathol.* **2007**, *8*, 713–722. [CrossRef]
2. Probst, C.; Schulthess, F.; Cotty, P.J. Impact of *Aspergillus* section *Flavi* community structure on the development of lethal levels of aflatoxins in Kenyan maize (*Zea mays*). *J. Appl. Microbiol.* **2010**, *108*, 600–610. [CrossRef] [PubMed]
3. Lewis, L.; Onsongo, M.; Njapau, H.; Schurz-Rogers, H.; Luber, G.; Kieszak, S.; Nyamongo, J.; Backer, L.; Dahiye, A.M.; Misore, A.; et al. Aflatoxin contamination of commercial maize products during an outbreak of acute aflatoxicosis in eastern and central Kenya. *Environ. Health Perspect.* **2005**, *113*, 1763–1767. [CrossRef] [PubMed]
4. Hedayati, M.T.; Pasqualotto, A.C.; Warn, P.A.; Bowyer, P.; Denning, D.W. *Aspergillus flavus*: Human pathogen, allergen and mycotoxin producer. *Microbiology* **2007**, *153*, 1677–1692. [CrossRef] [PubMed]
5. Rudramurthy, S.M.; Paul, R.A.; Chakrabarti, A.; Mouton, J.W.; Meis, J.F. Invasive *Aspergillosis* by *Aspergillus flavus*: Epidemiology, diagnosis, antifungal resistance, and management. *J. Fungi.* **2019**, *5*, 55. [CrossRef]
6. Caceres, I.; Khoury, A.A.; Khoury, R.E.; Lorber, S.; Oswald, I.P.; Khoury, A.E.; Atoui, A.; Puel, O.; Bailly, J.D. Aflatoxin biosynthesis and genetic regulation: A review. *Toxins* **2020**, *12*, 150. [CrossRef]
7. Latgé, J.P. The cell wall: A carbohydrate armour for the fungal cell. *Mol. Microbiol.* **2007**, *66*, 279–290. [CrossRef] [PubMed]
8. Bowman, S.M.; Free, S.J. The structure and synthesis of the fungal cell wall. *Bioessays* **2006**, *28*, 799–808. [CrossRef] [PubMed]
9. Gow, N.A.R.; Latge, J.; Munro, C.A. The fungal cell wall: Structure, biosynthesis, and function. *Microbiol. Spectr.* **2017**, *5*, 1128. [CrossRef]
10. Perlin, D.S. Cell wall-modifying antifungal drugs. *Curr. Top. Microbiol.* **2020**, *425*, 255–275. [CrossRef]
11. Munro, C.A.; Winter, K.; Buchan, A.; Henry, K.; Becker, J.M.; Brown, A.J.P.; Bulawa, C.E.; Gow, N.A.R. *Chs1* of *candida albicans* is an essential chitin synthase required for synthesis of the septum and for cell integrity. *Mol. Microbiol.* **2001**, *39*, 1414–1426. [CrossRef]
12. Tsigos, I.; Martinou, A.; Kafetzopoulos, D.; Bouriotis, V. Chitin deacetylases: New, versatile tools in biotechnology. *Trends Biotechnol.* **2000**, *18*, 305–312. [CrossRef]
13. Christodoulidou, A.; Briza, P.; Ellinger, A.; Bouriotis, V. Yeast ascospore wall assembly requires two chitin deacetylase isozymes. *FEBS Lett.* **1999**, *460*, 275–279. [CrossRef] [PubMed]
14. Christodoulidou, A.; Bouriotis, V.; Thireos, G. Two sporulation-specific chitin deacetylase-encoding genes are required for the ascospore wall rigidity of *Saccharomyces cerevisiae*. *J. Biol. Chem.* **1996**, *271*, 31420–31425. [CrossRef] [PubMed]
15. Baker, L.G.; Specht, C.A.; Donlin, M.J.; Lodge, J.K. Chitosan, the deacetylated form of chitin, is necessary for cell wall integrity in *Cryptococcus neoformans*. *Eukaryot. Cell.* **2007**, *6*, 855–867. [CrossRef]
16. Upadhya, R.; Baker, L.G.; Lam, W.C.; Specht, C.A.; Donlin, M.J.; Lodge, J.K. *Cryptococcus neoformans* Cda1 and its chitin deacetylase activity are required for fungal pathogenesis. *mBio* **2018**, *9*, e02087-18. [CrossRef]
17. Dai, M.D.; Wu, M.; Li, Y.; Su, Z.Z.; Lin, F.C.; Liu, X.H. The chitin deacetylase PoCda7 is involved in the pathogenicity of *Pyricularia oryzae*. *Microbiol. Res.* **2021**, *248*, 126749. [CrossRef] [PubMed]
18. Xie, M.; Zhao, X.; Lü, Y.; Jin, C. Chitin deacetylases Cod4 and Cod7 are involved in polar growth of *Aspergillus fumigatus*. *Microbiologyopen* **2020**, *9*, e00943. [CrossRef]
19. Mouyna, I.; Dellière, S.; Beauvais, A.; Gravelat, F.; Snarr, B.; Lehoux, M.; Zacharias, C.; Sun, Y.; de Jesus Carrion, S.; Pearlman, E.; et al. What are the functions of chitin deacetylases in *Aspergillus fumigatus*? *Front. Cell. Infect. Microbiol.* **2020**, *10*, 28. [CrossRef] [PubMed]
20. John, M.; Röhrig, H.; Schmidt, J.; Wieneke, U.; Schell, J. Rhizobium NodB protein involved in nodulation signal synthesis is a chitooligosaccharide deacetylase. *Proc. Natl. Acad. Sci. USA* **1993**, *90*, 625–629. [CrossRef]
21. Lee, S.; Kang, H.A.; Eyun, S.I. Evolutionary analysis and protein family classification of chitin deacetylases in *Cryptococcus neoformans*. *J. Microbiol.* **2020**, *58*, 805–811. [CrossRef]

22. Müller, G.A.; Tschöp, M.H.; Müller, T.D. Chip based sensing of the intercellular transfer of cell surface proteins: Regulation by the metabolic state. *Biomedicine* **2021**, *9*, 1452. [CrossRef]
23. Lima, S.L.; Colombo, A.L.; Junior, J.N.D.A. Fungal cell wall: Emerging antifungals and drug resistance. *Front. Microbiol.* **2019**, *10*, 2573. [CrossRef]
24. Kamakura, T.; Yamaguchi, S.; Saitoh, K.; Teraoka, T.; Yamaguchi, I. A novel gene, *CBP1*, encoding a putative extracellular chitin-binding protein, may play an important role in the hydrophobic surface sensing of *Magnaporthe grisea* during appressorium differentiation. *Mol. Plant Microbe Interact.* **2002**, *15*, 437–444. [CrossRef]
25. Rizzi, Y.S.; Happel, P.; Lenz, S.; Urs, M.J.; Bonin, M.; Cord-Landwehr, S.; Singh, R.; Moerschbacher, B.M.; Kahmann, R. Chitosan and chitin deacetylase activity are necessary for development and virulence of *Ustilago maydis*. *mBio* **2021**, *12*, e03419-20. [CrossRef] [PubMed]
26. Delgado, J.; Owens, R.A.; Doyle, S.; Asensio, M.A.; Núñez, F. Increased chitin biosynthesis contributes to the resistance of *Penicillium polonicum* against the antifungal protein PgAFP. *Appl. Microbiol. Biotechnol.* **2016**, *100*, 371–383. [CrossRef] [PubMed]
27. Yago, J.I.; Lin, C.H.; Chung, K.R. The SLT2 mitogen-activated protein kinase-mediated signalling pathway governs conidiation, morphogenesis, fungal virulence and production of toxin and melanin in the tangerine pathotype of *Alternaria alternata*. *Mol. Plant Pathol.* **2011**, *12*, 653–665. [CrossRef]
28. Yu, J.H. Regulation of development in *Aspergillus nidulans* and *Aspergillus fumigatus*. *Mycobiology* **2010**, *38*, 229–237. [CrossRef] [PubMed]
29. Matsuo, Y.; Tanaka, K.; Matsuda, H.; Kawamukai, M. *cda1+*, encoding chitin deacetylase is required for proper spore formation in *Schizosaccharomyces pombe*. *FEBS Lett.* **2005**, *579*, 2737–2743. [CrossRef]
30. Sun, X.; Liu, D.; Wang, Y.; Ma, A. Biogenesis of macrofungal sclerotia: Influencing factors and molecular mechanisms. *Appl. Microbiol. Biotechnol.* **2020**, *104*, 4227–4234. [CrossRef]
31. Dyer, P.S.; O'Gorman, C.M. Sexual development and cryptic sexuality in fungi: Insights from *Aspergillus* species. *FEMS Microbiol. Rev.* **2012**, *36*, 165–192. [CrossRef]
32. Gao, X.D.; Katsumoto, T.; Onodera, K. Purification and characterization of chitin deacetylase from *Absidia coerulea*. *J. Biochem.* **1995**, *117*, 257–263. [CrossRef]
33. Kafetzopoulos, D.; Martinou, A.; Bouriotis, V. Bioconversion of chitin to chitosan: Purification and characterization of chitin deacetylase from *Mucor rouxii*. *Proc. Natl. Acad. Sci. USA* **1993**, *90*, 2564–2568. [CrossRef] [PubMed]
34. Alfonso, C.; Nuero, O.M.; Santamaría, F.; Reyes, F. Purification of a heat-stable chitin deacetylase from *Aspergillus nidulans* and its role in cell wall degradation. *Curr. Microbiol.* **1995**, *30*, 49–54. [CrossRef]
35. Tokuyasu, K.; Mitsutomi, M.; Yamaguchi, I.; Hayashi, K.; Mori, Y. Recognition of chitooligosaccharides and their N-acetyl groups by putative subsites of chitin deacetylase from a deuteromycete, *Colletotrichum lindemuthianum*. *Biochemistry* **2000**, *39*, 8837–8843. [CrossRef]
36. Rushing, B.R.; Selim, M.I. Aflatoxin B1: A review on metabolism, toxicity, occurrence in food, occupational exposure, and detoxification methods. *Food Chem. Toxicol.* **2019**, *124*, 81–100. [CrossRef] [PubMed]
37. Amaike, S.; Keller, N.P. *Aspergillus flavus*. *Annu. Rev. Phytopathol.* **2011**, *49*, 107–133. [CrossRef] [PubMed]
38. Abbas, H.; Wilkinson, J.; Zablotowicz, R.; Accinelli, C.; Abel, C.; Bruns, H.; Weaver, M. Ecology of *Aspergillus flavus*, regulation of aflatoxin production, and management strategies to reduce aflatoxin contamination of corn. *Toxin Rev.* **2009**, *28*, 142–153. [CrossRef]
39. Prasad, R.; Shah, A.H.; Rawal, M.K. Antifungals: Mechanism of action and drug resistance. *Adv. Exp. Med. Biol.* **2016**, *892*, 327–349. [CrossRef]
40. Costa-de-Oliveira, S.; Silva, A.P.; Miranda, I.M.; Salvador, A.; Azevedo, M.M.; Munro, C.A.; Rodrigues, A.G.; Pina-Vaz, C. Determination of chitin content in fungal cell wall: An alternative flow cytometric method. *Cytom. A* **2013**, *83*, 324–328. [CrossRef]
41. Plaine, A.; Walker, L.; Da Costa, G.; Mora-Montes, H.M.; McKinnon, A.; Gow, N.A.; Gaillardin, C.; Munro, C.A.; Richard, M.L. Functional analysis of *Candida albicans* GPI-anchored proteins: Roles in cell wall integrity and caspofungin sensitivity. *Fungal Genet. Biol.* **2008**, *45*, 1404–1414. [CrossRef]
42. Greer, N.D. Voriconazole: The newest triazole antifungal agent. *Bayl. Univ. Med. Cent. Proc.* **2003**, *16*, 241–248. [CrossRef]
43. Richard, M.; Ibata-Ombetta, S.; Dromer, F.; Bordon-Pallier, F.; Jouault, T.; Gaillardin, C. Complete glycosylphosphatidylinositol anchors are required in *Candida albicans* for full morphogenesis, virulence and resistance to macrophages. *Mol. Microbiol.* **2002**, *44*, 841–853. [CrossRef] [PubMed]
44. Taira, T. Structures and antifungal activity of plant chitinases. *J. Appl. Glyosc.* **2010**, *57*, 167–176. [CrossRef]
45. Shimizu, K.; Keller, N.P. Genetic involvement of a cAMP-dependent protein kinase in a G protein signaling pathway regulating morphological and chemical transitions in *Aspergillus nidulans*. *Genetics* **2001**, *157*, 591–600. [CrossRef] [PubMed]
46. Yang, K.; Liu, Y.; Wang, S.; Wu, L.; Xie, R.; Lan, H.; Fasoyin, O.E.; Wang, Y.; Wang, S. Cyclase-associated protein cap with multiple domains contributes to mycotoxin biosynthesis and fungal virulence in *Aspergillus flavus*. *J. Agric. Food Chem.* **2019**, *67*, 4200–4213. [CrossRef] [PubMed]
47. Weidner, G.; d'Enfert, C.; Koch, A.; Mol, P.C.; Brakhage, A.A. Development of a homologous transformation system for the human pathogenic fungus *Aspergillus fumigatus* based on the *pyrG* gene encoding orotidine 5′-monophosphate decarboxylase. *Curr. Genet.* **1998**, *33*, 378–385. [CrossRef] [PubMed]

48. Skory, C.D.; Horng, J.S.; Pestka, J.J.; Linz, J.E. Transformation of *Aspergillus parasiticus* with a homologous gene (*pyrG*) involved in pyrimidine biosynthesis. *Appl. Environ. Microbiol.* **1990**, *56*, 3315–3320. [CrossRef]
49. Lan, H.; Sun, R.; Fan, K.; Yang, K.; Zhang, F.; Nie, X.; Wang, X.; Zhuang, Z.; Wang, S. The *Aspergillus flavus* histone acetyltransferase AflGcnE regulates morphogenesis, aflatoxin biosynthesis, and pathogenicity. *Front. Microbiol.* **2016**, *7*, 1324. [CrossRef]
50. Yang, K.; Liang, L.; Ran, F.; Liu, Y.; Li, Z.; Lan, H.; Gao, P.; Zhuang, Z.; Zhang, F.; Nie, X.; et al. The DmtA methyltransferase contributes to *Aspergillus flavus* conidiation, sclerotial production, aflatoxin biosynthesis and virulence. *Sci. Rep.* **2016**, *6*, 23259. [CrossRef]
51. Zhang, F.; Xu, G.; Geng, L.; Lu, X.; Yang, K.; Yuan, J.; Nie, X.; Zhuang, Z.; Wang, S. The stress response regulator AflSkn7 influences morphological development, stress response, and pathogenicity in the fungus *Aspergillus flavus*. *Toxins* **2016**, *8*, 202. [CrossRef]
52. Chang, P.K.; Scharfenstein, L.L.; Wei, Q.; Bhatnagar, D. Development and refinement of a high-efficiency gene-targeting system for *Aspergillus flavus*. *J. Microbiol. Methods* **2010**, *81*, 240–246. [CrossRef] [PubMed]

Disclaimer/Publisher's Note: The statements, opinions and data contained in all publications are solely those of the individual author(s) and contributor(s) and not of MDPI and/or the editor(s). MDPI and/or the editor(s) disclaim responsibility for any injury to people or property resulting from any ideas, methods, instructions or products referred to in the content.

toxins

MDPI

Article

The Effects of Aflatoxin B_1 on Liver Cholestasis and Its Nutritional Regulation in Ducks

Aimei Yu [1,†], Huanbin Wang [1,†], Qianhui Cheng [1], Shahid Ali Rajput [2] and Desheng Qi [1,*]

1 Department of Animal Nutrition and Feed Science, College of Animal Science and Technology, Huazhong Agricultural University, Wuhan 430070, China; yam@webmail.hzau.edu.cn (A.Y.); whbin@webmail.hzau.edu.cn (H.W.); cqh@webmail.hzau.edu.cn (Q.C.)

2 Faculty of Veterinary and Animal Science, Muhammad Nawaz Shareef University of Agriculture Multan, Multan 60000, Pakistan; shahid.ali@mnsuam.edu.pk

* Correspondence: qds@mail.hzau.edu.cn; Tel.: +86-27-8728-1793

† These authors contributed equally to this work.

Abstract: The aim of this study was to investigate the effects of aflatoxin B_1 (AFB_1) on cholestasis in duck liver and its nutritional regulation. Three hundred sixty 1-day-old ducks were randomly divided into six groups and fed for 4 weeks. The control group was fed a basic diet, while the experimental group diet contained 90 μg/kg of AFB_1. Cholestyramine, atorvastatin calcium, taurine, and emodin were added to the diets of four experimental groups. The results show that in the AFB_1 group, the growth properties, total bile acid (TBA) serum levels and total superoxide dismutase (T-SOD), glutathione peroxidase (GSH-Px), and glutathione (GSH) liver levels decreased, while the malondialdehyde (MDA) and TBA liver levels increased ($p < 0.05$). Moreover, AFB_1 caused cholestasis. Cholestyramine, atorvastatin calcium, taurine, and emodin could reduce the TBA serum and liver levels ($p < 0.05$), alleviating the symptoms of cholestasis. The qPCR results show that AFB_1 upregulated *cytochrome P450 family 7 subfamily A member 1* (*CYP7A1*) and *cytochrome P450 family 8 subfamily B member 1* (*CYP8B1*) gene expression and downregulated *ATP binding cassette subfamily B member 11* (*BSEP*) gene expression in the liver, and taurine and emodin downregulated *CYP7A1* and *CYP8B1* gene expression ($p < 0.05$). In summary, AFB_1 negatively affects health and alters the expression of genes related to liver bile acid metabolism, leading to cholestasis. Cholestyramine, atorvastatin calcium, taurine, and emodin can alleviate AFB_1-induced cholestasis.

Keywords: AFB_1; duck; cholestasis; cholestyramine; atorvastatin calcium; taurine; emodin; bile acid

Key Contribution: This study investigated the causes and nutritional regulation of AFB_1-induced liver cholestasis in ducks.

1. Introduction

Fungal toxins have caused significant harm to human health and livestock production [1]. Aflatoxin is a type of fungal toxin mainly produced by *Aspergillus flavus* [2]. It is widely present in various food crops and may cause contamination during crop maturation, harvesting, and storage [3]. There are various types of aflatoxin, among which is the B_1 type (AFB_1) has the most potent toxicity [4]. AFB_1 poisoning can lead to a decrease in animal growth and negatively affect health [5,6]. The liver is the main target organ for AFB_1 in animals, and ducklings are very sensitive to AFB_1 [5,6]. At present, there are no reports of AFB_1 causing liver cholestasis in ducklings.

Bile is synthesized in the liver; its main component is bile acids [7]. The liver synthesizes bile acids from cholesterol as a raw material [8]. There are two pathways for the synthesis of bile acids, namely the classic pathway mediated by cholesterol-7-α-hydroxylase (*CYP7A1*) and the alternative pathway mediated by *sterol-27-hydroxylase* (*CYP27A1*) [9]. The synthesized bile acids are discharged into the intestine through the bile duct, a process that is regulated by the bile salt efflux pump (BESP) [10]. After bile acids enter the gut, most of them are reabsorbed in the ileum and enter the bloodstream, while a small portion is excreted in the feces [9]. The reabsorbed bile acids are circulated through the bloodstream and returned to the liver for reuse [9].

Excessive synthesis of bile acids in the liver or obstruction of bile acid excretion can cause cholestasis [11]. Cholestyramine promotes the excretion of bile acids via feces, which can indirectly stimulate the efflux of bile acids from the liver [12]. Previously, it was reported that cholestyramine can reduce the concentration of bile acids and chenodeoxycholic acid in the serum of patients with cholestasis [13]. Atorvastatin calcium can inhibit liver cholesterol synthesis, reduce the raw materials available for bile acid synthesis, and indirectly inhibit bile acid synthesis [14]. Li et al. reported that atorvastatin calcium could further increase TBA levels in the ileum and feces of atherosclerosis model mice [15]. Taurine and emodin have antioxidant effects and may relieve AFB_1 poisoning [16,17]. In addition, taurine can promote the efflux of bile acids from the liver, and emodin inhibits the production of bile acids [18,19]. Moreover, taurine deficiency can lead to a decrease in the content of bile acids in cat bile [20]. A recent study revealed that emodin can upregulate the expression of *BSEP* genes and proteins, thereby promoting bile excretion [21]. At present, there are no reports on whether these four substances can alleviate the effects of AFB_1-induced liver cholestasis in ducklings.

This study explored the biological mechanism and nutritional regulation of AFB_1-induced liver cholestasis by evaluating the effects of AFB_1 on production performance, antioxidant indicators, total bile acid content in the liver, liver cholestasis, and the expression of genes related to bile acid metabolism in the liver of ducks. In addition, it investigated the ability of cholestyramine, atorvastatin calcium, taurine, and emodin to alleviate AFB_1 poisoning and cholestasis. The findings from this study can provide a reference for the prevention and treatment of AFB_1 poisoning in livestock and poultry.

2. Results

2.1. Production Performance and Organ Weight Changes in Ducks

The effects of the treatments on the production performance and organ weights of ducks are shown in Table 1. Compared with the control group, AFB_1 significantly reduced the body weight (BW), average daily weight gain (ADWG), and average daily feed intake (ADFI) of ducks at 1–2 and 3–4 weeks and increased the liver and spleen weights at week 4 ($p < 0.05$). Compared with the AFB_1 group, cholestyramine, taurine, and emodin significantly increased the BW and ADWG of ducks at 1–2 weeks ($p < 0.05$). At 3–4 weeks, compared with the AFB_1 group, cholestyramine and taurine significantly increased the BW and ADWG of ducks, while emodin significantly increased the BW of ducks ($p < 0.05$). In addition, cholestyramine, taurine, and emodin significantly reduced the liver weight ($p < 0.05$). Meanwhile, compared with the AFB_1 group, taurine and emodin significantly reduced the spleen weight ($p < 0.05$). Atorvastatin calcium did not significantly improve the production performance of ducks, but it did lessen the liver weight in week 4 ($p < 0.05$).

2.2. Changes in Liver Antioxidant Indicators

The effects of the treatments on the antioxidant indicators in duck liver are shown in Table 2. Compared with the control group, AFB_1 significantly increased the MDA content and decreased the T-SOD, GSH-Px, and GSH contents ($p < 0.05$). Compared with the AFB_1 group, in week 2, the MDA content in the AFB_1 + taurine and AFB_1 + emodin groups showed a downward trend, while the T-SOD, GSH-Px, and GSH contents showed an upward trend. In week 4, taurine significantly reduced the MDA content and increased the T-SOD, GSH-Px, and GSH contents ($p < 0.05$). Emodin significantly reduced the MDA content, increased the GSH content ($p < 0.05$), and showed an upward trend in the T-SOD and GSH-Px contents.

Table 1. Production performance and weight of the liver and spleen [1].

Items [2]	Control	AFB$_1$	AFB$_1$ + Cho	AFB$_1$ + Ato	AFB$_1$ + Tau	AFB$_1$ + Emo
IW (g)	54.26 ± 0.69	54.27 ± 0.60	54.18 ± 0.75	54.15 ± 0.57	54.24 ± 0.67	54.14 ± 0.60
			1–2 weeks			
BW (g)	438.86 ± 17.74 ^{a}	351.30 ± 8.70 ^{c}	392.52 ± 17.87 ^{b}	363.97 ± 8.86 ^{c}	404.31 ± 23.59 ^{b}	388.92 ± 10.86 ^{b}
ADWG (g)	27.47 ± 1.31 ^{a}	21.22 ± 0.63 ^{c}	24.17 ± 1.28 ^{b}	22.13 ± 0.62 ^{c}	25.01 ± 1.66 ^{b}	23.91 ± 0.78 ^{b}
ADFI (g)	50.71 ± 3.63 ^{a}	42.50 ± 1.43 ^{b}	46.67 ± 5.13 ^{ab}	41.53 ± 5.42 ^{b}	46.81 ± 5.71 ^{ab}	46.42 ± 5.25 ^{ab}
FCR	1.85 ± 0.12	2.00 ± 0.07	1.93 ± 0.18	1.88 ± 0.24	1.87 ± 0.17	1.94 ± 0.21
			3–4 weeks			
BW(g)	1365.58 ± 56.40 ^{a}	1082.15 ± 76.64 ^{d}	1220.16 ± 65.88 ^{b}	1113.13 ± 88.91 ^{cd}	1249.74 ± 89.56 ^{b}	1191.24 ± 65.51 ^{bc}
ADWG (g)	66.19 ± 4.59 ^{a}	52.20 ± 5.36 ^{d}	59.12 ± 5.18 ^{bc}	53.51 ± 6.54 ^{cd}	60.39 ± 5.33 ^{ab}	57.31 ± 4.64 ^{bcd}
ADFI (g)	145.47 ± 9.15 ^{a}	121.16 ± 21.29 ^{bc}	127.98 ± 8.87 ^{b}	112.00 ± 7.37 ^{c}	131.15 ± 11.21 ^{ab}	132.38 ± 11.80 ^{ab}
FCR	2.20 ± 0.06	2.33 ± 0.40	2.17 ± 0.06	2.12 ± 0.29	2.17 ± 0.10	2.18 ± 0.27
			Week 4			
Liver (g)	32.42 ± 4.79 ^{b}	43.53 ± 10.18 ^{a}	34.46 ± 5.46 ^{b}	31.41 ± 3.73 ^{b}	32.08 ± 4.45 ^{b}	27.93 ± 7.31 ^{b}
Spleen (g)	1.32 ± 0.35 ^{b}	2.09 ± 0.84 ^{a}	1.66 ± 0.26 ^{ab}	2.09 ± 0.78 ^{a}	1.36 ± 0.25 ^{b}	1.29 ± 0.39 ^{b}

[1] The values are expressed as the mean ± standard deviation (n = 6). Means with different superscripts differ significantly ($p < 0.05$). "AFB$_1$ + Cho" = AFB$_1$ + cholestyramine; "AFB$_1$ + Ato" = AFB$_1$ + atorvastatin calcium; "AFB$_1$ + Tau" = AFB$_1$ + taurine; "AFB$_1$ + Emo" = AFB$_1$ + emodin. [2] "IW" = initial weight; "BW" = body weight; "ADWG" = average daily weight gain; "ADFI" = average daily feed intake; "FCR" = feed conversion rate.

Table 2. Antioxidant indicators in the liver [1].

Items	Control	AFB$_1$	AFB$_1$ + Cho	AFB$_1$ + Ato	AFB$_1$ + Tau	AFB$_1$ + Emo
			Week 2			
T-SOD (U/mg protein)	100.41 ± 16.65 ^{a}	77.94 ± 5.65 ^{b}	74.36 ± 6.28 ^{b}	53.85 ± 7.39 ^{c}	84.78 ± 11.26 ^{b}	88.88 ± 19.70 ^{ab}
MDA (nmol/ mg protein)	1.90 ± 0.22 ^{b}	2.77 ± 0.60 ^{a}	2.41 ± 0.48 ^{ab}	2.41 ± 0.35 ^{ab}	2.26 ± 0.47 ^{ab}	1.94 ± 0.38 ^{b}
GSH-Px (U/mg protein)	57.34 ± 7.14 ^{a}	42.51 ± 7.65 ^{b}	46.00 ± 5.77 ^{b}	42.17 ± 6.08 ^{b}	50.69 ± 7.13 ^{ab}	51.08 ± 9.18 ^{ab}
GSH (U/mg protein)	37.33 ± 2.85 ^{a}	24.84 ± 4.09 ^{c}	30.12 ± 4.47 ^{bc}	33.06 ± 9.83 ^{ab}	26.95 ± 3.29 ^{bc}	28.13 ± 4.00 ^{bc}
			Week 4			
T-SOD (U/mg protein)	119.56 ± 6.06 ^{a}	90.70 ± 12.30 ^{b}	102.51 ± 13.70 ^{ab}	105.33 ± 17.25 ^{ab}	119.47 ± 12.21 ^{a}	98.88 ± 16.05 ^{b}
MDA (nmol/ mg protein)	1.70 ± 0.10 ^{b}	2.46 ± 0.48 ^{a}	1.96 ± 0.47 ^{b}	1.73 ± 0.46 ^{b}	1.70 ± 0.21 ^{b}	1.73 ± 0.58 ^{b}
GSH-Px (U/mg protein)	69.42 ± 10.79 ^{a}	44.80 ± 5.52 ^{c}	51.30 ± 17.29 ^{bc}	51.06 ± 5.60 ^{bc}	63.10 ± 13.37 ^{ab}	57.34 ± 7.63 ^{abc}
GSH (U/mg protein)	39.46 ± 2.03 ^{a}	29.68 ± 5.19 ^{b}	34.76 ± 4.28 ^{ab}	36.52 ± 6.21 ^{a}	36.97 ± 1.98 ^{a}	39.87 ± 5.96 ^{a}

[1] The values are expressed as the mean ± standard deviation (n = 6). Means with different superscripts differ significantly ($p < 0.05$). "AFB$_1$ + Cho" = AFB$_1$ + cholestyramine; "AFB$_1$ + Ato" = AFB$_1$ + atorvastatin calcium; "AFB$_1$ + Tau" = AFB$_1$ + taurine; "AFB$_1$ + Emo" = AFB$_1$ + emodin.

2.3. Histological Analysis of the Liver

The effects of the treatments on the morphology and steatosis of duck liver tissue are shown in Figure 1. Hematoxylin and eosin (H&E) staining showed that the liver morphology of the control group was normal. In contrast, the liver cells of the AFB$_1$ group showed swelling and an extremely uneven distribution of nuclei. Compared with the AFB$_1$ group, the liver cell swelling in the AFB$_1$ + taurine and AFB$_1$ + emodin groups was evidently decreased. Additionally, oil red O staining showed no abnormalities in the liver of the control group, while the liver of the AFB$_1$ group showed significant steatosis, denoted by large red areas of lipid staining. Strikingly, taurine and emodin treatment alleviated steatosis in the liver of ducks induced by AFB_1, with a reduction in the red area in the field of view. Compared with the AFB_1 group, there was no significant improvement in the liver of the AFB_1 + cholestyramine and AFB_1 + atorvastatin calcium groups.

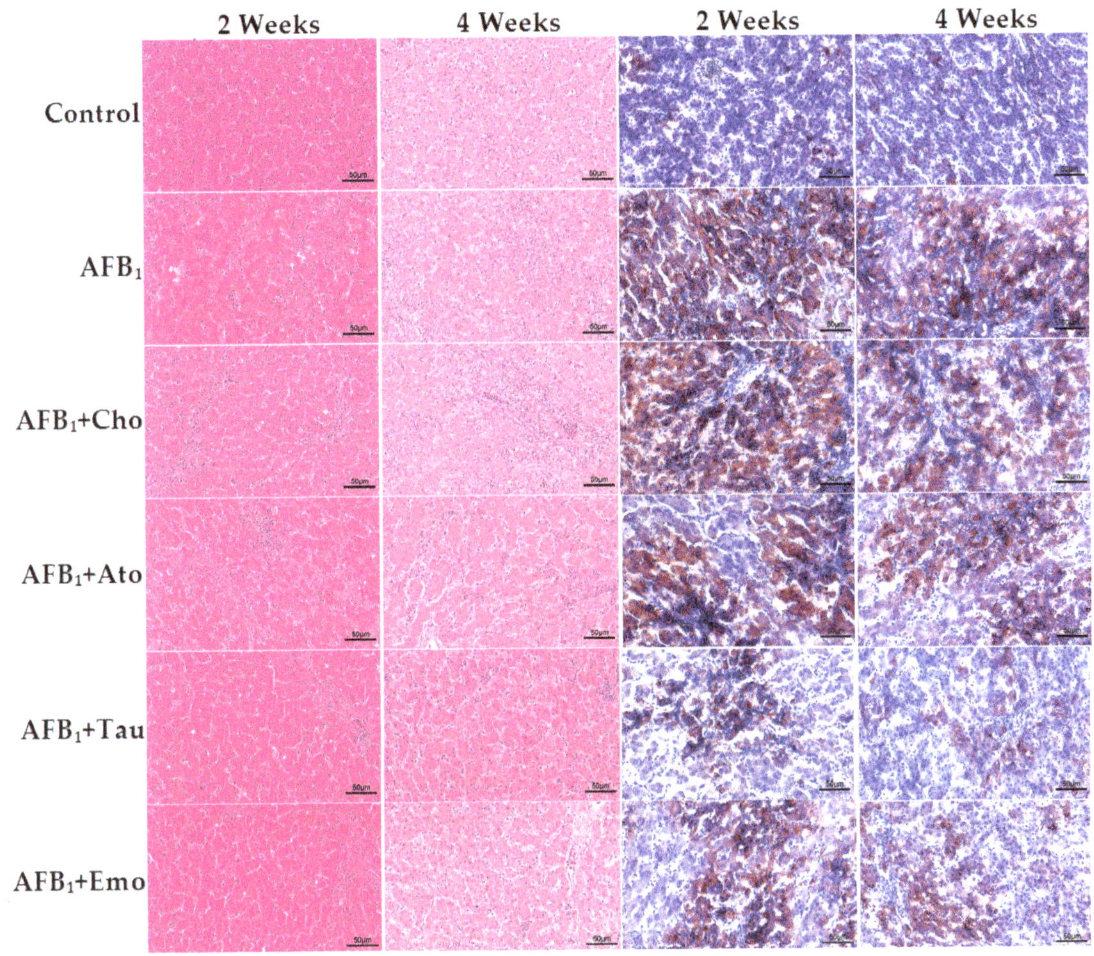

Figure 1. Hematoxylin and eosin (H&E) and oil red O staining of liver tissue (200× magnification). "AFB_1 + Cho" = AFB_1 + cholestyramine; "AFB_1 + Ato" = AFB_1 + atorvastatin calcium; "AFB_1 + Tau" = AFB_1 + taurine; "AFB_1 + Emo" = AFB_1 + emodin.

2.4. Changes in Biochemical Indicators in Duck Serum

The effects of the treatments on the biochemical indicators in duck serum are shown in Table 3. Compared with the control group, AFB_1 significantly increased the aspartate transaminase (AST), alanine transaminase (ALT), TBA, and total bilirubin (TBiL) levels and significantly reduced the total protein (TP) and albumin (ALB) levels ($p < 0.05$). In week 2, compared with the AFB_1 group, taurine and emodin treatments significantly reduced the AST and TBA levels ($p < 0.05$), and the TBiL levels also showed a downward trend. In addition, cholestyramine significantly reduced the TBA levels ($p < 0.05$). In week 4, compared with the AFB_1 group, cholestyramine and atorvastatin calcium significantly reduced the TBA and TBiL levels ($p < 0.05$). At the same time, taurine significantly increased the TP and ALB levels ($p < 0.05$), while the AST, ALT, TBA, and TBiL levels showed a descending trend. Emodin significantly reduced the TBA levels ($p < 0.05$), while the AST, ALT, and TBiL levels showed a decreasing trend.

Table 3. Biochemical indicators in the serum [1].

Items	Control	AFB_1	AFB_1 + Cho	AFB_1 + Ato	AFB_1 + Tau	AFB_1 + Emo
			Week 2			
AST (U/L)	33.25 ± 3.41 ^{d}	59.17 ± 7.01 ^{a}	53.58 ± 11.11 ^{ab}	46.22 ± 12.31 ^{bc}	43.37 ± 7.59 ^{c}	32.95 ± 2.52 ^{d}
ALT (U/L)	46.05 ± 8.83	53.77 ± 5.34	46.77 ± 9.49	49.95 ± 4.85	52.18 ± 9.88	43.82 ± 6.41
TP (g/L)	31.32 ± 2.80 ^{a}	20.37 ± 1.58 ^{c}	23.77 ± 4.86 ^{bc}	21.38 ± 1.73 ^{c}	23.18 ± 2.78 ^{bc}	25.13 ± 1.91 ^{b}
ALB (g/L)	9.88 ± 1.30 ^{a}	5.77 ± 1.05 ^{c}	6.27 ± 0.86 ^{bc}	6.13 ± 0.98 ^{bc}	7.35 ± 0.94 ^{b}	6.93 ± 1.17 ^{bc}
TBA (μmol/L)	12.65 ± 1.27 ^{c}	20.96 ± 3.64 ^{a}	12.66 ± 1.96 ^{c}	18.64 ± 1.22 ^{ab}	17.77 ± 1.75 ^{b}	16.78 ± 2.95 ^{b}
TbiL (μmol/L)	5.83 ± 1.15 ^{b}	7.43 ± 1.09 ^{a}	6.29 ± 1.01 ^{ab}	6.65 ± 0.87 ^{ab}	6.98 ± 1.58 ^{ab}	6.75 ± 1.27 ^{ab}
			Week 4			
AST (U/L)	33.12 ± 7.31 ^{b}	43.78 ± 8.73 ^{a}	38.48 ± 12.36 ^{ab}	38.90 ± 3.61 ^{ab}	37.85 ± 6.81 ^{ab}	39.07 ± 4.51 ^{ab}
ALT (U/L)	35.38 ± 7.74 ^{b}	45.58 ± 8.58 ^{a}	44.23 ± 6.16 ^{ab}	43.35 ± 6.59 ^{ab}	39.45 ± 4.56 ^{ab}	37.10 ± 7.19 ^{ab}
TP (g/L)	28.62 ± 1.58 ^{a}	22.98 ± 2.81 ^{c}	22.82 ± 1.61 ^{c}	25.40 ± 2.25 ^{bc}	27.67 ± 2.67 ^{ab}	24.83 ± 2.1 ^{bc}
ALB (g/L)	12.70 ± 0.89 ^{a}	9.70 ± 0.98 ^{b}	9.65 ± 0.94 ^{b}	9.82 ± 1.07 ^{b}	11.90 ± 1.68 ^{a}	11.10 ± 2.90 ^{ab}
TBA (μmol/L)	18.58 ± 2.04 ^{c}	29.03 ± 6.78 ^{a}	18.22 ± 5.81 ^{c}	21.63 ± 1.99 ^{bc}	24.92 ± 1.60 ^{ab}	22.83 ± 1.27 ^{bc}
TbiL (μmol/L)	5.88 ± 1.43 ^{bc}	9.99 ± 0.90 ^{a}	4.64 ± 1.43 ^{c}	6.78 ± 1.61 ^{b}	8.68 ± 0.99 ^{a}	8.59 ± 2.01 ^{a}

[1] The values are expressed as the mean ± standard deviation (n = 6). Means with different superscripts differ significantly ($p < 0.05$). "AFB_1 + Cho" = AFB_1 + cholestyramine; "AFB_1 + Ato" = AFB_1 + atorvastatin calcium; "AFB_1 + Tau" = AFB_1 + taurine; "AFB_1 + Emo" = AFB_1 + emodin.

2.5. The Degree of Liver Cholestasis in the Treatment Groups

The effect of the treatments on the cholestasis of duck liver in week 4 is shown in Figure 2. The control group presented a normal liver color, indicating no cholestasis. However, the liver of the AFB_1 group had a greenish color, indicative of cholestasis. The addition of cholestyramine, atorvastatin calcium, taurine, and emodin to the diet of the experimental group alleviated the liver cholestasis of duck to varying degrees induced by AFB_1.

2.6. Changes in the Total Bile Acid Content in the Liver

The effect of the treatments on the total bile acid content in the liver of the ducks is shown in Figure 3. Compared with the control group, AFB_1 significantly increased the TBA content in weeks 2 and 4 ($p < 0.05$). Compared with the AFB_1 group, cholestyramine, atorvastatin calcium, taurine, and emodin treatments significantly reduced the TBA content in the liver of ducks (2 and 4 weeks).

2.7. Changes in the Expression Levels of Liver-Bile-Acid-Metabolism-Related Genes

The effects of the treatments on the expression levels of genes related to liver bile acid metabolism are shown in Figure 4. AFB_1 significantly upregulated the expression of *CYP7A1* and *CYP8B1* in the liver and downregulated the expression of *BSEP*, *nuclear receptor subfamily 0 group B member 2* (*SHP*), and *solute carrier organic anion transporter family member 1A2* (*OATP*) ($p < 0.05$). Compared with the AFB_1 group, the expression of *CYP7A1* and *CYP8B1* in the AFB_1 + taurine and AFB_1 + emodin groups was significantly downregulated ($p < 0.05$), while the expression of *CYP7A1* and *CYP8B1* in the AFB_1 + cholestyramine and AFB_1 + atorvastatin calcium groups showed a downward trend. Compared with the AFB_1 group, the AFB_1 + atorvastatin calcium group showed significant upregulation of *BSEP* and *OATP* ($p < 0.05$), while the AFB_1 + emodin group showed a trend for increased *BSEP* expression. No significant gene expression changed in the AFB_1 + cholestyramine group.

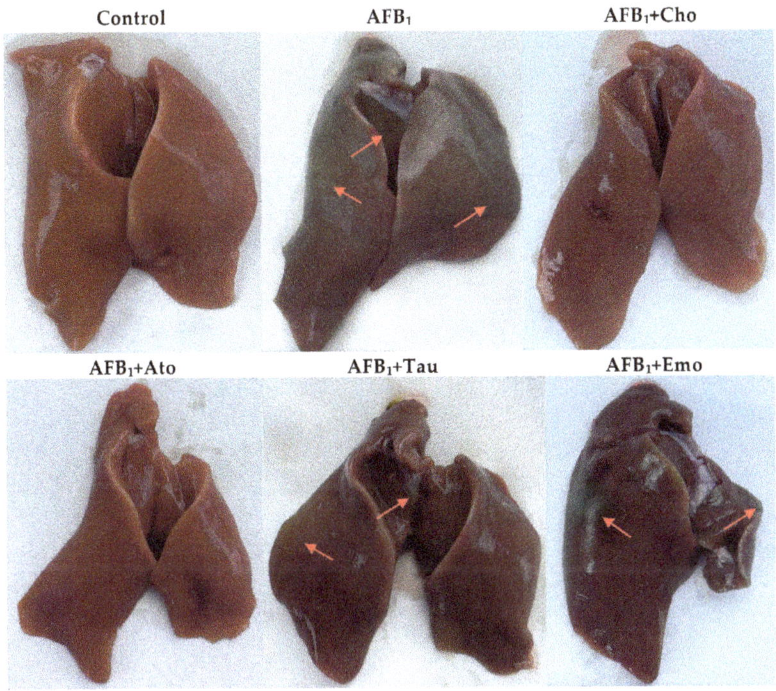

Figure 2. Liver anatomy map. Red arrow: location of cholestasis. "AFB_1 + Cho" = AFB_1 + cholestyramine; "AFB_1 + Ato" = AFB_1 + atorvastatin calcium; "AFB_1 + Tau" = AFB_1 + taurine; "AFB_1 + Emo" = AFB_1 + emodin.

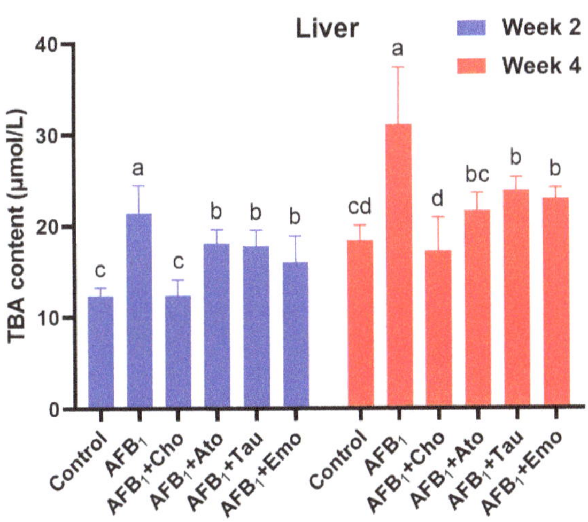

Figure 3. The total bile acid (TBA) content in the liver (n = 6). [a–d] Different lowercase letters indicate significant differences between the groups ($p < 0.05$). "AFB_1 + Cho" = AFB_1 + cholestyramine; "AFB_1 + Ato" = AFB_1 + atorvastatin calcium; "AFB_1 + Tau" = AFB_1 + taurine; "AFB_1 + Emo" = AFB_1 + emodin.

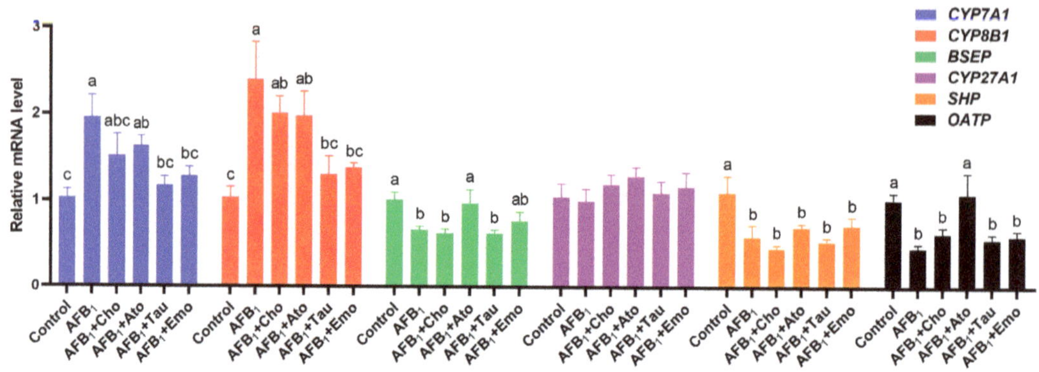

Figure 4. Changes in the expression of genes related to bile acid metabolism (n = 6). [a–c] Different lowercase letters indicate significant differences between the groups ($p < 0.05$). "AFB_1 + Cho" = AFB_1 + cholestyramine; "AFB_1 + Ato" = AFB_1 + atorvastatin calcium; "AFB_1 + Tau" = AFB_1 + taurine; "AFB_1 + Emo" = AFB_1 + emodin.

3. Discussion

Numerous studies have shown that AFB1 can cause growth retardation and liver damage in animals, with symptoms of poisoning mainly manifested as liver toxicity and oxidative stress [5,22,23]. Gao et al. [22] reported that AFB1 increased the levels of transaminases in chicken serum, decreased the antioxidant capacity of chickens, and caused swelling and necrosis of chicken liver cells. Wang et al. [5] reported that AFB1 not only leads to a decrease in growth performance and antioxidant levels in ducklings but also causes cellular swelling, steatosis, and nuclear shrinkage in the liver of ducklings. In addition, Liu et al. [23] reported that AFB1 reduced body weight and T-AOC and GSH-Px liver levels in mice. The present study showed that AFB1 can lead to a decrease in duck production performance and liver antioxidant levels, as well as symptoms such as liver cell swelling and steatosis, which is consistent with previous research results. AFB1 in feed poses a threat to the health of poultry and decreases the quality of their meat [24].

Bile acids are synthesized by the liver and secreted into the intestine to digest and absorb nutrients [7–9]. Bile acids also participate in the regulation of the structure of gut microbiota; thus, they are closely related to animal health [25]. There are currently no reports of AFB1 causing cholestasis in the duck liver. AFB_1 can upregulate *CYP7A1* gene expression in the mouse liver, promoting bile acid synthesis and thereby increasing bile acid content in the mouse serum, liver, and feces [26]. Furthermore, AFB_1 can cause the obstruction of bile secretion in rats [27]. In addition, rats orally administered 25 μg/kg of AFB_1 showed an increase in bile acids in feces, while oral administration of 5 μg/kg AFB_1 reduced the content of deoxycholic acid, a type of bile acid, in feces [28]. This result suggests that different doses of AFB_1 may have different effects on the metabolism of bile acids. The present study demonstrated that AFB_1 can cause liver cholestasis and increase the TBA serum and liver levels. *CYP7A1* and *CYP8B1* gene expression in the liver was upregulated, a finding similar to these reported results. Therefore, it is speculated that the AFB_1-mediated increase in liver bile acid synthesis leads to liver cholestasis.

BSEP is a key gene that controls the excretion of bile acids in liver cells [29–31]. This gene's absence or abnormal expression can lead to cholestasis [29–31]. Currently, there are no relevant reports on the regulation of the *BSEP* gene by AFB_1. The present study showed that AFB_1 downregulates *BSEP* gene expression in the liver. This downregulation can lead to obstruction of bile acid efflux in liver cells [29], and the significant increase in the TBA liver content in this study also supports this view. Therefore, abnormal *BSEP* gene expression may also be related to AFB_1-induced liver cholestasis.

Cholestyramine, as a bile acid chelator, can bind to bile acids in the intestine, promoting the excretion of bile acids via the feces and reducing the reabsorption of bile acids in the ileum [12]. In clinical practice, cholestyramine is mainly used to treat hypercholesterolemia and cholestasis [13,32]. The present study showed that cholestyramine can alleviate the symptoms of cholestasis caused by AFB_1 and reduce the bile acid content in duck liver and serum, which is similar to the reported results. In addition, this study revealed that cholestyramine improves the decline in duck growth performance caused by AFB_1, which may be related to its ability to alleviate cholestasis caused by AFB_1 and reduce the harmful effects of cholestasis. The specific reasons still need further research.

The liver synthesizes bile acids, and cholesterol is the raw material for bile acid synthesis [8]. Atorvastatin calcium is a lipid-lowering drug that can inhibit cholesterol synthesis in the liver [14]. Lastuvkova et al. reported that atorvastatin calcium increased the deoxycholic acid content in the plasma of nonalcoholic steatohepatitis model mice, promoting bile secretion and the excretion of bile acids in feces [33]. The present study demonstrated that atorvastatin calcium alleviates cholestasis and reduces the content of bile acids in the liver but does not change the TBA serum levels. The results of this study differ from previous reports, a phenomenon that may be related to the combined effect of atorvastatin calcium and AFB_1; the specific reasons need further investigation. In addition, atorvastatin calcium does not alleviate the decline in duck growth performance caused by AFB_1, which may be due to the involvement of cholesterol in duck growth and development [34]. The decrease in cholesterol synthesis affects the normal growth of ducklings.

Bile acids have two forms, free bile acids and conjugated bile acids, and taurine can combine with free bile acids in the liver to form taurine-conjugated bile acids, promoting the excretion of bile acids [9,18]. Taurine can promote the excretion of bile acids in rats and reduce cholesterol in serum [35]. The current study showed that taurine can reduce the total bile acid content in serum and liver and alleviate cholestasis caused by AFB_1, similar to the previous reports. In addition, this study revealed that taurine can enhance the antioxidant capacity of ducks and alleviate AFB_1-induced growth inhibition and liver damage in ducks. A previous study reported that taurine can alleviate AFB_1-induced liver injury in rats by enhancing antioxidant capacity and inhibiting mitochondrial-mediated cell apoptosis [36]; our results are similar to this report.

Emodin has antioxidant and anti-inflammatory effects and can promote bile acid excretion [17,19]. Emodin can alleviate α-naphthyl isothiocyanate-induced cholestasis in mice [37]. Similarly, in the present study, emodin can reduce the total bile acid content in serum and liver, alleviate cholestasis caused by AFB_1, and upregulate *BSEP* gene expression in the liver. There are currently no reports on the effect of emodin on AFB_1. Xia et al. reported that emodin can alleviate acute pancreatitis through its antioxidant and anti-inflammatory effects [38]. This study also demonstrated that emodin enhances the antioxidant capacity of ducks, which is consistent with our reported results. After the increase in antioxidant capacity, the poisoning symptoms caused by AFB_1 were alleviated, so the growth performance of ducks was also improved compared with the challenged group. AFB_1-induced liver cholestasis may be related to abnormal synthesis and excretion of bile acids. Reducing bile acid synthesis and promoting bile acid efflux are effective ways to alleviate cholestasis.

4. Conclusions

In summary, AFB_1 disrupts the normal growth of ducks, negatively affects the liver, and leads to cholestasis. The abnormal expression of bile acid metabolism-related genes in the liver may be an important reason for AFB_1-induced cholestasis in duck liver. Cholestyramine can promote the efflux of bile acids, while atorvastatin calcium, taurine, and emodin can reduce the synthesis of bile acids. These four compounds can alleviate AFB_1-mediated liver cholestasis. This study can provide a reference for the prevention and treatment of AFB_1 poisoning in livestock and poultry.

5. Materials and Methods

5.1. Duck, Treatments, and Sample Collection

This experiment was approved by the Animal Ethics and Use Committee of Huazhong Agricultural University (approval number HZAUDU-2024-0005). Three-hundred sixty 1-day-old cherry valley meat ducks were randomly divided into 6 groups: control, AFB_1, AFB_1 + cholestyramine, AFB_1 + atorvastatin calcium, AFB_1 + taurine, and AFB_1 + emodin. Each group had 6 replicates, with 10 ducks per replicate. All ducks are kept in stainless steel cages and can freely obtain a basic diet (Supplementary Table S1) and water. Before the formal experiment began, the duck house was fumigated and disinfected with potassium permanganate and formalin. AFB_1 is derived from the fermentation products of *Aspergillus flavus* cultured in the laboratory. Cholestyramine, atorvastatin calcium, taurine, and emodin were purchased from Shanghai Macklin Biochemical Co., Ltd. (Shanghai, China; purity $\geq$ 98%). The doses added to their diet were AFB_1, 90 μg/kg; cholestyramine, 6000 mg/kg; atorvastatin calcium, 75 mg/kg; taurine, 1500 mg/kg; and emodin, 400 mg/kg. After adding AFB1 to the basic diet, the actual content of AFB1 was determined using a commercial reagent kit (Romer Labs, Tulln an der Donau, Austria). The actual content of AFB1 in a uniformly mixed diet is 87 μg/kg. The ducks were raised for a total of 4 weeks, and their production performance was recorded during the feeding period. Blood samples were collected from the subwing vein at weeks 2 and 4, and serum was separated using the method reported by Wang et al. [13]. The serum was stored in a −80 °C freezer for further analysis. At the end of the experiment, the ducks were euthanized using the acute blood loss method, and the liver and spleen weights were recorded before taking photos of the liver.

5.2. Determination of Production Performance

At the beginning of the experiment, the average weight of each group of ducks was determined and recorded as the initial weight. Then, the average weight of each group of ducks was measured at weeks 2 and 4. The average daily weight gain was calculated based on the initial weight, the average weight at week 2, the average weight at week 4, and the number of days of feeding. The weight of feed fed each week and the remaining weight of feed were recorded to calculate the average daily feed intake. The ratio of the average daily feed intake to the average daily weight gain was recorded as the feed conversion rate.

5.3. Determination of Liver Antioxidant Indicators

The liver was lysed using RIPA lysis buffer (Beijing Solarbio Science & Technology Co., Ltd., Beijing, China) to obtain liver tissue homogenate. The protein concentration was measured by using the BCA Protein Assay Kit (Beyotime Biotechnology Co., Ltd., Wuhan, China). The T-SOD, MDA, GSH Px, GSH, and TBA contents in the homogenate were measured by using a commercial kit (Nanjing Jiancheng Biotechnology Co., Ltd., Nanjing, China), following the manufacturer's instructions.

5.4. H&E Staining

After fixing liver tissue blocks with paraformaldehyde and embedding them in paraffin, sections were cut and stained with H&E.

5.5. Oil Red O Staining

The liver tissue blocks fixed with paraformaldehyde were dehydrated with sucrose solution and then frozen. Sections were cut and stained with oil red O and hematoxylin.

5.6. Determination of Serum Biochemical Indicators

The AST, ALT, TP, ALB, TBA, and TBiL serum levels were measured using a fully automated biochemical analyzer (Beckman, CA, USA).

5.7. Determination of the TBA Content in the Liver

A commercially available reagent kit (Nanjing Jiancheng Biotechnology Co., Ltd.) was used to determine the TBA content in the liver, following the manufacturer's guidelines.

5.8. RNA Extraction, cDNA Synthesis, and Real-Time qPCR

Trizol (ABclonal Technology Co., Ltd., Wuhan, China) was used to extract RNA from the liver. ABScript III RT Master Mix for qPCR with gDNA Remover and 2X Universal SYBR Green Fast qPCR Mix (ABclonal Technology Co., Ltd.) was used to synthesize cDNA and perform real-time qPCR, following the manufacturer's instructions. The primers for real-time qPCR are listed in Supplementary Table S2. All genes' Cq values were normalized using the GAPDH Cq value as a reference, and the results were analyzed using the $2^{-\Delta\Delta Ct}$ method.

5.9. Statistical Analysis

The experimental results were analyzed using SPSS 27.0 (IBM Corp., Armonk, NY, USA) and are presented as the mean $\pm$ standard deviation. One-way analysis of variance and the t-test were to evaluate intergroup differences, where $p < 0.05$ indicated a significant difference.

Supplementary Materials: The following supporting information can be downloaded at https://www.mdpi.com/article/10.3390/toxins16060239/s1, Table S1. Composition of the basal diets and nutrient levels; Table S2. Primers used for real-time qPCR.

Author Contributions: Experimental research, writing—original draft preparation and data curation, A.Y.; methodology, H.W.; data curation, Q.C.; investigation, S.A.R.; supervision, project administration, and funding acquisition, D.Q. All authors have read and agreed to the published version of the manuscript.

Funding: This research was funded by the National Key R&D Program of China (Grant No.: 2016YFD0501207).

Institutional Review Board Statement: All animal procedures used in this study were approved by the Institutional Animal Care and Use Committee of Huazhong Agricultural University, China (approval number ID: HZAUDU-2024-0005).

Informed Consent Statement: Not applicable.

Data Availability Statement: The data presented in this study are available on request from the corresponding author.

Acknowledgments: We thank the National Key Research and Development Program for the support of this research. We would also like to thank Shuai Wang, associate researcher in the laboratory, for his guidance on this experimental protocol.

Conflicts of Interest: The authors declare no conflicts of interest.

References

1. Cimbalo, A.; Alonso-Garrido, M.; Font, G.; Manyes, L. Toxicity of mycotoxins in vivo on vertebrate organisms: A review. *Food Chem. Toxicol.* **2020**, *137*, 111161. [CrossRef] [PubMed]
2. Shabeer, S.; Asad, S.; Jamal, A.; Ali, A. Aflatoxin Contamination, Its Impact and Management Strategies: An Updated Review. *Toxins* **2022**, *14*, 307. [CrossRef] [PubMed]
3. Ortega-Beltran, A.; Bandyopadhyay, R. Contributions of integrated aflatoxin management strategies to achieve the sustainable development goals in various African countries. *Glob. Food Secur.* **2021**, *30*, 100559. [CrossRef]
4. Cao, W.; Yu, P.; Yang, K.; Cao, D. Aflatoxin B_1: Metabolism, toxicology, and its involvement in oxidative stress and cancer development. *Toxicol. Mech. Methods* **2022**, *32*, 395–419. [CrossRef] [PubMed]
5. Wang, Y.; Wu, J.; Wang, L.; Yang, P.; Liu, Z.; Rajput, S.A.; Hassan, M.; Qi, D. Epigallocatechin Gallate and Glutathione Attenuate Aflatoxin B_1-Induced Acute Liver Injury in Ducklings via Mitochondria-Mediated Apoptosis and the Nrf2 Signalling Pathway. *Toxins* **2022**, *14*, 876. [CrossRef] [PubMed]

6. Liu, F.; Wang, Y.; Zhou, X.; Liu, M.; Jin, S.; Shan, A.; Feng, X. Resveratrol Relieved Acute Liver Damage in Ducks (*Anas platyrhynchos*) Induced by AFB_1 via Modulation of Apoptosis and Nrf2 Signaling Pathways. *Animals* **2021**, *11*, 3516. [CrossRef] [PubMed]
7. Ahmed, M. Functional, Diagnostic and Therapeutic Aspects of Bile. *Clin. Exp. Gastroenterol.* **2022**, *15*, 105–120. [CrossRef] [PubMed]
8. Wang, Y.; Yutuc, E.; Griffiths, W.J. Cholesterol metabolism pathways—Are the intermediates more important than the products? *FEBS J.* **2021**, *288*, 3727–3745. [CrossRef] [PubMed]
9. Jia, W.; Wei, M.; Rajani, C.; Zheng, X. Targeting the alternative bile acid synthetic pathway for metabolic diseases. *Protein Cell* **2021**, *12*, 411–425. [CrossRef] [PubMed]
10. Garzel, B.; Hu, T.; Li, L.; Lu, Y.; Heyward, S.; Polli, J.; Zhang, L.; Huang, S.M.; Raufman, J.P.; Wang, H. Metformin Disrupts Bile Acid Efflux by Repressing Bile Salt Export Pump Expression. *Pharm. Res.* **2020**, *37*, 26. [CrossRef] [PubMed]
11. Arab, J.P.; Cabrera, D.; Arrese, M. Bile Acids in Cholestasis and its Treatment. *Ann. Hepatol.* **2017**, *16*, s53–s57. [CrossRef] [PubMed]
12. Nishida, S.; Horinouchi, A.; Higashimura, Y.; Akahori, R.; Matsumoto, K. Cholestyramine, a Bile Acid Sequestrant, Increases Cecal Short Chain Fatty Acids and Intestinal Immunoglobulin A in Mice. *Biol. Pharm. Bull.* **2020**, *43*, 565–568. [CrossRef] [PubMed]
13. Laatikainen, T. Effect of cholestyramine and phenobarbital on pruritus and serum bile acid levels in cholestasis of pregnancy. *Am. J. Obstet. Gynecol.* **1978**, *132*, 501–506. [CrossRef] [PubMed]
14. Wang, H.; Wu, K.; Mi, X.; Rajput, S.A.; Qi, D. Effects of 3-Hydroxy-3-methylglutaryl-CoA Reductase Inhibitors on Cholesterol Metabolism in Laying Hens. *Animals* **2023**, *13*, 1868. [CrossRef] [PubMed]
15. Li, W.; Liu, H.; Liang, J.; Wang, T.; Liu, J.; Pi, X.; Zou, W.; Qu, L. Effects of Atorvastatin on Bile Acid Metabolism in High-fat Diet-fed ApoE -/- Mice. *J. Cardiovasc. Pharmacol.* **2023**, *81*, 454–462. [CrossRef] [PubMed]
16. Jong, C.J.; Sandal, P.; Schaffer, S.W. The Role of Taurine in Mitochondria Health: More Than Just an Antioxidant. *Molecules* **2021**, *26*, 4913. [CrossRef]
17. Cui, Y.; Chen, L.J.; Huang, T.; Ying, J.Q.; Li, J. The pharmacology, toxicology and therapeutic potential of anthraquinone derivative emodin. *Chin. J. Nat. Med.* **2020**, *18*, 425–435. [CrossRef] [PubMed]
18. Guertin, F.; Roy, C.C.; Lepage, G.; Perea, A.; Giguere, R.; Yousef, I.; Tuchweber, B. Effect of taurine on total parenteral nutrition-associated cholestasis. *JPEN J. Parenter. Enter. Nutr.* **1991**, *15*, 247–251. [CrossRef] [PubMed]
19. Hu, N.; Liu, J.; Xue, X.; Li, Y. The effect of emodin on liver disease—Comprehensive advances in molecular mechanisms. *Eur. J. Pharmacol.* **2020**, *882*, 173269. [CrossRef]
20. Miyazaki, T.; Sasaki, S.I.; Toyoda, A.; Shirai, M.; Ikegami, T.; Matsuzaki, Y.; Honda, A. Influences of Taurine Deficiency on Bile Acids of the Bile in the Cat Model. *Adv. Exp. Med. Biol.* **2019**, *1155*, 35–44. [CrossRef] [PubMed]
21. Xiong, X.L.; Ding, Y.; Chen, Z.L.; Wang, Y.; Liu, P.; Qin, H.; Zhou, L.S.; Zhang, L.L.; Huang, J.; Zhao, L. Emodin Rescues Intrahepatic Cholestasis via Stimulating FXR/BSEP Pathway in Promoting the Canalicular Export of Accumulated Bile. *Front. Pharmacol.* **2019**, *10*, 522. [CrossRef] [PubMed]
22. Gao, X.; Jiang, L.; Xu, J.; Liu, W.; Li, S.; Huang, W.; Zhao, H.; Yang, Z.; Yu, X.; Wei, Z. Aflatoxin B_1-activated heterophil extracellular traps result in the immunotoxicity to liver and kidney in chickens. *Dev. Comp. Immunol.* **2022**, *128*, 104325. [CrossRef] [PubMed]
23. Liu, L.; Chen, F.; Qin, S.; Ma, J.; Li, L.; Jin, T.; Zhao, R. Effects of Selenium-Enriched Yeast Improved Aflatoxin B_1-Induced Changes in Growth Performance, Antioxidation Capacity, IL-2 and IFN-gamma Contents, and Gene Expression in Mice. *Biol. Trace Elem. Res.* **2019**, *191*, 183–188. [CrossRef] [PubMed]
24. Hussain, Z.; Khan, M.Z.; Khan, A.; Javed, I.; Saleemi, M.K.; Mahmood, S.; Asi, M.R. Residues of aflatoxin B_1 in broiler meat: Effect of age and dietary aflatoxin B_1 levels. *Food Chem. Toxicol.* **2010**, *48*, 3304–3307. [CrossRef] [PubMed]
25. Collins, S.L.; Stine, J.G.; Bisanz, J.E.; Okafor, C.D.; Patterson, A.D. Bile acids and the gut microbiota: Metabolic interactions and impacts on disease. *Nat. Rev. Microbiol.* **2023**, *21*, 236–247. [CrossRef] [PubMed]
26. Liu, Y.; Li, J.; Kang, W.; Liu, S.; Liu, J.; Shi, M.; Wang, Y.; Liu, X.; Chen, X.; Huang, K. Aflatoxin B_1 induces liver injury by disturbing gut microbiota-bile acid-FXR axis in mice. *Food Chem. Toxicol.* **2023**, *176*, 113751. [CrossRef] [PubMed]
27. Carrillo, M.C.; Rodriguez, J.V.; Monti, J.A.; Pellegrino, J.M.; Rodriguez, G.E. Impairment of bile secretion induced by aflatoxin B_1 in the rat. *Res. Commun. Chem. Pathol. Pharmacol.* **1982**, *38*, 521–524. [PubMed]
28. Zhou, J.; Tang, L.; Wang, J.; Wang, J.S. Aflatoxin B_1 Disrupts Gut-Microbial Metabolisms of Short-Chain Fatty Acids, Long-Chain Fatty Acids, and Bile Acids in Male F344 Rats. *Toxicol. Sci.* **2018**, *164*, 453–464. [CrossRef] [PubMed]
29. Stieger, B.; Beuers, U. The canalicular bile salt export pump BSEP (ABCB11) as a potential therapeutic target. *Curr. Drug Targets* **2011**, *12*, 661–670. [CrossRef] [PubMed]
30. Sohail, M.I.; Donmez-Cakil, Y.; Szollosi, D.; Stockner, T.; Chiba, P. The Bile Salt Export Pump: Molecular Structure, Study Models and Small-Molecule Drugs for the Treatment of Inherited BSEP Deficiencies. *Int. J. Mol. Sci.* **2021**, *22*, 784. [CrossRef] [PubMed]
31. Kis, E.; Ioja, E.; Rajnai, Z.; Jani, M.; Mehn, D.; Heredi-Szabo, K.; Krajcsi, P. BSEP inhibition: In vitro screens to assess cholestatic potential of drugs. *Toxicol. In Vitro* **2012**, *26*, 1294–1299. [CrossRef] [PubMed]
32. Stanciu, M.C.; Nichifor, M.; Teaca, C.A. Bile Acid Sequestrants Based on Natural and Synthetic Gels. *Gels* **2023**, *9*, 500. [CrossRef] [PubMed]

33. Lastuvkova, H.; Faradonbeh, F.A.; Schreiberova, J.; Hroch, M.; Mokry, J.; Faistova, H.; Nova, Z.; Hyspler, R.; Igreja Sa, I.C.; Nachtigal, P.; et al. Atorvastatin Modulates Bile Acid Homeostasis in Mice with Diet-Induced Nonalcoholic Steatohepatitis. *Int. J. Mol. Sci.* **2021**, *22*, 6468. [CrossRef]
34. Schade, D.S.; Shey, L.; Eaton, R.P. Cholesterol Review: A Metabolically Important Molecule. *Endocr. Pract.* **2020**, *26*, 1514–1523. [CrossRef] [PubMed]
35. Masuda, M.; Horisaka, K. Effect of taurine and homotaurine on bile acid metabolism in dietary hyperlipidemic rats. *J. Pharmacobiodyn.* **1986**, *9*, 934–940. [CrossRef] [PubMed]
36. Wu, G.; San, J.; Pang, H.; Du, Y.; Li, W.; Zhou, X.; Yang, X.; Hu, J.; Yang, J. Taurine attenuates AFB_1-induced liver injury by alleviating oxidative stress and regulating mitochondria-mediated apoptosis. *Toxicon* **2022**, *215*, 17–27. [CrossRef] [PubMed]
37. Wang, X.; Han, L.; Bi, Y.; Li, C.; Gao, X.; Fan, G.; Zhang, Y. Paradoxical Effects of Emodin on ANIT-Induced Intrahepatic Cholestasis and Herb-Induced Hepatotoxicity in Mice. *Toxicol. Sci.* **2019**, *168*, 264–278. [CrossRef] [PubMed]
38. Xia, S.; Ni, Y.; Zhou, Q.; Liu, H.; Xiang, H.; Sui, H.; Shang, D. Emodin Attenuates Severe Acute Pancreatitis via Antioxidant and Anti-inflammatory Activity. *Inflammation* **2019**, *42*, 2129–2138. [CrossRef] [PubMed]

Disclaimer/Publisher's Note: The statements, opinions and data contained in all publications are solely those of the individual author(s) and contributor(s) and not of MDPI and/or the editor(s). MDPI and/or the editor(s) disclaim responsibility for any injury to people or property resulting from any ideas, methods, instructions or products referred to in the content.

toxins

MDPI

Article

Rhein Inhibits Cell Development and Aflatoxin Biosynthesis via Energy Supply Disruption and ROS Accumulation in *Aspergillus flavus*

Xiaoyan Wang [1], Kashif Iqbal Sahibzada [1,2], Ruibo Du [1], Yang Lei [1], Shan Wei [1], Na Li [1], Yuansen Hu [1] and Yangyong Lv [1,*]

1 College of Biological Engineering, Henan University of Technology, Zhengzhou 450001, China; wangxiaoyan990416@163.com (X.W.); kashif.iqbal@dhpt.uol.edu.pk (K.I.S.); drb17719229855@163.com (R.D.); leiyang@haut.edu.cn (Y.L.); weishansd2014@163.com (S.W.); nalinwsuaf@haut.edu.cn (N.L.); hys308@126.com (Y.H.)

2 Department of Health Professional Technologies, Faculty of Allied Health Sciences, The University of Lahore, Lahore 54570, Pakistan

* Correspondence: lvyangyong@haut.edu.cn

Abstract: *Aspergillus flavus* and its carcinogenic secondary metabolites, aflatoxins, not only cause serious losses in the agricultural economy, but also endanger human health. Rhein, a compound extracted from the Chinese herbal medicine *Rheum palmatum* L. (Dahuang), exhibits good anti-inflammatory, anti-tumor, and anti-oxidative effects. However, its effect and underlying mechanisms against *Aspergillus flavus* have not yet been fully illustrated. In this study, we characterized the inhibition effect of rhein on *A. flavus* mycelial growth, sporulation, and aflatoxin B_1 (AFB_1) biosynthesis and the potential mechanism using RNA-seq analysis. The results indicate that *A. flavus* mycelial growth and AFB_1 biosynthesis were significantly inhibited by 50 μM rhein, with a 43.83% reduction in colony diameter and 87.2% reduction in AFB_1 production. The RNA-seq findings demonstrated that the differentially expressed genes primarily participated in processes such as spore formation and development, the maintenance of cell wall and membrane integrity, management of oxidative stress, the regulation of the citric acid cycle, and the biosynthesis of aflatoxin. Biochemical verification experiments further confirmed that 50 μM rhein effectively disrupted cell wall and membrane integrity and caused mitochondrial dysfunction through disrupting energy metabolism pathways, leading to decreased ATP synthesis and ROS accumulation, resulting in impaired aflatoxin biosynthesis. In addition, a pathogenicity test showed that 50 μM rhein inhibited *A. flavus* spore growth in peanut and maize seeds by 34.1% and 90.4%, while AFB_1 biosynthesis was inhibited by 60.52% and 99.43%, respectively. In conclusion, this research expands the knowledge regarding the antifungal activity of rhein and provides a new strategy to mitigate *A. flavus* contamination.

Keywords: rhein; *Aspergillus flavus*; AFB_1 biosynthesis; transcriptomics

Key Contribution: Rhein affects the growth and spore formation of *A. flavus* and blocks aflatoxin biosynthesis.

1. Introduction

Aspergillus flavus has been found to contaminate peanuts, maize, and wheat during cultivation, storage, processing, and transportation [1,2]. *A. flavus* also produces highly carcinogenic secondary metabolites called aflatoxins, of which aflatoxin B_1 is classified as a grade 1 carcinogen [3,4]. The consumption of aflatoxin-contaminated food can cause poisoning, which may lead to hepatocellular carcinoma, seriously endangering the life and health of both humans and livestock [5]. Studies have shown that over 120 different countries have established maximum permissible levels of aflatoxin contamination in food products [6]. It is therefore important to develop rapid, efficient, and safe measures to address the problems associated with AFB_1 contamination, given the serious threat it poses to human health.

Biological, physical, and chemical methods have been employed to prevent the contamination of *A. flavus* in food [7]. Biological measures, such as endophytic *Aspergillus fumigatus* and lactic acid bacteria, are efficient in inhibiting the growth of *A. flavus* and lessening mycotoxin production [8,9]. Physical methods, such as γ-irradiation of maize artificially inoculated with *A. flavus*, also resulted in a reduction in aflatoxins. In recent years, compounds from natural plants, including cinnamaldehyde, citral, and eugenol, have shown excellent antifungal performance against *A. flavus* and aflatoxin biosynthesis, owing to their eco-friendly properties [10]. Therefore, it is imperative to develop natural compounds to effectively prevent *A. flavus* growth and aflatoxin contamination.

Rhein, an anthraquinone compound derived from the natural plant Rhubard, also known as a kind of traditional Chinese herbal medicine, exhibits various effects, including anti-inflammatory, anti-tumor, antioxidant, anti-diabetic, cardioprotective, neuroprotective, hepatoprotective, and nephroprotective properties [11–13]. For example, rhein inhibits colon cancer cell growth by targeting the mammalian target of rapamycin (mTOR) pathway and human kidney cancer cells through the mitogen-activated protein kinase (MAPK)/nuclear factor kappa-light-chain enhancer of activated B cells (NF-κB) signaling pathway [14,15]. Additionally, studies have demonstrated that rhein can inhibit *Propionibacterium acnes* growth through inhibiting NADH dehydrogenase-2 activity [16]. Previous reports have also shown that rhein and its derivatives exhibit insecticidal and antifungal activity against *Saprolegnia* sp., *Phytophthora capsici*, *Candida albicans*, *Cryptococcus neoformans*, and *Trichophyton mentagrophytes* [17–23]. However, the effects and underlying mechanisms of rhein against *A. flavus* mycelium growth and aflatoxin biosynthesis have not yet been investigated.

In the current study, we investigated the effects of rhein on *A. flavus* mycelium growth, spore formation, and AFB_1 biosynthesis. Rhein was found to significantly diminish the pathogenicity of *A. flavus* on both maize and peanut seeds. An analysis of the transcriptomic data showed that differential gene expression was involved in conidium formation and development (*wetA*, *BrlA*, *AbaA*), cell wall and membrane structure, oxidative stress response, the citrate cycle, and aflatoxin biosynthesis (*aflA*, *aflC*, *aflO*, *aflQ*). Our study provides new strategies for the prevention of aflatoxin contamination.

2. Results

2.1. Effect of Rhein on A. flavus Growth, Spore Formation, and AFB_1 Production

To examine the antifungal effects of rhein on mycelium growth and AFB_1 production, 2 μL of *A. flavus* spore solution (1×10^6 spores/mL) was inoculated into PDA medium with varying concentrations of rhein (0, 25, or 50 μM), then incubated at 30 °C for 5 days. The results demonstrate that *A. flavus* mycelium growth was significantly inhibited by treatment with 25 and 50 μM rhein. Additionally, the mycelium became dense and the colony diameter was reduced by 32.34% and 43.83%, respectively. Regarding the formation of *A. flavus* spores, it was observed that the number of spores and conidiophores decreased significantly with increasing rhein concentration, where the number of spores was significantly reduced by 41.32% and 63.18% under the 25 and 50 μM rhein treatments, respectively. Electron microscopy analysis revealed that the surface of the conidia without rhein treatment was regular and the shape was plentiful, while the morphology of rhein-treated *A. flavus* spores appeared shrunken and irregular (Figure 1A–C). TLC and HPLC analysis showed that, compared to the control, AFB_1 production decreased by 76.5% and 87.2% after treatment with 25 and 50 μM rhein, respectively, suggesting a dose-dependent inhibition of AFB_1 biosynthesis (Figure 1D,E). Additionally, the dry weight of mycelia also decreased significantly with increasing rhein concentration (Figure 1F). In summary, rhein showed significant inhibitory effects against *A. flavus* mycelium growth, spore formation, and AFB_1 yield; moreover, 50 μM rhein showed better inhibition of *A. flavus* mycelium growth, sporulation, and AFB_1. Based on these results, the *A. flavus* treated with 50 µM rhein was selected for RNA-seq analysis.

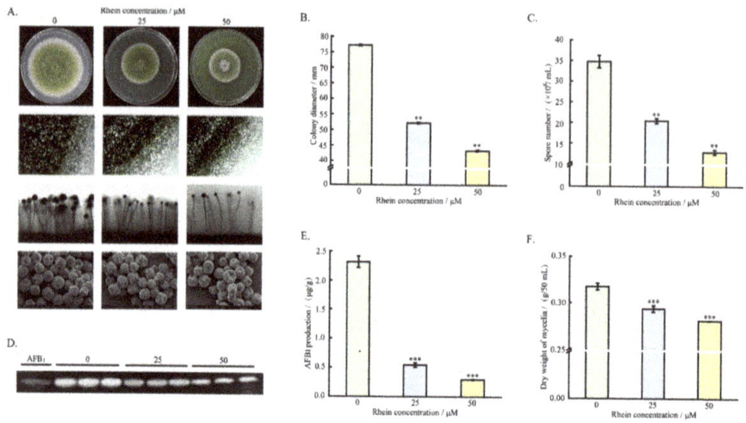

Figure 1. Effects of rhein (0, 25, 50 µM) on *A. flavus* morphology and AFB_1 biosynthesis. (**A**) Observation of colony and conidial morphology; (**B**) colony diameter; (**C**) the number of spores; (**D**) AFB_1 production according to TLC analysis; (**E**) AFB_1 production according to HPLC analysis; and (**F**) dry weight of mycelia. ** represents $p < 0.01$, *** represents $p < 0.001$.

2.2. Transcriptomic Analysis

The *A. flavus* treated with 50 µM rhein was selected for RNA-seq analysis with triple biological replicates in order to explicate the underlying mechanism through which rhein affects *A. flavus* mycelium growth and aflatoxin biosynthesis. The results show 633 differentially expressed genes (DEGs) in rhein-treated *A. flavus*, with 150 DEGs upregulated and 483 DEGs downregulated compared to the control. To further elucidate the function of the DEGs, enrichment analysis was conducted. The results reveal that the DEGs were primarily linked to the biosynthetic pathway of secondary metabolites, metabolic pathways, and the citric acid cycle, as indicated by the KEGG enrichment analysis (Figure 2A). GO enrichment analysis revealed that DEGs were associated with conidium formation and development, the cell wall and membrane, the hydrogen peroxide metabolic process, and oxidoreductase activity (Figure 2B–D).

To better elucidate the underlying mechanism, the representative DEGs were classified into four categories, including conidium formation and development, cell wall and cell membrane, glycolysis, citrate cycle, oxidative stress, and aflatoxin biosynthesis (Table 1).

Table 1. The representative DEGs in the control compared to the rhein treatment.

Gene ID	log_2(FC)	Gene Functional Description
	Conidium formation and development	
AFLA_052030	−1.55	Developmental regulatory protein wetA
AFLA_082850	−2.47	C2H2 type conidiation transcription factor BrlA
AFLA_029620	−2.06	Transcription factor AbaA
AFLA_098380	−1.85	Conidial hydrophobin RodA
AFLA_014260	−1.44	Conidial hydrophobin RodB
AFLA_044790	−1.63	Conidiation-specific family protein
	Cell wall and membrane	
AFLA_052780	−1.60	Cell wall glucanase (Scw4)
AFLA_061960	−1.94	Cell wall protein
AFLA_137200	1.56	Chitin synthase
AFLA_025900	−2.71	Chitin deacetylase
AFLA_007630	−1.52	Integral membrane protein
AFLA_039810	−1.23	DUF1212 domain membrane protein

Table 1. *Cont.*

Gene ID	$\log_2$(FC)	Gene Functional Description
	Oxidative stress	
AFLA_096210	−1.79	Catalase
AFLA_034380	−1.28	Catalase
AFLA_071780	−2.42	Peroxidase family protein
AFLA_001980	−1.45	Glutamate cysteine ligase
	Glycolysis, citrate cycle, and aflatoxin biosynthesis	
AFLA_101470	1.05	Glyceraldehyde-3-phosphate dehydrogenase
AFLA_036370	−1.29	Phosphoenolpyruvate carboxykinase
AFLA_107660	−2.13	Succinyl-CoA synthetase
AFLA_117410	−4.26	Citrate synthase
AFLA_117420	−1.75	Fatty acid synthase subunit alpha (aflA)
AFLA_081850	−1.23	Norsolorinic acid synthase (aflC)
AFLA_065570	−1.35	O-methyltransferase (aflO)
AFLA_073130	1.28	O-methylsterigmatocystin oxidoreductase ordA-like (aflQ)

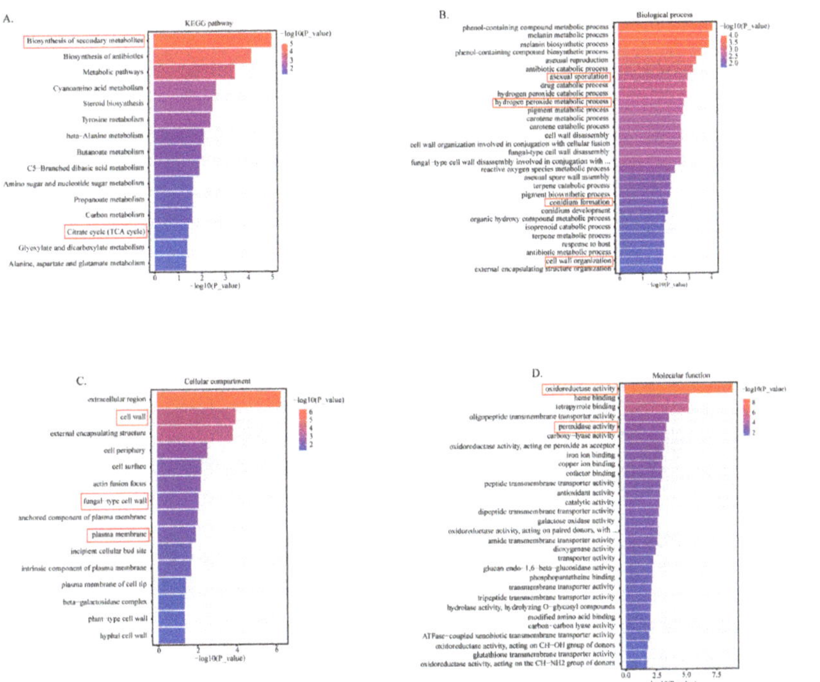

Figure 2. Transcriptome analysis of rhein-treated and untreated *A. flavus* mycelia. (**A**) KEGG pathway enrichment; (**B**) enrichment of biological processes; (**C**) analysis of the enrichment of cellular components in DEGs; and (**D**) analysis of DEGs for molecular function enrichment. The red boxes indicated the related items in Table 1 and the subsequent biochemical verification experiments.

2.3. Effect of Rhein on the Hydrophobicity of *A. flavus*

To verify the differential expression of conidial hydrophobin genes, the effect of rhein on the hydrophobicity of *A. flavus* was investigated. The results demonstrate that bromophenol blue solution and sterile water fully penetrated the surface of the rhein-treated *A. flavus* mycelium into the medium as compared to the control. The mycelial surface was observed using a stereoscopic microscope, and at the same magnification, the sterile water droplets on the surface of *A. flavus* colonies treated with rhein were found to be smaller (Figure 3). These results show that treatment with rhein inhibits the hydrophobicity of the cell surface of the mycelia.

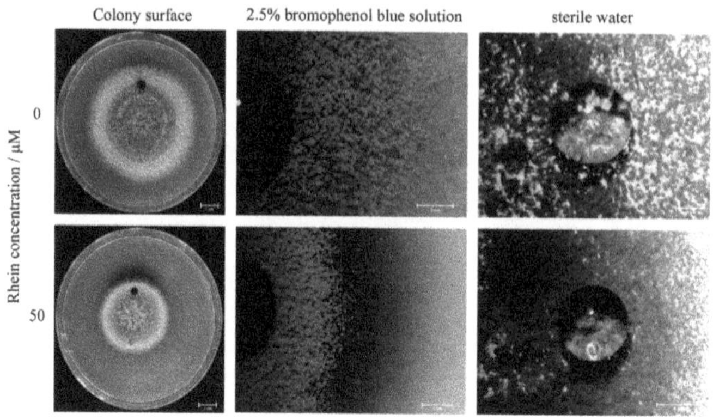

Figure 3. Determination of the hydrophobicity of *A. flavus* treated with or without rhein.

2.4. Effect of the Rhein on the Integrity of A. flavus Cell Wall

To study the effect of rhein on *A. flavus* cell wall integrity, the strains were cultured for 4 days under the reagent CFW. The results unveil that *A. flavus* strains were more sensitive to CFW after rhein treatment than the controls (Figure 4A,B). Additionally, the integrity of the cell wall was examined through CFW staining, showing a decrease in CFW fluorescence intensity with increasing rhein concentration (Figure 4C). These results suggest that rhein disrupts *A. flavus* cell wall synthesis.

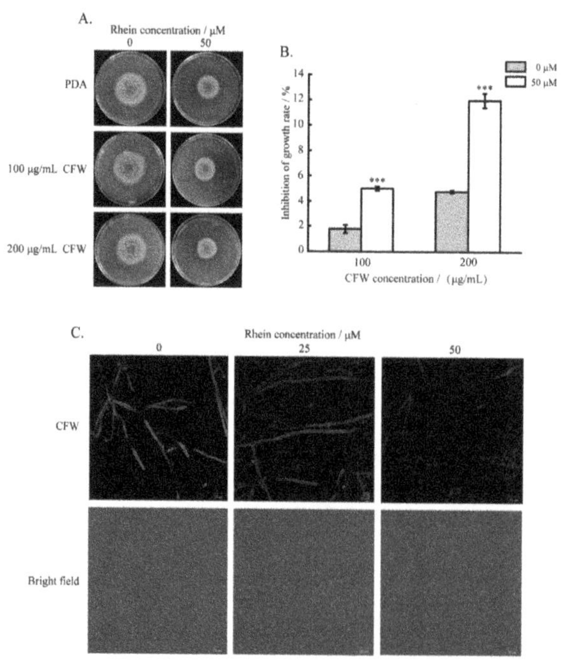

Figure 4. Effects of rhein on *A. flavus* cell wall. (**A**) The sensitivity of *A. flavus* to CFW after treatment with rhein; (**B**) the CFW inhibition of *A. flavus* mycelial growth; (**C**) *A. flavus* mycelia treated with rhein (0, 25, 50 μM) stained with CFW by CLSM. *** represents $p < 0.001$.

2.5. Effect of Rhein on the Integrity of Cell Membrane

Different concentrations of SDS were introduced into PDA medium to examine the effect of rhein on the integrity of *A. flavus* cell membranes. The results demonstrate that the inhibition rate of rhein-treated *A. flavus* reached 11.2% and 39.1% under the conditions of 100 and 200 μg/mL SDS, respectively. Rhein-treated *A. flavus* was more sensitive than the control group (Figure 5A,B). Additionally, PI staining revealed that the red fluorescence intensity gradually increased with the concentration of rhein (Figure 5C). These results reveal that the addition of rhein disrupts the cell membrane of *A. flavus*.

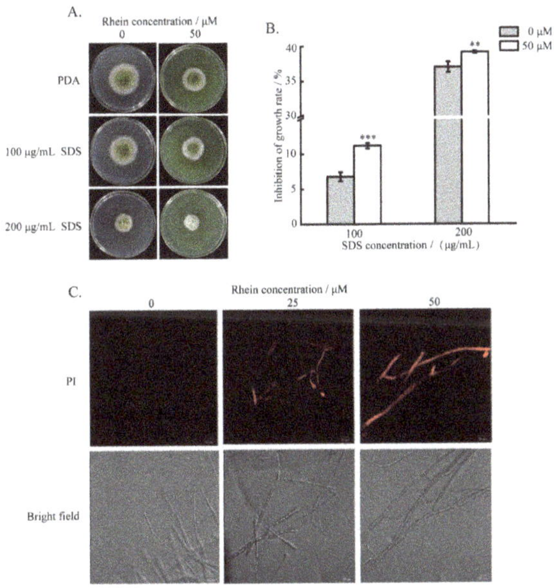

Figure 5. Effects of rhein on A. flavus cell membrane. (**A**) The sensitivity to SDS of A. flavus treated with rhein; (**B**) the SDS inhibition of A. flavus mycelial growth; (**C**) A. flavus mycelial treated with rhein (0, 25, 50 μM) stained with PI by CLSM. ** represents $p < 0.01$, *** represents $p < 0.001$.2.6. Effects of Rhein on Intracellular ROS, H_2O_2, and ATP in A. flavus.

To validate the analysis of RNA-seq data, the intracellular ROS accumulation, H_2O_2, and ATP contents were measured. The results demonstrate that, accompanied by the downregulated expression of catalase, peroxidase, and glutamate cysteine ligase-related genes (Figure 6A), the intracellular H_2O_2 content of *A. flavus* increased (Figure 6B), which was in agreement with the results of the transcriptome analysis. The intracellular ATP content of *A. flavus* also decreased slightly after rhein treatment (Figure 6C). In addition, the results of DCFH-DA staining demonstrate that *A. flavus* mycelia treated with rhein exhibited stronger green fluorescence than those untreated with rhein (Figure 6D), indicating that rhein induced the accumulation of ROS.

2.6. Effect of the Rhein on A. flavus Pathogenicity on Crop Seeds

To investigate the effect of rhein on the sporogenesis and the AFB_1 yield of *A. flavus* on crop seeds, different concentrations of rhein were incubated for 5 days with maize and peanut seeds infected with *A. flavus* spore suspension. The results indicate that the colonization ability of *A. flavus* on peanut and maize seeds was reduced after treatment with rhein (Figure 7A). The number of spores on the surface of the seeds was reduced by 90.4% and 34.1%, respectively, in comparison to the control group, following treatment with 50 μM of rhein (Figure 7B). Additionally, the TLC and HPLC results demonstrate the inhibition of AFB_1 production. With the 50 μM rhein treatment, the AFB_1 yield of *A. flavus* on peanuts and maize decreased by 60.52% and 99.43%, respectively (Figure 7C,D). These results clearly demonstrate that rhein significantly inhibits the pathogenicity of *A. flavus* on maize and peanut seeds.

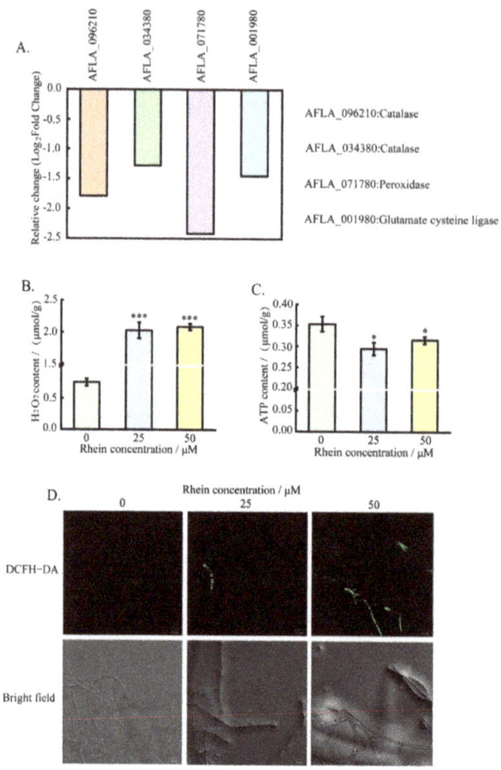

Figure 6. Effects of rhein on *A. flavus* intracellular ROS. (**A**) Expression levels of differential genes associated with oxidative stress in *A. flavus* after rhein treatment; (**B**) H_2O_2 content; (**C**) ATP content; (**D**) *A. flavus* mycelial treated with rhein (0, 25, 50 μM) stained with DCFH-DA by CLSM. * represents $p < 0.05$, *** represents $p < 0.001$.

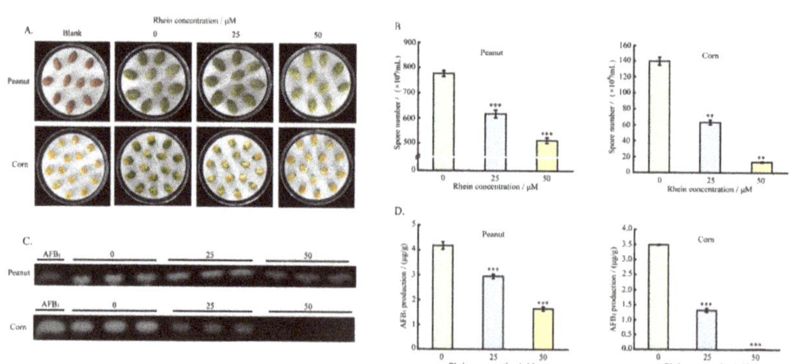

Figure 7. Effect of rhein on *A. flavus* pathogenicity to crop seeds. (**A**) Growth of *A. flavus* on seeds with or without rhein treatment; (**B**) number of *A. flavus* spores on seeds; (**C**) TLC; and (**D**) HPLC analysis of AFB_1 yield. ** represents $p < 0.01$, *** represents $p < 0.001$.

3. Discussion

Developing safe and effective antifungal drugs is urgently needed to reduce the threat posed by *A. flavus* to cereals and food. In our study, the effects of rhein, a compound from traditional Chinese medicine, on *A. flavus* mycelium growth, sporulation, and AFB_1 biosynthesis were characterized. The results demonstrate that rhein disrupted *A. flavus* cell wall and membrane integrity, thereby inhibiting its growth and sporulation. Additionally, an RNA-seq analysis and subsequent validation demonstrated that mitochondrial dysfunction and the accumulation of intracellular ROS might lead to the inhibition of aflatoxin biosynthesis.

Rhein treatment significantly inhibited cell growth and development in *A. flavus*. Various regulators and transcription factors regulate the *A. flavus* spore-forming process. It has been reported that conidia formation and maturation in *A. flavus* are mainly governed by BrlA, AbaA, and WetA [24]. The *AbaA* gene plays a role in conidial cell differentiation and is activated by BrlA at metaphase [25,26], while the *wetA* gene, activated by *AbaA* in anaphase, participates in synthesizing key conidial cell wall components [27]. In our study, rhein treatment substantially reduced the *A. flavus* spore count, and the expression of genes related to conidial growth, such as *AFLA_052030* (*wetA*), *AFLA_082850* (*BrlA*), *AFLA_029620* (*AbaA*), and *AFLA_044790*, was significantly downregulated (Table 1). These findings suggest that rhein may inhibit spore formation in *A. flavus*. In addition, RodA and RodB are conidial surface proteins, with RodA playing a role in the surface rod structure and RodB contributing to the conidial cell wall structure in *Aspergillus fumigatus* [28,29]. In our study, the bromophenol blue solution diffused rapidly and was absorbed on the surface of rhein-treated mycelium (Figure 3). Furthermore, the expression of genes related to hydrophobic proteins (*AFLA_098380*, *AFLA_014260*) was significantly downregulated. This is in agreement with the results of the hydrophobicity assay in the *Afper1* gene deletion strain studied by Lv et al. [30]. These results suggest that rhein treatment may have reduced the hydrophobicity of the mycelium surface.

The cell wall and membrane of the fungi are dynamic structures which are essential for maintaining cell viability and integrity, making them promising targets for antifungal treatment [31–35]. In this study, CFW staining showed that rhein disrupted cell wall synthesis, and genes involved in cell wall protein synthesis (*AFLA_061960*) were also downregulated. However, transcriptome data analysis demonstrated that gene expression of chitin synthetase was upregulated, suggesting that upregulated chitin synthetase might contribute to compensating for the destroyed cell wall after rhein treatment. In addition, PI staining demonstrated that the red fluorescence intensity was increased, suggesting cell membrane damage. Similar phenomena have been observed in *A. flavus* cell membranes when treated with honokiol [36], (E)-2-hexenal [37], and phenyllactic acid [38]. Moreover, the gene expression levels of *AFLA_007630* and *AFLA_039810*, which are involved in membrane protein synthesis, were significantly downregulated. This suggests that rhein may induce impaired synthesis of the cell wall and membrane, leading to the disruption of cell components. These results also indicate that the antifungal activity of rhein may target the integrity of the cell membrane and cell wall.

It is well known that mitochondria are organelles that generate ATP through oxidative phosphorylation reactions, providing energy to the cell [39]. Oxidative phosphorylation occurs during the citric acid cycle (TCA) in the mitochondria [40]. Previous studies have indicated that mitochondrial dysfunction leads to the accumulation of ROS by reducing ATP levels [41]. In our study, genes associated with the TCA cycle, including citrate synthetase and succinyl-CoA synthetase, were significantly downregulated after rhein treatment (Table 1). This finding mirrors the genetic changes observed in *A. flavus* treated with estragole [42]. In addition, *A. flavus* treated with rhein exhibited a remarkable decrease in intracellular ATP levels, which aligns with the findings of Duan et al. [43] regarding *A. flavus* treated with 1-octanol. These results reveal that rhein affects the changes in enzymes in other metabolic pathways, such as the TCA cycle, leading to mitochondrial dysfunction and ultimately the accumulation of ROS.

Peroxisomes are an important component of eukaryotic cells, participating in various cellular physiological processes and being capable of rapidly producing and removing H_2O_2 and O_2, thereby regulating the dynamic changes in ROS levels [44–46]. In this study, the genes encoding catalase and peroxidase (*AFLA_096210, AFLA_034380*) were significantly downregulated after treatment with rhein (Figure 6A). Subsequently, experimental verification showed that the intracellular H_2O_2 content was increased in *A. flavus* treated with rhein. These results suggest that H_2O_2 accumulation disrupts oxidation reduction (redox) homeostasis. In addition, glutathione biosynthesis is also involved in oxidative stress, with glutathione ligase reported as the key enzyme catalyzing glutathione biosynthesis [47]. Elevated levels of ROS also promote mitochondrial and metabolic dysfunction, resulting in oxidative stress [48]. Our transcriptome data show that the transcriptional level of the glutathione ligase gene (*AFLA_001980*) was remarkably downregulated (Table 1). We also assessed the intracellular ROS level in *A. flavus* using DCFH-DA staining and noted the dose-dependent increase in fluorescence intensity with rhein treatment. These results suggest that rhein triggered oxidative stress, resulting in a significant accumulation of ROS, consistent with the intracellular ROS level observed in *A. flavus* treated with methyl 2-methylbutyrate [49]. Therefore, we speculate that rhein treatment might cause changes in the intracellular peroxisomes of *A. flavus*, which disrupted the redox balance and subsequently led to ROS accumulation.

Aflatoxin biosynthesis involves a complex regulatory network composed of multiple genes and enzymatic reactions, regulated by the pathway-specific transcription factor aflR and several structural genes, including *aflA, aflB, aflC, aflO, aflQ, aflR,* and *aflS*. Additionally, it is also governed by environmental conditions, such as light, nutrient sources, and oxidative stress [50–53]. According to previous reports, three proteins, aflA, aflB, and aflC, are jointly involved in the synthesis of norsolorinic acid (NOR) in the biosynthesis pathway of aflatoxin [5]. AflO, which encodes O-methyltransferase, converts demethylsterigmatocystin (DMST) to a new compound, sterigmatocystin (ST), and aflQ is a P-450 monooxygenase that converts O-methylsterigmatocystin (OMST) to AFB_1 [3]. In our study, the production of AFB_1 by rhein-treated *A. flavus* growing on both PDA medium and grain was significantly decreased. Analysis of transcriptome data revealed that the genes involved in the pathway for the aflatoxin synthesis (*aflA, aflC, aflO,* and *aflQ*) were all remarkedly downregulated after rhein treatment. In *A. flavus*, deletion of catalase CTA1 contributes to the intracellular ROS accumulation and downregulation of *aflC, aflD,* and *aflQ*, consequently leading to impaired AFB_1 biosynthesis [54]. Additionally, the accumulated intracellular ROS in *A. flavus* treated with curcumin also leads to a reduction in aflatoxin production [55]. In our study, the downregulated expression levels of *aflA, aflC, aflO,* and *aflQ* and decreased AFB_1 biosynthesis may be attributed to the ROS accumulation triggered by rhein treatment.

4. Conclusions

In summary, rhein inhibited the growth and development of *A. flavus* by interfering with cell wall and membrane synthesis. Additionally, rhein might disrupt energy metabolism, cause mitochondrial dysfunction, and decrease ATP synthesis, leading to ROS accumulation, thereby inhibiting AFB_1 biosynthesis (Figure 8). Furthermore, rhein treatment effectively inhibited mycelial growth and the production of AFB_1 in both peanut and maize seeds. Our findings indicate that rhein can be considered as a potential novel agent for preventing fungal infections in food and feedstuffs.

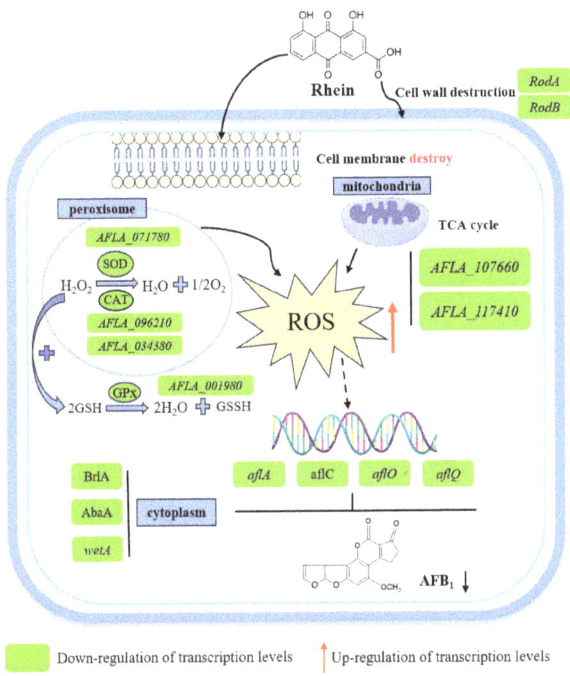

Figure 8. Mechanistic diagram of rhein inhibiting *A. flavus* growth and AFB_1 biosynthesis.

5. Materials and Methods

5.1. Strains, Chemicals, and Medium

The model strain *A. flavus* NRRL 3357, a kind gift from Prof. Zhumei He (Sun Yat-sen University, China), was used in this study. Rhein (CAS:478-43-3, 98.5% purity) was obtained from J&K Scientific Ltd. (Beijing, China). Peanut and maize seeds were sourced from local markets. Potato dextrose agar medium (PDA) was selected for fungal growth, conidial production, and aflatoxin B_1 (AFB_1) biosynthesis. Rhein was dissolved in DMSO to form a solution with a final concentration of 10 mg/mL.

5.2. Effect of Rhein on Phenotype, Spore, and Aflatoxin B_1 Production of A. flavus

The effects of rhein on *A. flavus* colony size, spore morphology, and aflatoxin B_1 yield were examined with PDA medium. Rhein at different concentrations (25 and 50 μM) was added to PDA medium to create treatment groups according to previously described methods [56], while the same volume of DMSO was added to the control group. The spore solution (1×10^6 spores/mL) was inoculated, and the colony diameter was recorded after culturing at 30 °C for 5 days. The morphology of spores and conidial pedicles was observed using a light microscope, stereoscopic microscope, and electron microscope.

AFB_1 was extracted and detected using the methods described in a previous study [42]. In brief, 25 μM or 50 μM of rhein were introduced into PDA medium, and 2 μL of spore solution was inoculated followed by incubation for 5 days at 30 °C. AFB_1 was then extracted by adding chloroform and distilled water (volume ratio of 1:1). Finally, AFB_1 was analyzed through HPLC and TLC.

To evaluate the impact of rhein on mycelium production, 25 or 50 μM of rhein was introduced into liquid medium containing 50 mL DPY. Subsequently, *A. flavus* spores were inoculated to achieve a concentration of 1×10^6 spores/mL. After shaking for 3 days at 30 °C, the mycelium was filtered and dried for 6 h, then weighed.

5.3. RNA-seq Analysis

The *A. flavus* spore suspension and 50 μM rhein were grown in PDA medium for 60 h at 30 °C, with DMSO replacing rhein in the control group. The TRIzol method by Thermo Fisher, Shanghai, China, was employed to extract total RNA from the mycelia with three biological replicates, and the quality and integrity of RNA were assessed. Subsequently, Shanghai Yuanxin Biotechnology Co., Ltd. (Shanghai, China) constructed the cDNA library for RNA-seq analysis. Each sample with three biological replicates, $|\log_2$fold change$| > 1$, and p value < 0.05 was defined as a differentially expressed gene. KEGG enrichment analysis and GO were performed using Goatools and KOBAS software, respectively. Raw transcriptome data were entered into the CNCB database under CRA accession number CRA015824.

5.4. Effect of Rhein on Hydrophobicity of A. flavus

Hydrophobicity was determined using the method of a previous study with slight modifications [57]. *A. flavus* spores were point-inoculated on PDA medium supplemented with rhein (50 μM), then incubated for a duration of 4 days at 30 °C. Subsequently, the colony surface was treated with sterile water and 2.5% bromophenol blue solution for 12 h and then photographed.

5.5. Effect of Rhein on A. flavus Cell Wall

A. flavus spore suspension was inoculated into PDA medium supplemented with rhein (50 μM) and fluorescent brightener 28 (CFW, 100 and 200 μg/mL) and incubated at 30 °C for 3 days. Moreover, the *A. flavus* colony diameter was recorded and photographed to evaluate the effect of rhein on *A. flavus* cell walls.

A. flavus spore suspension and rhein (25 and 50 μM) were added to 50 mL DPY medium and incubated for 48 h. Following the staining method described in a previous study [58], the mycelia were collected, washed, and then stained with Calcofluor white stain (CFW) for 1 min in the dark on slides. Finally, confocal laser scanning microscopy (CLSM) was utilized to observe the mycelia.

5.6. Effect of Rhein on A. flavus Cell Membrane

A. flavus spore was point-inoculated onto PDA supplemented with 50 μM rhein and different concentrations of sodium dodecyl sulphate (SDS, 100, and 200 μg/mL), and then it was incubated at 30 °C for 4 days. Then, the colony diameters were assessed and the relative inhibition rate was calculated to evaluate the effect of rhein on inhibition of *A. flavus* cell membranes.

Similarly, *A. flavus* mycelia were treated as described above. *A. flavus* spore suspension and rhein (25 and 50 μM) were added to 50 mL DPY medium and allowed to grow for 48 h. The mycelia were then collected and washed. Subsequently, propidium iodide (PI) staining was carried out at 37 °C for 30 min. Finally, the mycelia were collected and examined through CLSM.

5.7. Determination of A. flavus Intracellular ROS, H_2O_2, and ATP Content

A. flavus spore suspension and various rhein concentrations (25 and 50 μM) were added to 50 mL DPY medium and allowed to grow for 48 h at 30 °C. The mycelia were treated as described above, then stained with DCFH-DA and incubated in the dark at 37 °C for 30 min. Finally, samples were visualized using CLSM.

A. flavus spore was point-inoculated on PDA medium with cellophane with or without rhein (50 μM) and incubated at 30 °C for 60 h. Subsequently, *A. flavus* mycelia were ground into powder and collected. The H_2O_2 and ATP contents were determined with reference to the biochemical kit (Solarbio Science, Beijing, China).

5.8. Effect of Rhein on *A. flavus* Mycelium Growth and AFB_1 Yield on Peanuts and Maize

Different concentrations of rhein (25 and 50 μM) and *A. flavus* spore suspension were added into 50 mL sterile water until a final concentration of 1×10^6 spores/mL was achieved. Then, sterilized peanuts and corn were placed in the water and shaken for 1 h. The treated peanuts and corn were then evenly placed in a dish covered with filter paper and cultured at 30 °C in the dark for 5 days. To maintain appropriate moisture levels, sterile water was added periodically during this incubation period. The infected peanut and corn seeds were placed in triangular flasks after 5 days of incubation, and the spores were counted after 30 min of thorough agitation after adding 20 mL of sterile water. AFB_1 was extracted and determined as described above.

5.9. Statistical Analysis

In this study, all experiments were performed in triplicate and presented as mean ± SD, followed by significance analysis using SAS software (version 9.2), with statistical significance considered at $p < 0.05$.

Author Contributions: Methodology, X.W.; validation, R.D.; data curation, Y.L. (Yang Lei) and S.W.; writing—original draft preparation, X.W.; writing—review and editing, K.I.S., N.L. and Y.L. (Yangyong Lv); project administration, Y.H. All authors have read and agreed to the published version of the manuscript.

Funding: This work was funded by the Natural Science Foundation of Henan Province (222300420037), the Natural Science Foundation of China (31972176), and the Cultivation Programme for Young Backbone Teachers in Henan University of Technology (21420114).

Institutional Review Board Statement: Not applicable.

Informed Consent Statement: Not applicable.

Data Availability Statement: The raw transcriptome read data are available in the CNCB database under accession number CRA015824.

Conflicts of Interest: The authors declare no conflicts of interest.

References

1. Li, D.; Qin, L.; Wang, Y.; Xie, Q.; Li, N.; Wang, S.; Yuan, J. AflSte20 regulates morphogenesis, stress response, and aflatoxin biosynthesis of *Aspergillus flavus*. *Toxins* **2019**, *11*, 730. [CrossRef] [PubMed]
2. Zhang, F.; Zhong, H.; Han, X.; Guo, Z.; Yang, W.; Liu, Y.; Yang, K.; Zhuang, Z.; Wang, S. Proteomic profile of *Aspergillus flavus* in response to water activity. *Fungal Biol.* **2015**, *119*, 114–124. [CrossRef] [PubMed]
3. Caceres, I.; Khoury, A.A.; Khoury, R.E.; Lorber, S.; Oswald, I.P.; Khoury, A.E.; Atoui, A.; Puel, O.; Bailly, J.D. Aflatoxin biosynthesis and genetic regulation: A review. *Toxins* **2020**, *12*, 150. [CrossRef]
4. Hu, Y.; Yang, G.; Zhang, D.; Liu, Y.; Li, Y.; Lin, G.; Guo, Z.; Wang, S.; Zhuang, Z. The PHD transcription factor Rum1 regulates morphogenesis and aflatoxin biosynthesis in *Aspergillus flavus*. *Toxins* **2018**, *10*, 301. [CrossRef] [PubMed]
5. Khan, R.; Ghazali, F.M.; Mahyudin, N.A.; Samsudin, N.I.P. Aflatoxin biosynthesis, genetic regulation, toxicity, and control strategies: A review. *J Fungi* **2021**, *7*, 606. [CrossRef]
6. Eom, T.J.; Moon, H.; Yu, J.H.; Park, H.S. Characterization of the velvet regulators in *Aspergillus flavus*. *J. Microbiol.* **2018**, *56*, 893–901. [CrossRef] [PubMed]
7. Gong, A.; Song, M.; Zhang, J. Current strategies in controlling *Aspergillus flavus* and aflatoxins in grains during storage: A review. *Sustainability* **2024**, *16*, 3171. [CrossRef]
8. Abdelaziz, A.M.; El-Wakil, D.A.; Attia, M.S.; Ali, O.M.; AbdElgawad, H.; Hashem, A.H. Inhibition of *Aspergillus flavus* growth and aflatoxin production in Zea mays L. using Endophytic *Aspergillus fumigatus*. *J Fungi* **2022**, *8*, 482. [CrossRef]
9. Ben Taheur, F.; Mansour, C.; Kouidhi, B.; Chaieb, K. Use of lactic acid bacteria for the inhibition of *Aspergillus flavus* and *Aspergillus carbonarius* growth and mycotoxin production. *Toxicon* **2019**, *166*, 15–23. [CrossRef]
10. Liang, D.; Xing, F.; Selvaraj, J.N.; Liu, X.; Wang, L.; Hua, H.; Zhou, L.; Zhao, Y.; Wang, Y.; Liu, Y. Inhibitory effect of cinnamaldehyde, citral, and eugenol on aflatoxin biosynthetic gene expression and aflatoxin B_1 biosynthesis in *Aspergillus flavus*. *J. Food Sci.* **2015**, *80*, M2917–M2924. [CrossRef]
11. Wu, C.; Cao, H.; Zhou, H.; Sun, L.; Xue, J.; Li, J.; Bian, Y.; Sun, R.; Dong, S.; Liu, P.; et al. Research progress on the antitumor effects of rhein: Literature review. *Anticancer Agents Med. Chem.* **2017**, *17*, 1624–1632. [CrossRef] [PubMed]
12. Li, G.M.; Chen, J.R.; Zhang, H.Q.; Cao, X.Y.; Sun, C.; Peng, F.; Yin, Y.P.; Lin, Z.; Yu, L.; Chen, Y.; et al. Update on pharmacological activities, security, and pharmacokinetics of rhein. *Evid. Based Complement Altern. Med.* **2021**, *2021*, 4582412. [CrossRef] [PubMed]

13. Sun, H.; Luo, G.; Chen, D.; Xiang, Z. A comprehensive and system review for the pharmacological mechanism of action of rhein, an active anthraquinone ingredient. *Front. Pharmacol.* **2016**, *7*, 247. [CrossRef] [PubMed]
14. Zhang, H.; Yi, J.K.; Huang, H.; Park, S.; Park, S.; Kwon, W.; Kim, E.; Jang, S.; Kim, S.Y.; Choi, S.K.; et al. Rhein suppresses colorectal cancer cell growth by inhibiting the mTOR pathway in vitro and in vivo. *Cancers* **2021**, *13*, 2176. [CrossRef] [PubMed]
15. Ma, Y.L.; Chen, F.; Shi, J. Rhein inhibits malignant phenotypes of human renal cell carcinoma by impacting on MAPK/NF-κB signaling pathways. *Onco Targets Ther.* **2018**, *11*, 1385–1394. [CrossRef] [PubMed]
16. Nguyen, A.T.; Kim, K.Y. Rhein inhibits the growth of Propionibacterium acnes by blocking NADH dehydrogenase-2 activity. *J. Med. Microbiol.* **2020**, *69*, 689–696. [CrossRef] [PubMed]
17. Yao, J.Y.; Lin, L.Y.; Yuan, X.M.; Ying, W.L.; Xu, Y.; Pan, X.Y.; Ge, P.H. Antifungal activity of Rhein and aloe-emodin from *Rheum palmatum* on fish pathogenic *Saprolegnia* sp. *J. World Aquacult. Soc.* **2017**, *48*, 137–144. [CrossRef]
18. Wang, B.; Yang, J.; Zhao, X.; Feng, X.; Xu, S.; Li, P.; Li, L.; Chen, Y. Antifungal activity of the botanical compound rhein against *Phytophthora capsici* and the underlying mechanisms. *Pest Manag. Sci.* **2024**, *80*, 1228–1239. [CrossRef] [PubMed]
19. Agarwal, S.K.; Singh, S.S.; Verma, S.; Kumar, S. Antifungal activity of anthraquinone derivatives from *Rheum emodi*. *J. Ethnopharmacol.* **2000**, *72*, 43–46. [CrossRef]
20. Chewchinda, S.; Wuthi-udomlert, M.; Gritsanapan, W. HPLC quantitative analysis of rhein and antidermatophytic activity of cassia fistula pod pulp extracts of various storage conditions. *Biomed. Res. Int.* **2013**, *2013*, 821295. [CrossRef]
21. Zhu, X.; Chen, S.; Zheng, Y.; Zhang, Y.; Hsiang, T.; Huang, R.; Qi, J.; Gan, T.; Chang, Y.; Li, J. Antifungal and insecticidal activities of rhein derivatives: Synthesis, characterization and preliminary structure-activity relationship studies. *Nat. Prod. Res.* **2022**, *36*, 4140–4146. [CrossRef] [PubMed]
22. Shang, X.F.; Zhao, Z.M.; Li, J.C.; Yang, G.Z.; Liu, Y.Q.; Dai, L.X.; Zhang, J.Y. Insecticidal and antifungal activities of *Rheum palmatum* L. anthraquinones and structurally related compounds. *Ind. Crop. Prod.* **2019**, *137*, 508–520. [CrossRef]
23. Chen, S.; Wang, M.; Yu, L.; Shi, J.; Zhang, Y.; Tian, Y.; Li, L.; Zhu, X.; Li, J. Rhein-amino acid easter caonjugates as potential antifungal agents: Synthesis and biological evaluation. *Molecules* **2023**, *28*, 2074. [CrossRef] [PubMed]
24. Cho, H.J.; Son, S.H.; Chen, W.; Son, Y.E.; Lee, I.; Yu, J.H.; Park, H.S. Regulation of conidiogenesis in *Aspergillus flavus*. *Cells* **2022**, *11*, 2796. [CrossRef]
25. Andrianopoulos, A.; Timberlake, W.E. The *Aspergillus nidulans abaA* gene encodes a transcriptional activator that acts as a genetic switch to control development. *Mol. Cell Biol.* **1994**, *14*, 2503–2515. [CrossRef] [PubMed]
26. Mead, M.E.; Borowsky, A.T.; Joehnk, B.; Steenwyk, J.L.; Shen, X.X.; Sil, A.; Rokas, A. Recurrent loss of abaA, a master regulator of asexual development in filamentous fungi, correlates with changes in genomic and morphological traits. *Genome Biol. Evol.* **2020**, *12*, 1119–1130. [CrossRef] [PubMed]
27. Wu, M.Y.; Mead, M.E.; Kim, S.C.; Rokas, A.; Yu, J.H. WetA bridges cellular and chemical development in *Aspergillus flavus*. *PLoS ONE* **2017**, *12*, e0179571. [CrossRef]
28. Pedersen, M.H.; Borodina, I.; Moresco, J.L.; Svendsen, W.E.; Frisvad, J.C.; Søndergaard, I. High-yield production of hydrophobins RodA and RodB from *Aspergillus fumigatus* in Pichia pastoris. *Appl. Microbiol. Biotechnol.* **2011**, *90*, 1923–1932. [CrossRef]
29. Grünbacher, A.; Throm, T.; Seidel, C.; Gutt, B.; Röhrig, J.; Strunk, T.; Vincze, P.; Walheim, S.; Schimmel, T.; Wenzel, W.; et al. Six hydrophobins are involved in hydrophobin rodlet formation in *Aspergillus nidulans* and contribute to hydrophobicity of the spore surface. *PLoS ONE* **2014**, *9*, e94546. [CrossRef]
30. Lv, Y.; Yang, H.; Wang, J.; Wei, S.; Zhai, H.; Zhang, S.; Hu, Y. Afper1 contributes to cell development and aflatoxin biosynthesis in *Aspergillus flavus*. *Int. J. Food Microbiol.* **2022**, *377*, 109828. [CrossRef]
31. Gow, N.A.R.; Latge, J.P.; Munro, C.A. The fungal cell wall: Structure, biosynthesis, and function. *Microbiol. Spectr.* **2017**, *5*. [CrossRef] [PubMed]
32. Garcia-Rubio, R.; de Oliveira, H.C.; Rivera, J.; Trevijano-Contador, N. The fungal cell wall: Candida, cryptococcus, and *Aspergillus* Species. *Front. Microbiol.* **2020**, *10*, 2993. [CrossRef]
33. Sant, D.G.; Tupe, S.G.; Ramana, C.V.; Deshpande, M.V. Fungal cell membrane-promising drug target for antifungal therapy. *J. Appl. Microbiol.* **2016**, *121*, 1498–1510. [CrossRef] [PubMed]
34. Li, Q.; Zhu, X.; Xie, Y.; Liang, J. Antifungal properties and mechanisms of three volatile aldehydes (octanal, nonanal and decanal) on *Aspergillus flavus*. *Grain Oil Sci. Technol.* **2021**, *4*, 131–140. [CrossRef]
35. Bowman, S.M.; Free, S.J. The structure and synthesis of the fungal cell wall. *Bioessays* **2006**, *28*, 799–808. [CrossRef] [PubMed]
36. Zhang, W.; Li, B.; Lv, Y.; Wei, S.; Zhang, S.; Hu, Y. Transcriptomic analysis shows the antifungal mechanism of honokiol against *Aspergillus flavus*. *Int. J. Food Microbiol.* **2023**, *384*, 109972. [CrossRef] [PubMed]
37. Ma, W.; Zhao, L.; Zhao, W.; Xie, Y. (E)-2-Hexenal, as a potential natural antifungal compound, inhibits *Aspergillus flavus* spore germination by disrupting mitochondrial energy metabolism. *J. Agric. Food Chem.* **2019**, *67*, 1138–1145. [CrossRef] [PubMed]
38. Li, M.; Yao, B.; Meng, X. Inhibitory effect and possible mechanism of phenyllactic acid on *Aspergillus flavus* spore germination. *J. Basic Microbiol.* **2022**, *62*, 1457–1466. [CrossRef]
39. Kowalczyk, P.; Sulejczak, D.; Kleczkowska, P.; Bukowska-Ośko, I.; Kucia, M.; Popiel, M.; Wietrak, E.; Kramkowski, K.; Wrzosek, K.; Kaczyńska, K. Mitochondrial oxidative stress-a causative factor and therapeutic target in many diseases. *Int. J. Mol. Sci.* **2021**, *22*, 13384. [CrossRef]
40. Chen, Y.; Zhou, Z.; Min, W. Mitochondria, oxidative stress and innate immunity. *Front. Physiol.* **2018**, *9*, 1487. [CrossRef]

41. van Hameren, G.; Campbell, G.; Deck, M.; Berthelot, J.; Gautier, B.; Quintana, P.; Chrast, R.; Tricaud, N. In vivo real-time dynamics of ATP and ROS production in axonal mitochondria show decoupling in mouse models of peripheral neuropathies. *Acta Neuropathol. Commun.* **2019**, *7*, 86. [CrossRef] [PubMed]
42. Liang, L.; Zhang, W.; Hao, J.; Wang, Y.; Wei, S.; Zhang, S.; Hu, Y.; Lv, Y. Estragole inhibits growth and aflatoxin biosynthesis of *Aspergillus flavus* by affecting reactive oxygen species homeostasis. *Microbiol. Spectr.* **2023**, *11*, e0134823. [CrossRef] [PubMed]
43. Duan, W.Y.; Qin, Y.L.; Zhang, S.B.; Zhai, H.C.; Lv, Y.Y.; Wei, S.; Hu, Y.S. Inhibitory mechanisms of plant volatile 1-octanol on the germination of *Aspergillus flavus* spores. *Food Biophys.* **2024**, *19*, 96–108. [CrossRef]
44. Del Río, L.A.; López-Huertas, E. ROS Generation in Peroxisomes and its Role in Cell Signaling. *Plant Cell Physiol.* **2016**, *57*, 1364–1376. [CrossRef] [PubMed]
45. Sandalio, L.M.; Rodríguez-Serrano, M.; Romero-Puertas, M.C.; del Río, L.A. role of peroxisomes as a source of reactive oxygen species (ROS) signaling molecules. *Subcell Biochem.* **2013**, *69*, 231–255. [CrossRef]
46. Pellegrino, E.; Aylan, B.; Bussi, C.; Fearns, A.; Bernard, E.M.; Athanasiadi, N.; Santucci, P.; Botella, L.; Gutierrez, M.G. Peroxisomal ROS control cytosolic Mycobacterium tuberculosis replication in human macrophages. *J. Cell Biol.* **2023**, *222*, e202303066. [CrossRef] [PubMed]
47. Krzywanski, D.M.; Dickinson, D.A.; Iles, K.E.; Wigley, A.F.; Franklin, C.C.; Liu, R.M.; Kavanagh, T.J.; Forman, H.J. Variable regulation of glutamate cysteine ligase subunit proteins affects glutathione biosynthesis in response to oxidative stress. *Arch. Biochem. Biophys.* **2004**, *423*, 116–125. [CrossRef] [PubMed]
48. Newsholme, P.; Cruzat, V.F.; Keane, K.N.; Carlessi, R.; de Bittencourt, P.I., Jr. Molecular mechanisms of ROS production and oxidative stress in diabetes. *Biochem. J.* **2016**, *473*, 4527–4550. [CrossRef]
49. Wei, S.; Zhang, Y.; Wu, M.; Lv, Y.; Zhang, S.; Zhai, H.; Hu, Y. Mechanisms of methyl 2-methylbutyrate suppression on *Aspergillus flavus* growth and aflatoxin B_1 biosynthesis. *Int. J. Food Microbiol.* **2024**, *409*, 110462. [CrossRef]
50. Kenne, G.J.; Gummadidala, P.M.; Omebeyinje, M.H.; Mondal, A.M.; Bett, D.K.; McFadden, S.; Bromfield, S.; Banaszek, N.; Velez-Martinez, M.; Mitra, C.; et al. Activation of aflatoxin biosynthesis alleviates total ROS in *Aspergillus parasiticus*. *Toxins* **2018**, *10*, 57. [CrossRef]
51. Tai, B.; Chang, J.; Liu, Y.; Xing, F. Recent progress of the effect of environmental factors on *Aspergillus flavus* growth and aflatoxins production on foods. *Food Qual. Safety* **2020**, *4*, 21–28. [CrossRef]
52. Meneely, J.P.; Kolawole, O.; Haughey, S.A.; Miller, S.J.; Krska, R.; Elliott, C.T. The challenge of global aflatoxins legislation with a focus on peanuts and peanut products: A systematic review. *Expo Health* **2023**, *15*, 467–487. [CrossRef]
53. Tumukunde, E.; Xie, R.; Wang, S. Updates on the functions and molecular mechanisms of the genes involved in *Aspergillus flavus* development and biosynthesis of aflatoxins. *J Fungi* **2021**, *7*, 666. [CrossRef] [PubMed]
54. Zhu, Z.; Yang, M.; Bai, Y.; Ge, F.; Wang, S. Antioxidant-related catalase CTA1 regulates development, aflatoxin biosynthesis, and virulence in pathogenic fungus *Aspergillus flavus*. *Environ. Microbiol.* **2020**, *22*, 2792–2810. [CrossRef] [PubMed]
55. Zhang, D.; Yang, Y.; Yao, B.; Hu, T.; Ma, Z.; Shi, W.; Ye, Y. Curcumin inhibits *Aspergillus flavus* infection and aflatoxin production possibly by inducing ROS burst. *Food Res. Int.* **2023**, *167*, 112646. [CrossRef]
56. Chen, B.; Ye, F.; Yu, L.; Jia, G.; Huang, X.; Zhang, X.; Peng, S.; Chen, K.; Wang, M.; Gong, S.; et al. Development of cell-active N^6-methyladenosine RNA demethylase FTO inhibitor. *J. Am. Chem. Soc.* **2012**, *134*, 17963–17971. [CrossRef]
57. Fasoyin, O.E.; Wang, B.; Qiu, M.; Han, X.; Chung, K.R.; Wang, S. Carbon catabolite repression gene *creA* regulates morphology, aflatoxin biosynthesis and virulence in *Aspergillus flavus*. *Fungal Genet. Biol.* **2018**, *115*, 41–51. [CrossRef]
58. Zhang, W.; Li, B.; Lv, Y.; Wei, S.; Zhang, S.; Hu, Y. Synergistic effects of combined cinnamaldehyde and nonanal vapors against *Aspergillus flavus*. *Int. J. Food Microbiol.* **2023**, *402*, 110277. [CrossRef]

Disclaimer/Publisher's Note: The statements, opinions and data contained in all publications are solely those of the individual author(s) and contributor(s) and not of MDPI and/or the editor(s). MDPI and/or the editor(s) disclaim responsibility for any injury to people or property resulting from any ideas, methods, instructions or products referred to in the content.

toxins

MDPI

Article

X-ray Irradiation Reduces Live *Aspergillus flavus* Viability but Not Aflatoxin B1 in Naturally Contaminated Maize

Hannah Glesener [1,2], Darya Abdollahzadeh [1], Christopher Muse [1], Rosa Krajmalnik-Brown [1,3], Mark A. Weaver [4] and Lee E. Voth-Gaeddert [1,5,*]

1 Biodesign Center for Health Through Microbiomes, Arizona State University, Tempe, AZ 85281, USA; glesener@asu.edu (H.G.); dabdoll1@asu.edu (D.A.); christopher.l.muse@asu.edu (C.M.); dr.rosy@asu.edu (R.K.-B.)
2 School for Engineering of Matter, Transport, and Energy, Arizona State University, Tempe, AZ 85281, USA
3 School of Sustainable Engineering and the Built Environment, Arizona State University, Tempe, AZ 85281, USA
4 USDA-ARS National Biological Control Laboratory, Stoneville, MS 38776, USA; mark.weaver@usda.gov
5 Center for Indigenous Health Research, Wuqu' Kawoq | Maya Health Alliance, Chimaltenango 04006, Guatemala
* Correspondence: lee.gaeddert@asu.edu

Abstract: Food crops around the world are commonly contaminated with *Aspergillus flavus*, which can produce the carcinogenic mycotoxin aflatoxin B1 (AFB1). The objective of this study is to test an X-ray irradiation sterilization method for studying AFB1 in contaminated maize samples in the laboratory. Maize that had been naturally contaminated with 300 ppb AFB1 by the growth of aflatoxigenic *A. flavus* was ground and then irradiated at 0.0, 1.0, 1.5, 2.0, 2.5, and 3.0 kGy. *A. flavus* was quantified by dilution plating on potato dextrose agar (PDA) and modified Rose Bengal media (MDRB) for viability and qPCR for gene presence. AFB1 was quantified by HPLC and ELISA. *A. flavus* viability, but not gene copies, significantly decreased with increasing doses of radiation (PDA: $p < 0.001$; MDRB: $p < 0.001$; qPCR: $p = 0.026$). AFB1 concentration did not significantly change with increasing doses of radiation (HPLC: $p = 0.153$; ELISA: $p = 0.567$). Our results imply that X-ray irradiation is an effective means of reducing viable *A. flavus* without affecting AFB1 concentrations. Reducing the hazard of fungal spores and halting AFB1 production at the targeted dose are important steps to safely and reproducibly move forward research on the global mycotoxin challenge.

Keywords: corn; aflatoxin; irradiation; *A. flavus*; food crops; laboratory safety

Key Contribution: *Aspergillus flavus* is not viable in maize after receiving 3.0 kGy of X-ray irradiation. X-ray irradiation is a promising sterilization technique to improve safety during laboratory aflatoxin studies in crops by eliminating viable fungal spores and halting the growth of fungi for research.

1. Introduction

Fungal toxins (mycotoxins) are a burden on the global food supply. In high-income countries, the burden is mainly economic due to crop loss [1]. In low- and middle-income countries, the burden includes poor health attributed to the consumption of contaminated food [1,2]. Research to understand the effects of chronic exposure on human health is ongoing [3,4]. Meanwhile, the global spread of mycotoxins in food crops is increasing and becoming increasingly unpredictable due to climate change conditions (e.g., water stress, elevated CO_2, and temperature) [5]. When mycotoxins were first discovered in the 1960s, it quickly became clear that full eradication was not possible and humans would have to manage exposure through engineering controls in food processing systems [6]. The control points can generally be categorized as pre-harvest (e.g., protecting crops in the field) or post-harvest (e.g., proper storage of grains). Despite sophisticated technologies and successful interventions at both control points, mycotoxins are still pervasive in stored crops, particularly in communities that do not have the economic resources to implement industrial prevention systems [4]. It is paramount for scientists to innovate and implement solutions applicable across all control points within the food system.

Maize—the most abundantly grown crop by volume in the world and the third most consumed crop—is commonly contaminated by the carcinogenic aflatoxin B1 (AFB1) produced by *Aspergillus flavus* [7,8]. Scientists study maize in research labs to understand the complex relationships between food and fungi, yet there can be a disconnect between the characteristics of maize after lab processing compared to maize in the field. For example, sterilized maize can be artificially contaminated with AFB1 by spiking toxic doses into maize in the lab, yet this may not fully capture the critical morphological and biochemical processes that occur between the maize, fungus, and toxin [9]. While important insights about maize have been discovered through artificial contamination of maize with AFB1, it is important to study naturally occurring toxin formation by fungi for translational applications. However, natural contamination can introduce hazards to a lab environment [10,11].

When maize is brought into the lab, it can be sterilized to prevent the growth of microbes and to limit hazards from toxins in the lab. Sterilization is normally either physical (e.g., steam) or chemical (e.g., ethylene oxide) [12,13]. Both physical and chemical sterilization techniques affect the structure of the food product, interfering with certain applied science experiments such as nixtamalization [14]. An alternative sterilization technique is irradiation, a cold pasteurization technique used in industrial food supply systems because it limits structural changes to food, does not use heat, and effectively limits microbial growth. However, this technique is rarely used for research purposes in labs studying mycotoxin-contaminated food.

Researchers have mostly studied gamma irradiation using the radioactive decay of ^{60}Co or ^{137}Cs. Gamma irradiation is slowly being phased out because it is dangerous to operate, it is difficult to dispose of used materials, and the dose uniformity is highly variable. Electron beam (e-beam) irradiation is another option; however, this technique does not have good depth penetration and can mainly reduce surface contamination of foods. The third option is X-ray irradiation, which is estimated to be widely used in the future for food safety [15]. X-ray irradiation can be adapted on a smaller scale in the food industry without complex operating facilities, which means that irradiation units can be directly incorporated into continuous local food processes, saving the transportation and handling costs associated with larger, remote facilities used for irradiation [16]. Companies like Rayfresh Foods are working towards building this technology, though adoption is a hurdle due to the cost of equipment implementation and maintenance. Electricity costs and usage are also high, which is a downside to X-ray irradiation, yet the market for smaller, safer, and easily operated irradiation technology is increasing at the same time as research into the improved efficiency of renewable electricity [17]. Therefore, we chose X-ray irradiation as the cold-pasteurization method for this study [15].

Previous research on the irradiation of *A. flavus* and AFB1 on food products has mostly focused on reducing the fungal load for storage or decreasing the AFB1 content. Gamma irradiation (^{60}Co) of ground pistachio nuts at increasing doses (0.0, 0.5, 1.0, 1.5, 2.0, 4.0, and 6.0 kGy) showed decreased *A. flavus* viability, with a 99.99% reduction at 4 and 6 kGy [18]. AFB1 degradation was observed in corn at increasing doses (4.0, 6.0, and 10.0 kGy) and found to have no significant reduction at 4 kGy but 72.11% and 84.23% at the respective higher doses [19]. Another study looking at *A. flavus* growth during grain storage showed that gamma irradiation (^{60}Co) of whole maize (0, 2.5, 3.5, 4.5, and 6.0 kGy) decreased the viability of *A. flavus* at a proportional rate, with no growth seen at 6.0 kGy over 8 weeks [20]. X-ray irradiation (Rad Source 420M) of cannabis flowers (2.0, 2.5, and 5.0 kGy) showed no growth on potato dextrose agar (PDA) at or above 2.5 kGy [21]. It is unknown whether the X-ray irradiation results seen in cannabis flowers will translate to contaminated maize because of the different material matrix, though *A. flavus* reduction in maize gamma irradiation studies point to the likelihood that X-ray irradiation will also reduce *A. flavus* viability in maize.

This study tests the feasibility of X-ray irradiation as a sterilization technique for a lab-based, aflatoxigenic *A. flavus*-contaminated maize study. The objective is to test a sterilization method for studying toxins in food products in the laboratory using AFB1 produced by *A. flavus* on maize. The aim of this research is to understand the irradiation dose necessary to kill *A. flavus* on the maize while maintaining AFB1 concentrations for subsequent detoxification studies.

2. Results

2.1. X-ray Irradiation Decreases Aspergillus flavus Viability

Maize contaminated with AFB1-producing *A. flavus* was X-ray irradiated at increasing doses—0.0, 1.0, 1.5, 2.0, 2.5, and 3.0 kGy. As seen in Figure 1, dilution plating of irradiated samples on a modified Rose Bengal medium (MDRB; designed for *A. flavus*) shows no observed growth at or above 2.5 kGy. There was a consistent decline in observed colony-forming units (CFU) in samples as the irradiation dose increased.

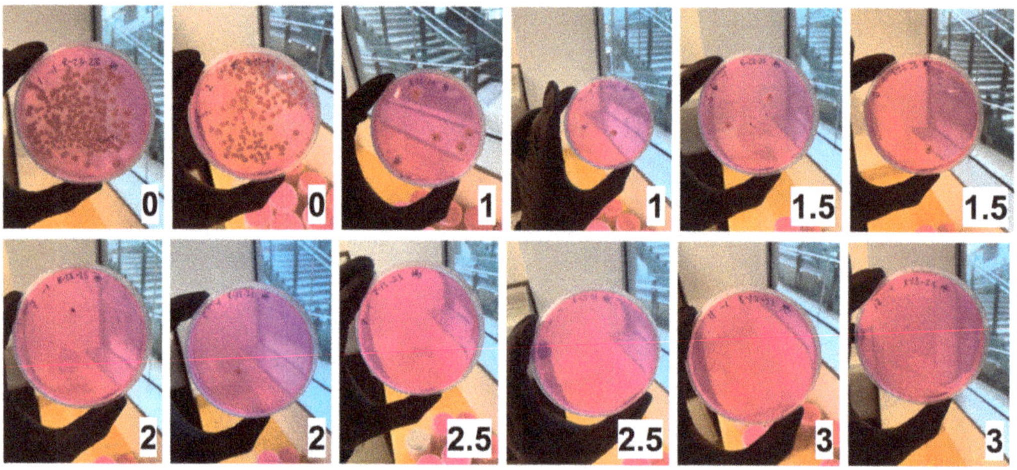

Figure 1. Duplicate irradiated maize samples plated on modified Rose Bengal agar (MDRB) in serial dilutions. A one-hundred-fold dilution is shown. Top left to bottom right images display increasing irradiation doses from 0.0 kilogray (kGy) to 3.0 kGy, indicated in the bottom right of each image by a black number in a white box.

For samples plated on MDRB, CFU per gram of maize decreased from an average of four and a half observed CFU at 0.0 kGy to no observed CFU at 2.5 and 3.0 kGy (Figure 2a). Dilution plating of irradiated samples on potato dextrose agar (PDA; designed for general fungi) showed similar results with no observed CFU at 3.0 kGy (Figure 2b). Each plate that contained at least one CFU was included in the log CFU calculations (Supplemental Tables S1 and S2). Some treatments contained multiple plates with observed colonies (five observations for PDA irradiated at 1.0 kGy), while others only contained one plate with one observed colony (one CFU was observed on one plate at 2.0 kGy for MDRB and 2.5 kGy for PDA).

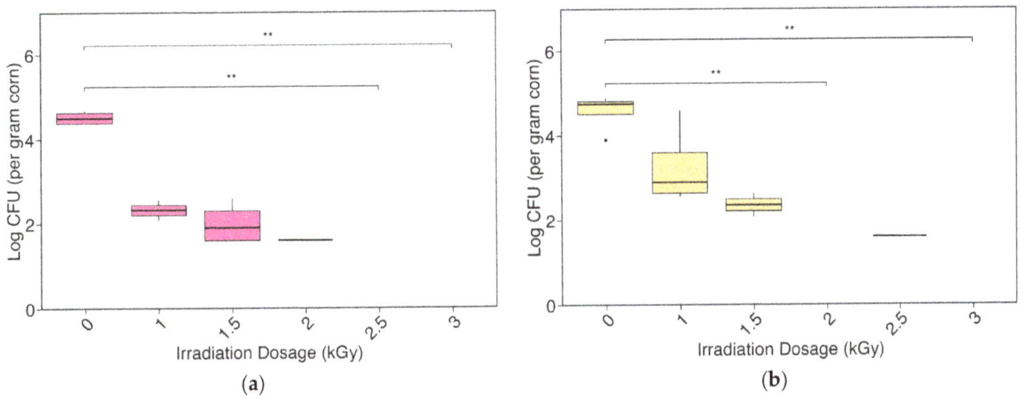

Figure 2. Box plots represent serial dilution plating (tenfold) of X-ray-irradiated maize samples: (**a**) modified Rose Bengal agar (MDRB); (**b**) potato dextrose agar (PDA). All plates with observed samples are reported. Significant results between treatments are marked on plots (** for $p < 0.01$).

DNA extracts from the irradiated samples were amplified via a commercial qPCR kit, targeting the single-copy *A. flavus* MDR1 gene. qPCR showed consistent *A. flavus* presence across doses, with no detectable pattern of higher or lower concentrations between irradiation doses. The maize samples had a mean of 7.04 ± 0.34 log gene copies of MDR1 (Figure 3).

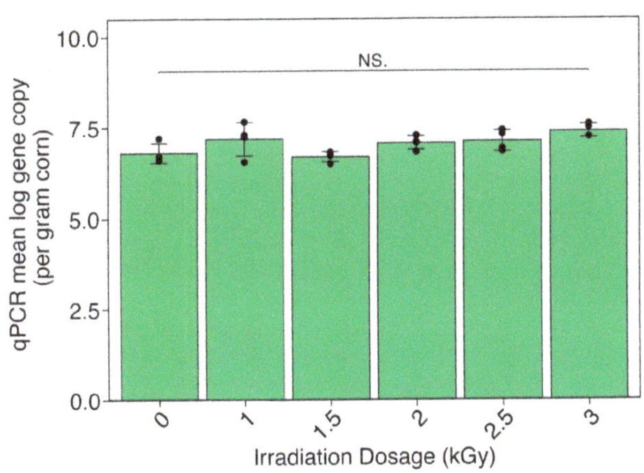

Figure 3. Quantitative polymerase chain reaction (qPCR) results for the MDR1 gene in X-ray-irradiated maize samples. Data points represent logarithmic values of duplicate qPCR measurements for duplicate samples. Error bars represent the mean ± the standard deviation. No significant differences were noted between any of the irradiation doses (NS. for $p \geq 0.05$).

2.2. X-ray Irradiation Did Not Decrease Aflatoxin B1 Concentration in Maize

The irradiation dosage did not significantly reduce the AFB1 concentration as measured by HPLC (ANOVA: $p = 0.153$; Figure 4a). The mean AFB1 concentration of samples on the HPLC was 240.22 ± 17.66 ppb. The irradiation dosage did not significantly reduce the AFB1 concentration as measured by ELISA (ANOVA: $p = 0.566$; Figure 4b). ELISA tests showed a mean AFB1 concentration of 267.08 ± 53.59 ppb.

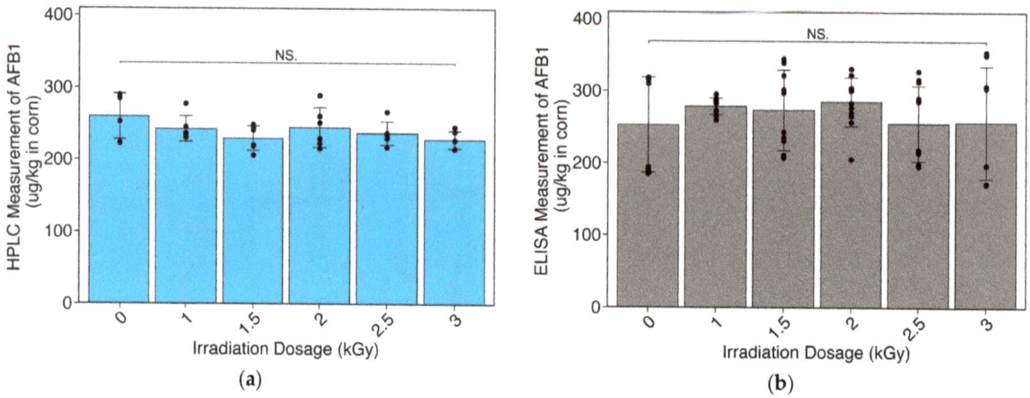

Figure 4. Aflatoxin B1 (AFB1) quantification in X-ray-irradiated samples using high-performance liquid chromatography (HPLC) and an enzyme-linked immunosorbent assay (ELISA): (**a**) HPLC data represent triplicate readings of duplicate samples; (**b**) ELISA data represent triplicate readings of duplicate samples that were diluted twice at a 1:1 ratio to ensure samples were detected by kit detection limits (2–50 ppb). AFB1 concentration is reported in μg/kg (parts per billion). Error bars represent the mean ± the standard deviation. No significant differences are noted between any of the irradiation doses for HPLC or ELISA (NS. for $p \geq 0.05$).

3. Discussion

We tested the feasibility of X-ray irradiation for sterilization of aflatoxigenic *Aspergillus flavus*-contaminated maize for laboratory-based aflatoxin B1 (AFB1) degradation work. We found that viable *A. flavus* was not detected in ground maize samples at an X-ray irradiation dose of 3.0 kGy. AFB1 concentrations were not significantly changed at a radiation dose up to 3.0 kGy. These findings build upon previous research showing that gamma irradiation and electron beam irradiation can be effective means of reducing the fungal concentration in contaminated foods [22,23]. It also builds on previous research demonstrating that AFB1 degradation did not occur below 4.0 kGy when treated with irradiation [19]. This is important for researchers who intend to study AFB1 in naturally contaminated samples while maintaining the integrity of the structure (physical/chemical properties) of the food product through cold pasteurization.

Irradiation acts in two primary ways to kill organisms—directly or indirectly. Indirect death occurs when high-energy particles turn cellular water into hydrogen peroxide and other cytotoxic radicals that can lead to cell death [24]. Direct death occurs when an organism's reproductive DNA is destroyed. Because the MDR1 gene—the qPCR target gene in our experiment—is about 4.6 kbp long and the *A. flavus* genome is about 36.5 Mbp long, it is likely that a large number of copies of the target gene remained intact even if the genome was disrupted through irradiation. This could be a reason why qPCR results show that MDR1 gene copies remained high with increasing irradiation doses despite a reduction in viability. In addition, our results showed a more than two-log difference between fungal concentrations in plated samples compared to qPCR samples before radiation treatment. This could be due to dormant or dead *A. flavus* spores that are picked up by qPCR but not represented through plating. Previous research has shown that qPCR leads to higher counts than dilution plating, although both techniques are common and complementary practices in fungal microbiology [25].

Some limitations of this research are that we did not test any nutritional or physical characteristics of the maize product, which means we cannot draw conclusions about the efficacy of X-ray irradiation on food quality. However, the U.S. Food and Drug Administration (FDA) has approved irradiation for treatment of sprouting seeds up to 8.0 kGy, and our treatment was far below this, so it is likely that the organoleptic (sensory appeal) and nutritional quality of the maize remained acceptable [26]. Another limitation of this study is that we did not use whole maize kernels because we were restricted in size due to the size of the irradiator (15 g samples). The size limitation led to an uneven distribution of AFB1 in the maize, so ground maize was necessary to homogenize the contamination concentration. Whole maize kernels contaminated with AFB1 should be tested for food safety studies to ensure irradiation penetration; however, for the purpose of this study, ground maize sufficed.

These findings are important because they show the effective sterilization of a food product using non-radioactive, non-thermal, and non-chemical X-ray irradiation, which has rarely been performed on food and can be tested more broadly using other food matrices. Because mycotoxins are prevalent in the global food supply, it is critical that scientists study food products in the laboratory while keeping exposure to the hazards of mycotoxin-contaminated materials to a minimum. X-ray irradiation can help neutralize two out of three of the following primary hazards of working with AFB1-contaminated materials: inhalation of spores, contamination of other research materials, and toxic/carcinogenic quality. The carcinogenic hazard of AFB1 remains present and can be mitigated by proper handling. X-ray irradiation could be a good alternative to current sterilization techniques, especially given the safety profile of the technique and the expected increase in adoption and availability of equipment in the future.

4. Materials and Methods

4.1. Chemicals and Media

The following chemicals were used for the extraction, purification, and quantification of aflatoxin B1 (AFB1) and *Aspergillus flavus*. All the following chemicals were purchased from Sigma-Aldrich (St. Louis, MO, USA): methanol (≥99.9%), acetonitrile (≥99.9%), acetone (≥99.5%), sodium chloride (≥99.0%, CAS: 7647-14-5), peptone (CAS: 91079-38-8), yeast extract (CAS: 8013-01-2), magnesium sulfate heptahydrate (≥99.0%, CAS: 10034-99-8), potassium phosphate monobasic (≥99.0%, CAS: 7778-77-0), Rose Bengal (≥80%, CAS: 632-69-9), streptomycin sulfate salt (CAS: 3810-74-0), chlortetracycline hydrochloride (≥75%, CAS: 64-72-2), 2, 6-dichloro-4-nitroaniline (96%, CAS: 99-30-9), agar powder (CAS: 9002-18-0), triton-X 100, potato dextrose agar (PDA) and aflatoxin mix (certified reference material: CRM46304, 1000 ug/kg AFB1). Ethanol (99.5%) was purchased from Decon Labs (Prussia, PA, USA). Glucose (≥99.5%, CAS: 50-99-7) was purchased from HiMedia Laboratories (Mumbai, Maharashtra, India). The modified Rose Bengal media (MDRB) contained (per liter) 30 g of sodium chloride, 20 g of agar, 10 g of glucose, 2.5 g of peptone, 0.5 g of yeast extract, 1 g of potassium phosphate monobasic, 0.5 g of magnesium sulfate heptahydrate, 25 mg of Rose Bengal (stored at 4 °C in 16.5 mg/mL ethanol), 4 mg of 2, 6-dichloro-4-nitroaniline (stored at 4 °C in 0.2 g/100 mL acetone), 187.5 mg of streptomycin (stored at 4 °C in 3.75 g/200 mL of 0.2 μm filtered water), and 15 mg of chlortetracycline (stored at 4 °C in 1.5 g/100 mL methanol). The solution was adjusted to pH 5.0 before adding agar, and the media was cooled to 60 °C before adding antibiotics [27].

4.2. Maize Inoculation and Sample Preparation

A composite grain sample was sourced from Guatemala in 2023 [2]. Intact white maize kernels were dried and stored at 4 °C before use in the present experiments. Maize was heated at 90 °C for 30 min to attenuate some of the microflora but was not sterilized to minimize chemical or structural changes. A portion of the grain sample was aliquoted as a negative control for the X-ray irradiation experiment, while the rest of this maize was moistened and inoculated in a biosafety cabinet with a mix of spores of aflatoxigenic *A. flavus* from Guatemala and representing common genotypes, as verified by DNA microsatellite analysis [2]. The maize was incubated, with samples collected and analyzed for AFB1 every 8–12 h, until the target concentration of 300 ppb AFB1 was reached. Contaminated grain was then dried, milled, and stored at 4 °C with minimal exposure to light.

4.3. X-ray Irradiation of Maize

Aflatoxin-contaminated ground maize was prepared for irradiation by dispensing 16 g in 58 × 70 mm plastic bags (2 mil). A subsample of six bags was irradiated by an X-ray using a 160KV RS2000 Series Biological Irradiator (Rad Source Technologies Inc., Buford, GA, USA) at level 5 (Figure 5). The equipment was operated for 35 cycles of 25 min of run time with a 30-minute rest period to preserve the X-ray bulb. Bags were removed from the irradiator at the target levels of 0.0, 1.0, 1.5, 2.0, 2.5, and 3.0 kGy occurring at cycles 0, 10, 15, 20, 25, 30, and 35. Immediately upon removal, the irradiated bags were labeled and stored at 4 °C.

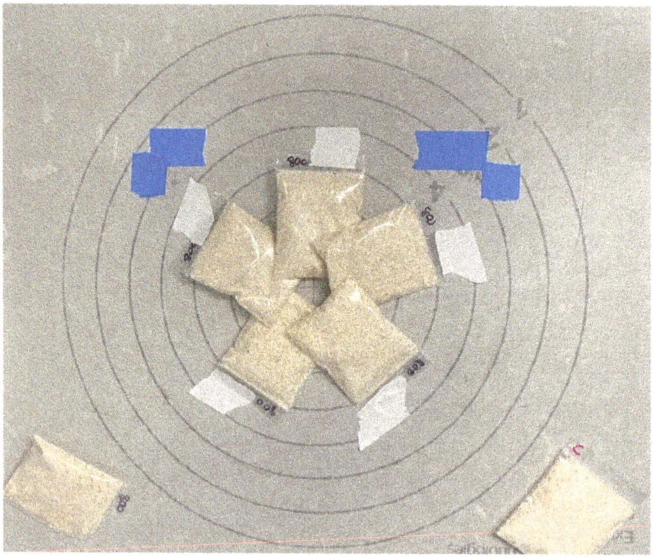

Figure 5. Alignment of the bags of maize at level 5 on the X-ray irradiator plate after the no-dosage bag was removed. Control bags of corn for AFB1 contamination and non-contamination were kept on the plate outside of the zone of irradiation.

4.4. Sample Aliquoting

X-ray-irradiated samples were aliquoted in a Labconco biosafety cabinet (Kansas City, MO, USA). Each bag of ground maize was emptied into a 50 mL Falcon tube and vortexed for 30 sec to homogenize the sample. Each 50 mL tube was then aliquoted for the following assays: 250 mg ± 10 mg directly into bead beating tubes for DNA extractions, 2 g ± 20 mg into a 10 mL tube for dilution plating, and 5 g ± 20 mg into a new 50 mL falcon tube for aflatoxin extraction. Experimental controls were also aliquoted in the biosafety cabinet in the same portions as experimental samples. The negative control was a portion of the composite grain sample before *A. flavus* inoculation. The positive control was a certified reference material with a known AFB1 concentration (200 ppb) purchased from Trilogy Analytical Laboratory (Washington, MO, USA). The samples were then stored at 4 °C in the dark until further analysis.

4.5. Aflatoxin B1 Extraction

Aflatoxin B1 (AFB1) was extracted from the maize samples based on previously published methods based on AOAC official method 993.16 [17]. In a chemical fume hood, 25 mL of 70% methanol was added with a serological pipette to each 5 g maize sample in a 50 mL conical vial. Each sample was vigorously shaken for 3 min, followed by centrifugation for 10 min at 4000 rpm. The liquid phase was filtered with a #1 Whatman filter. Each sample yielded between 12 and 15 mL of filtrate.

4.6. Aflatoxin B1 Quantification

The sample filtrate was aliquoted for ELISA (10 mL) and HPLC (2 mL) AFB1 analysis. ELISA samples were quantified with the AgraQuant® Aflatoxin B1 ELISA kit (Romer Labs, Tulln, Austria). The kit range of quantification was 2–50 ppb, with a limit of detection (LOD) and limit of quantification (LOQ) of 2 ppb. Each maize filtrate was diluted twice to ensure AFB1 quantification, given the small range of AFB1 detection. Briefly, 10 mL of each sample filtrate was diluted with 40 mL of 70% methanol. Half of the resulting filtrate (25 mL) was further diluted with a 1:1 ratio of 25 mL 70% methanol (1:9 ratio of the original filtrate). The two diluted sample filtrates (100 µL of 1:4 and 1:9) were used to quantify AFB1 following the AgraQuant kit instructions. ELISA 96-well plates were read on a BioTek Synergy HTX Multimode Reader (Agilent Technologies, Santa Clara, CA, USA) with the monochromator set at 450 nm. Using the kit-provided AFB1 solutions, a five-point standard curve (0, 2, 5, 20, and 50 ppb) with triplicate readings had an R^2 value of 0.9908.

For HPLC, each 2 mL sample was centrifuged at 10,000 rpm for 10 min, then filtered with a 0.2 µm PTFE filter into a 2 mL amber glass vial. Closely following Agilent Technologies technical application method, samples were injected (20 µL) and ran for 8 min in triplicate on an Agilent 1260 Infinity II HPLC using an Agilent Eclipse XDB-C18 4.6 × 150, 3.5 µm Rapid Resolution column (Agilent Technologies, Santa Clara, CA, USA) and photo diode array (PDA) detector set at a wavelength of 365 nm with a reference wavelength of 380 nm to observe AFB1 [27]. The mobile phase consisted of 50:40:10 water, methanol, and acetonitrile (v/v). The column temperature was 40 °C. The pump flow rate was 0.8 mL/min.

To prepare the HPLC standard curve, a certified reference material aflatoxin mix was obtained from Sigma-Aldrich (CRM46304; St. Louis, MO, USA). The aflatoxin mix contained 1000 µg/kg AFB1, 1000 µg/kg AFG1, 300 µg/kg AFB2, and 300 µg/kg AFB2. The standards were calibrated on the HPLC using the aflatoxin mix detection established by the column manufacturer [27] with a higher injection volume (20 µL instead of 10 µL) to ensure detection. LOD was 5 ppb using a minimum signal-to-noise ratio of 2:1, and LOQ was 20 ppb using a signal-to-noise ratio of 10:1 combined with recovery calculations ± 20% (0.8 to 1.2; Supplementary Table S3 and Supplementary Figure S1). A seven-point standard curve (5, 10, 20, 50, 100, 200, and 500 ppb) with triplicate readings had an R^2 value of 0.9989.

4.7. Aspergillus flavus Quantification

DNA was extracted from 250 mg of each irradiated sample using a Qiagen PowerSoil kit (Germantown, MD, USA) with the modification of a 50 µL final elution. The extracted DNA was stored at −80 °C. Moreover, 8 µL of DNA was used for qPCR with a Youseq *A. flavus* qPCR Test Kit (CAT No.: YSL-qP-IC-A.flavu-100, Youseq Ltd., Winchester, UK), which targets the MDR1 gene and uses Taqman chemistry. PCR was performed with an Applied Biosystems QuantStudio 3 Real-Time PCR (Thermo Fisher Scientific Corp., Waltham, MA, USA).

Colony-forming units (CFU) were determined after suspending 2 g of each sample in 8 mL of sterile 0.1% Triton water and mixing on a rotary shaker for 30 min. Ten-fold serial dilutions were performed to a final dilution of 10^{-6} and 100 uL aliquots from each dilution spread on MDRB and PDA plates with a flame-sterilized spreader. Plates were sealed with parafilm and stored in an incubator at 37 °C for 3 days. Colony-forming units were counted for each plate at 66 h.

4.8. Statistics

Descriptive statistics and a one-way ANOVA test for statistical differences were performed using R Studio Software (version 2023.06.2 + 561). Significance is noted by $p < 0.05$, with additional notations for significance (NS. for $p \geq 0.05$, * for $p < 0.05$, ** for $p < 0.01$, *** for $p < 0.001$).

Supplementary Materials: The following supporting information can be downloaded at: https://www.mdpi.com/article/10.3390/toxins16080329/s1. Table S1: Dilution plating counts for modified Rose Bengal media; Table S2: Dilution plating counts for potato dextrose agar; Table S3: HPLC standard curve, LOD and LOQ calculations; Figure S1: HPLC AFB1 detection limits using certified reference standard diluted with mobile phase.

Author Contributions: Conceptualization, H.G., M.A.W. and L.E.V.-G.; methodology, H.G., C.M. and L.E.V.-G.; validation, M.A.W. and R.K.-B.; formal analysis, H.G. and D.A.; investigation, H.G. and L.E.V.-G.; data curation, H.G.; writing—original draft preparation, H.G; writing—review and editing, H.G., D.A., C.M., M.A.W., R.K.-B. and L.E.V.-G.; visualization, H.G. and D.A.; supervision, R.K.-B. and L.E.V.-G.; funding acquisition, R.K.-B. and L.E.V.-G. All authors have read and agreed to the published version of the manuscript.

Funding: This research was funded by the National Institute of Environmental Health Sciences.

Data Availability Statement: The original contributions presented in the study are included in the article and Supplementary Materials. Further inquiries can be directed to the corresponding authors. The raw data supporting the conclusions of this article will be made available by the authors on request.

Acknowledgments: Thank you to Qiwen Cheng, Ann Miller, Olga Torres, and the Maya Health Alliance Team: Peter Rohloff, Wendy Rodriquez, and Gabriela Montenegro. A special thanks to Douglas Lake at Arizona State University for lending time on the RS2000 X-ray Irradiator.

Conflicts of Interest: The authors declare no conflicts of interest. The funders had no role in the design of the study, in the collection, analysis, or interpretation of data, in the writing of the manuscript, or in the decision to publish the results.

References

1. Rushing, B.R.; Selim, M.I. Aflatoxin B1: A Review on Metabolism, Toxicity, Occurrence in Food, Occupational Exposure, and Detoxification Methods. *Food Chem. Toxicol.* **2019**, *124*, 81–100. [CrossRef]
2. Weaver, M.A.; Bowen, C.; Park, L.C.; Bastidas, A.; Drewry, S.G.; Mandel, J.R. Genetic Diversity of *Aspergillus flavus* on Maize in Guatemala. *Foods* **2023**, *12*, 3864. [CrossRef]
3. Cheng, Q.; Glesener, H.; Montenegro, G.; Torres, O.; Miller, A.C.; Krajmalnik-Brown, R.; Rohloff, P.; Voth-Gaeddert, L.E. Assessment of Aflatoxin Exposure, Growth Faltering and the Gut Microbiome among Children in Rural Guatemala: Protocol for an Observational Prospective Cohort and Bioreactor Simulations. *BMJ Paediatr. Open* **2023**, *7*, e001960. [CrossRef]
4. Voth-Gaeddert, L.; Torres, O.; Maldonado, J.; Krajmalnik-Brown, R.; Rittmann, B.; Oerther, D. Aflatoxin Exposure, Child Stunting, and Dysbiosis in the Intestinal Microbiome Among Children in Guatemala. *Environ. Eng. Sci.* **2019**, *36*, 958–968. [CrossRef]
5. Battilani, P.; Toscano, P.; Van Der Fels-Klerx, H.J.; Moretti, A.; Camardo Leggieri, M.; Brera, C.; Rortais, A.; Goumperis, T.; Robinson, T. Aflatoxin B1 Contamination in Maize in Europe Increases Due to Climate Change. *Sci. Rep.* **2016**, *6*, 24328. [CrossRef]
6. Abbas, H.; Wilkinson, J.; Zablotowicz, R.; Accinelli, C.; Abel, C.; Bruns, H.; Weaver, M. Ecology of *Aspergillus flavus*, Regulation of Aflatoxin Production, and Management Strategies to Reduce Aflatoxin Contamination of Corn. *Toxin Rev.* **2009**, *28*, 142–153. [CrossRef]
7. Erenstein, O.; Jaleta, M.; Sonder, K.; Mottaleb, K.; Prasanna, B.M. Global Maize Production, Consumption and Trade: Trends and R&D Implications. *Food Secur.* **2022**, *14*, 1295–1319. [CrossRef]
8. Caceres, I.; Al Khoury, A.; El Khoury, R.; Lorber, S.; Oswald, I.P.; El Khoury, A.; Atoui, A.; Puel, O.; Bailly, J.-D. Aflatoxin Biosynthesis and Genetic Regulation: A Review. *Toxins* **2020**, *12*, 150. [CrossRef] [PubMed]
9. Bertuzzi, T.; Rastelli, S.; Mulazzi, A.; Pietri, A. Evaluation and Improvement of Extraction Methods for the Analysis of Aflatoxins B1, B2, G1 and G2 from Naturally Contaminated Maize. *Food Anal. Methods* **2012**, *5*, 512–519. [CrossRef]
10. Hedayati, M.T.; Pasqualotto, A.C.; Warn, P.A.; Bowyer, P.; Denning, D.W. *Aspergillus flavus*: Human Pathogen, Allergen and Mycotoxin Producer. *Microbiology* **2007**, *153*, 1677–1692. [CrossRef] [PubMed]
11. Al Hallak, M.; Verdier, T.; Bertron, A.; Roques, C.; Bailly, J.-D. Fungal Contamination of Building Materials and the Aerosolization of Particles and Toxins in Indoor Air and Their Associated Risks to Health: A Review. *Toxins* **2023**, *15*, 175. [CrossRef] [PubMed]
12. Méndez-Albores, J.A.; Arámbula-Villa, G.; Loarca-Piña, M.G.; González-Hernández, J.; Castaño-Tostado, E.; Moreno-Martínez, E. Aflatoxins' Fate during the Nixtamalization of Contaminated Maize by Two Tortilla-Making Processes. *J. Stored Prod. Res.* **2004**, *40*, 87–94. [CrossRef]
13. Weaver, M.A.; Callicott, K.A.; Mehl, H.L.; Opoku, J.; Park, L.C.; Fields, K.S.; Mandel, J.R. Characterization of the *Aspergillus flavus* Population from Highly Aflatoxin-Contaminated Corn in the United States. *Toxins* **2022**, *14*, 755. [CrossRef] [PubMed]
14. Schaarschmidt, S.; Fauhl-Hassek, C. Mycotoxins during the Processes of Nixtamalization and Tortilla Production. *Toxins* **2019**, *11*, 227. [CrossRef] [PubMed]

15. Indiarto, R.; Pratama, A.W.; Sari, T.I.; Theodora, H.C. Food Irradiation Technology: A Review of the Uses and Their Capabilities. *Int. J. Eng. Trends Technol.* **2020**, *68*, 91–98. [CrossRef]
16. Mshelia, R.D.; Dibal, N.I.; Chiroma, S.M. Food irradiation: An effective but under-utilized technique for food preservations. *J. Food Sci. Technol.* **2023**, *60*, 2517–2525. [CrossRef] [PubMed]
17. Mills, A.D.; Levin, T.; Wiser, R.; Seel, J.; Botterud, A. Impacts of variable renewable energy on wholesale markets and generating assets in the United States: A review of expectations and evidence. *Renew. Sustain. Energy Rev.* **2020**, *120*, 109670. [CrossRef]
18. Makari, M.; Hojjati, M.; Shahbazi, S.; Askari, H. Effect of Co-60 Gamma Irradiation on *Aspergillus flavus*, Aflatoxin B1 and Qualitative Characteristics of Pistachio Nuts (*Pistacia vera* L.). *J. Food Meas. Charact.* **2021**, *15*, 5256–5265. [CrossRef]
19. Ghanem, I.; Orfi, M.; Shamma, M. Effect of Gamma Radiation on the Inactivation of Aflatoxin B1 in Food and Feed Crops. *Braz. J. Microbiol.* **2008**, *39*, 787–791. [CrossRef]
20. Khalil, O.A.A.; Hammad, A.A.; Sebaei, A.S. *Aspergillus flavus* and Aspergillus Ochraceus Inhibition and Reduction of Aflatoxins and Ochratoxin A in Maize by Irradiation. *Toxicon* **2021**, *198*, 111–120. [CrossRef]
21. Frink, S.; Marjanovic, O.; Tran, P.; Wang, Y.; Guo, W.; Encarnacion, N.; Alcantara, D.; Moezzi, B.; Vrdoljak, G. Use of X-ray Irradiation for Inactivation of Aspergillus in Cannabis Flower. *PLoS ONE* **2022**, *17*, e0277649. [CrossRef] [PubMed]
22. Aquino, S.; Ferreira, F.; Ribeiro, D.H.B.; Corrêa, B.; Greiner, R.; Villavicencio, A.L.C.H. Evaluation of Viability of *Aspergillus flavus* and Aflatoxins Degradation in Irradiated Samples of Maize. *Braz. J. Microbiol.* **2005**, *36*, 352–356. [CrossRef]
23. Nemţanu, M.R.; Braşoveanu, M.; Karaca, G.; Erper, İ. Inactivation Effect of Electron Beam Irradiation on Fungal Load of Naturally Contaminated Maize Seeds: Fungal Inactivation of Contaminated Maize Seeds by Irradiation. *J. Sci. Food Agric.* **2014**, *94*, 2668–2673. [CrossRef] [PubMed]
24. Markov, K.; Mihaljević, B.; Domijan, A.-M.; Pleadin, J.; Delaš, F.; Frece, J. Inactivation of Aflatoxigenic Fungi and the Reduction of Aflatoxin B1 in Vitro and in Situ Using Gamma Irradiation. *Food Control* **2015**, *54*, 79–85. [CrossRef]
25. Pietarinen, V.-M.; Rintala, H.; Hyvärinen, A.; Lignell, U.; Kärkkäinen, P.; Nevalainen, A. Quantitative PCR Analysis of Fungi and Bacteria in Building Materials and Comparison to Culture-Based Analysis. *J. Environ. Monit.* **2008**, *10*, 655. [CrossRef]
26. 21 CFR Part 179—Irradiation in the Production, Processing and Handling of Food. Available online: https://www.ecfr.gov/current/title-21/part-179 (accessed on 13 May 2024).
27. Barbas, C.; Dams, A.; Majors, R.E. Analysis of Organic Acids Using High-Performance Liquid Chromatography. Agilent Technologies. Available online: https://www.agilent.com/Library/applications/5989-3634EN.pdf (accessed on 12 March 2023).

Disclaimer/Publisher's Note: The statements, opinions and data contained in all publications are solely those of the individual author(s) and contributor(s) and not of MDPI and/or the editor(s). MDPI and/or the editor(s) disclaim responsibility for any injury to people or property resulting from any ideas, methods, instructions or products referred to in the content.

toxins

MDPI

Article

Combining Multiple Omics with Molecular Dynamics Reveals SCP2-Mediated Cytotoxicity Effects of Aflatoxin B1 in SW480 Cells

Mengting Chen [1,†], Jiaxin Wen [1,†], Yiyan Qiu [1,2,†], Xinyue Gao [1], Jian Zhang [1,3], Yifan Lin [1], Zekai Wu [1], Xiaohuang Lin [1] and An Zhu [1,3,*]

1 Key Laboratory of Gastrointestinal Cancer (Fujian Medical University), Ministry of Education, Fuzhou 350108, China
2 Jimei District Center for Disease Control and Prevention, Xiamen 361022, China
3 School of Public Health, Fujian Medical University, Fuzhou 350108, China
* Correspondence: zhuan@fjmu.edu.cn
† These authors contributed equally to this work.

Abstract: Aflatoxins belong to a class of mycotoxins, among which aflatoxin B1 (AFB1) has detrimental effects on the health of both animals and humans. It is associated with long-term exposure-induced carcinogenicity, hepatotoxicity, renal toxicity, neurotoxicity, and immunosuppressive properties, resulting in a variety of diseases. The intestine is the first barrier for human exposure to AFB1, but limited investigations have been conducted to clarify the underlying mechanisms of intestinal cytotoxicity. The mechanism of AFB1-induced cytotoxicity was investigated in this study using an integrated approach combining transcriptome, proteome, and metabolome analysis along with molecular dynamics simulation. After exposing SW480 cells to 50 μM AFB1 for 72 h, the transcriptome, proteome, and metabolome exhibited significant enrichment in pathways associated with oxidative stress, fatty acid and lipid metabolism, and glutathione metabolism. The experimental results demonstrated that AFB1 significantly reduces SW480 cells viability, and induces oxidative stress, calcium overload, mitochondrial damage, and lipid metabolism disorders.

Keywords: aflatoxin B1; multiple omics; intestinal toxicity; oxidative stress; mitochondrial damage; lipid metabolism disorder

Key Contribution: Transcriptome, proteome, and metabolome sequencing combined with molecular dynamics were used to explore the mechanism of AFB1-induced SW480 cell cytotoxicity. AFB1 resulted in the decreased viability of SW480 cells, altered levels of oxidative stress and lipid metabolism, and mitochondrial damage.

1. Introduction

Aflatoxins are secondary metabolites produced by *Aspergillus flavus* and *Aspergillus parasitic*. Among them, aflatoxin B1 (AFB1) is one of the most widely distributed and toxic variants, and was classified as a Group I carcinogen by the World Health Organization (WHO) in 1993 [1]. The presence of AFB1 is commonly detected in various crops, including cereals, peanuts, maize, and their by-products, throughout the entire growth cycle as well as during transportation and storage processes [2]. The acceptable limits of AFB1 in food for human consumption and animal feed have been reported to be 40 parts per billion (ppb) and 300 ppb, respectively [3]. The consumption of feed contaminated with AFB1 significantly impairs animal performance and health [4,5]. Therefore, AFB1 contamination poses one of the most critical problems threatening food safety.

Transcriptomics is the technology that focuses on the types, abundance, and spatiotemporal expression patterns of mRNA, which is transcribed from DNA [6]. In 2001, the human genome sequence, covering more than 96% of the euchromatic regions, was completed roughly through the cooperation of 20 groups [7]. Since then, model organisms such as mice [8], rats [9], and chimpanzees [10] were sequenced to build genome sequences, which have laid the foundation for subsequent research. Recently, Wu et al. [11] proved that the reproductive toxicity of oridonin was induced by the tight junction, mitochondrial damage, and the Wnt signaling pathway, utilizing high-throughput sequencing and various genome databases. Therefore, transcriptomics has not only deepened the understanding of gene function and regulatory mechanisms but also provided insights into the molecular-level responses of organisms to environmental changes, developmental processes, and disease.

Proteomics has made remarkable progress in recent years, which has greatly expanded its applications in biomedicine [12]. The technology makes it possible to specifically identify and measure the expression levels of multiple proteins in biological samples, providing valuable information for the recognition of disease mechanisms and biomarkers. Furthermore, proteomics has been increasingly utilized in the study of fungal toxicity mechanisms. Previous research has revealed the antifungal mechanism of copper through proteomic analysis; namely, copper stress down-regulated protein synthesis and up-regulated protein degradation in *Cryptococcus neoformans*, thereby inhibiting its growth [13]. In another study, the potential mechanism underlying the damage to intestinal epithelial integrity caused by the combination of aflatoxin M1 (AFM1) and ochratoxin A was analyzed using integrated transcriptomics and proteomics [14].

Metabolomics has received considerable attention as a novel experimental approach for monitoring dynamic changes in metabolites and identifying key targets to elucidate the mechanisms underlying metabolic processes [15]. Metabolomics analysis has been widely applied in the study of diverse diseases, like cardiovascular diseases [16], neurodegenerative diseases [17], and digestive diseases [18]. The metabolomics results of male F344 rats gavaged with 75 μg/kg AFB1 for 5 weeks demonstrate that AFB1 disrupts the production of fatty acids, amino acids, and carbohydrates, leading to intestinal toxicity [19]. In addition, the metabolomics analysis of NCM460 cells treated with 2.5 μM AFB1 for 24 h revealed that AFB1 disrupted intestinal homeostasis by significantly increasing the content of glycerophospholipids, thereby exerting enteric toxicity [20].

The intestinal tract not only acts as the main location for nutrient digestion and absorption, but also functions as a vital barrier against toxins and detrimental substances. Consequently, it is common to observe higher concentrations of mycotoxins within intestinal cells compared to other tissues [21]. The presence of AFB1 induces intestinal injury across different species, as it alters the structure of intestinal epithelial cells and compromises the integrity of the intestinal barrier, ultimately resulting in a loss of barrier function [4]. Prolonged exposure to a diet with a low level of AFB1 resulted in a reduction in the digestibility of the total intestine and destroyed the integrity of the pig intestinal barrier by diminishing its antioxidant capacity and promoting proinflammatory cytokine production [22]. In this investigation, AFB1 was used to treat SW480 cells to study potential mechanisms of cytotoxicity via multi-omics and molecular dynamics.

2. Result

2.1. Cell Viability

The cell viability was determined via the MTT assay following the culture of SW480 cells with diverse doses of AFB1 for 72 h (Figure 1B). Compared to the control group, cell viability was reduced to 86.8%, 67.0%, 47.5%, and 16.9% with an increase of 5, 10, 50, and 100 μM AFB1.

2.2. Transcriptomic Analysis Based on Differentially Expressed Genes (DEGs)

The transcriptomic analysis was performed to investigate the alterations in gene function associated with AFB1 treatment in SW480 cells. The control group can be distinguished from the AFB1 group based on principal component analysis (PCA) score plots (Figure 2A). A total of 13,850 genes were detected and matched based on gene symbols through mRNA-seq. The differential expression analysis comparing the control group with the AFB1 group revealed a total of 1037 DEGs meeting the criteria of a fold change >2 or <0.5 and a p value < 0.05. The expression levels of 320 genes were found to be up-regulated, while those of 717 genes were down-regulated among these DEGs (Figure 2B). We performed a gene ontology (GO) and Kyoto Encyclopedia of Genes and Genomes (KEGG) enrichment analysis of the DEGs. The DEGs in the control group and AFB1 group were mainly enriched in response to oxidative stress, long-chain fatty acid metabolic process, endoplasmic reticulum lumen, triglyceride lipase activity, calcium signaling pathway, and ferroptosis (Figure 2C). The Sankey diagram presented in Figure 2D illustrates the specific genes that exhibit enrichment under the GO and KEGG entries mentioned above. We analyzed the *NQO1* and *CAT* genes enriched in response to oxidative stress and the *CYP4F3* and *CYP4F12* genes enriched in the long-chain fatty acid metabolic process to explore their specific changes. The integrative genomics viewer (IGV) plots (Figure 2E) showed the up-regulation of *CAT* and *NQO1*, while a down-regulation of *CYP4F12* and *CYP4F3* was observed following AFB1 treatment.

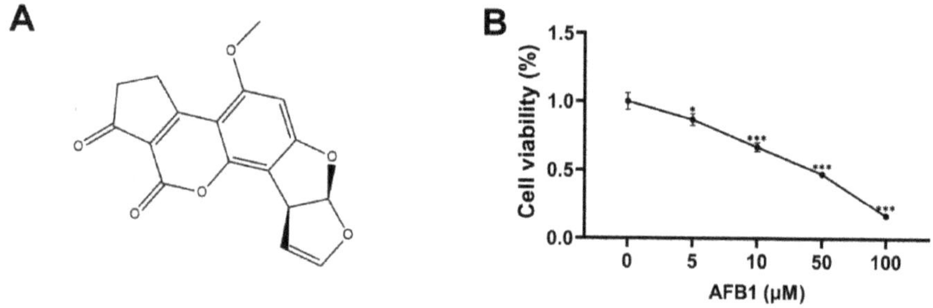

Figure 1. (**A**) The chemical structure of AFB1. (**B**) The viability of SW480 cells following treatment with 0, 5, 10, 50, and 100 μM AFB1 for 72 h. The data were presented as the mean ± SD. * $p < 0.05$, *** $p < 0.001$, n = 3.

The gene set enrichment analysis (GSEA) enrichment plots (Figure 3A–D) displayed the up-regulation of pathways associated with antioxidant activity and cellular oxidant detoxification, while down-regulation was observed in the pathways related to calcium ion homeostasis and lipase activity. The Circos heatmaps visualize the expression of genes associated with antioxidant activity and lipase activity, as illustrated in Figure 3E and 3F, respectively. After 72 h of AFB1 treatment of SW480 cells, it was observed that the gene expression related to antioxidant activity and lipase activity exhibited an inverse alteration compared to the control group.

2.3. Proteomics Analysis Based on Differentially Expressed Proteins (DEPs)

The distinction in proteome samples between the untreated and treated AFB1 groups can be observed through the PCA score plots (Figure 4A). Correlation analysis of transcriptomic and proteomic results using nine quadrants revealed a total of 69 genes and proteins exhibiting concordant changes, with 41 showing up-regulation and 28 displaying down-regulation (Figure 4B). With $p < 0.05$ and a fold change >1.2 or <0.83 as the threshold, a total of 1764 DEPs were identified, including 1354 up-regulated proteins and 410 down-regulated proteins (Figure 4C). In order to further clarify the functional roles of the 1764 DEPs, GO and KEGG pathway enrichment analyses were conducted. The GO analysis demonstrated significant enrichment in fatty acid beta-oxidation, cellular response to oxidative stress, endoplasmic reticulum localization, and oxidoreductase activity (Figure 4D). The KEGG pathway is primarily enriched in metabolic pathways, fatty acid metabolism, and chemical carcinogenesis–reactive oxygen species (Figure 4E). GSEA enrichment analysis (Figure 5A–D) showed that calcium ion binding was down-regulated and endoplasmic reticulum organization, fatty acid metabolic process, and lipid modification were up-regulated. The PPI interaction network maps (Figure 5E) were constructed based on the proteins enriched in the aforementioned four pathways, and the corresponding protein expression levels were visualized in Figure 5F using heat maps.

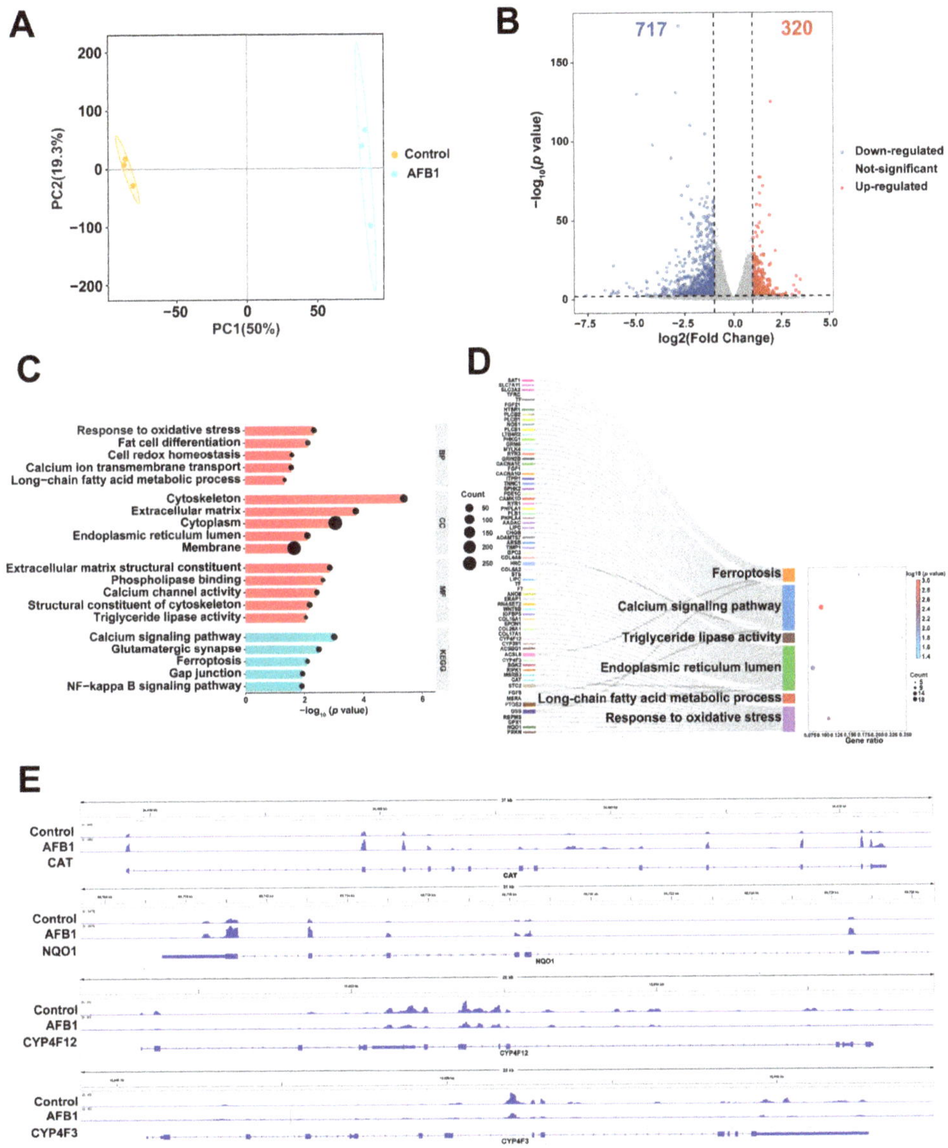

Figure 2. (**A**) Performing PCA on mRNA expression in the control and AFB1 groups. (**B**) Volcano plot shows a total of 1037 differential genes, 320 up-regulated and 717 down-regulated between the two groups, were screened. (**C**) The GO and KEGG enrichment of DEGs. (**D**) The Sankey diagram of ferroptosis, calcium signaling pathway, triglyceride lipase activity, endoplasmic reticulum lumen, long-chain fatty acid metabolic process, and response to oxidative stress pathways. (**E**) The level of *CAT*, *NQO1*, *CYP4F12*, and *CYP4F3* mRNA transcripts observed by IGV.

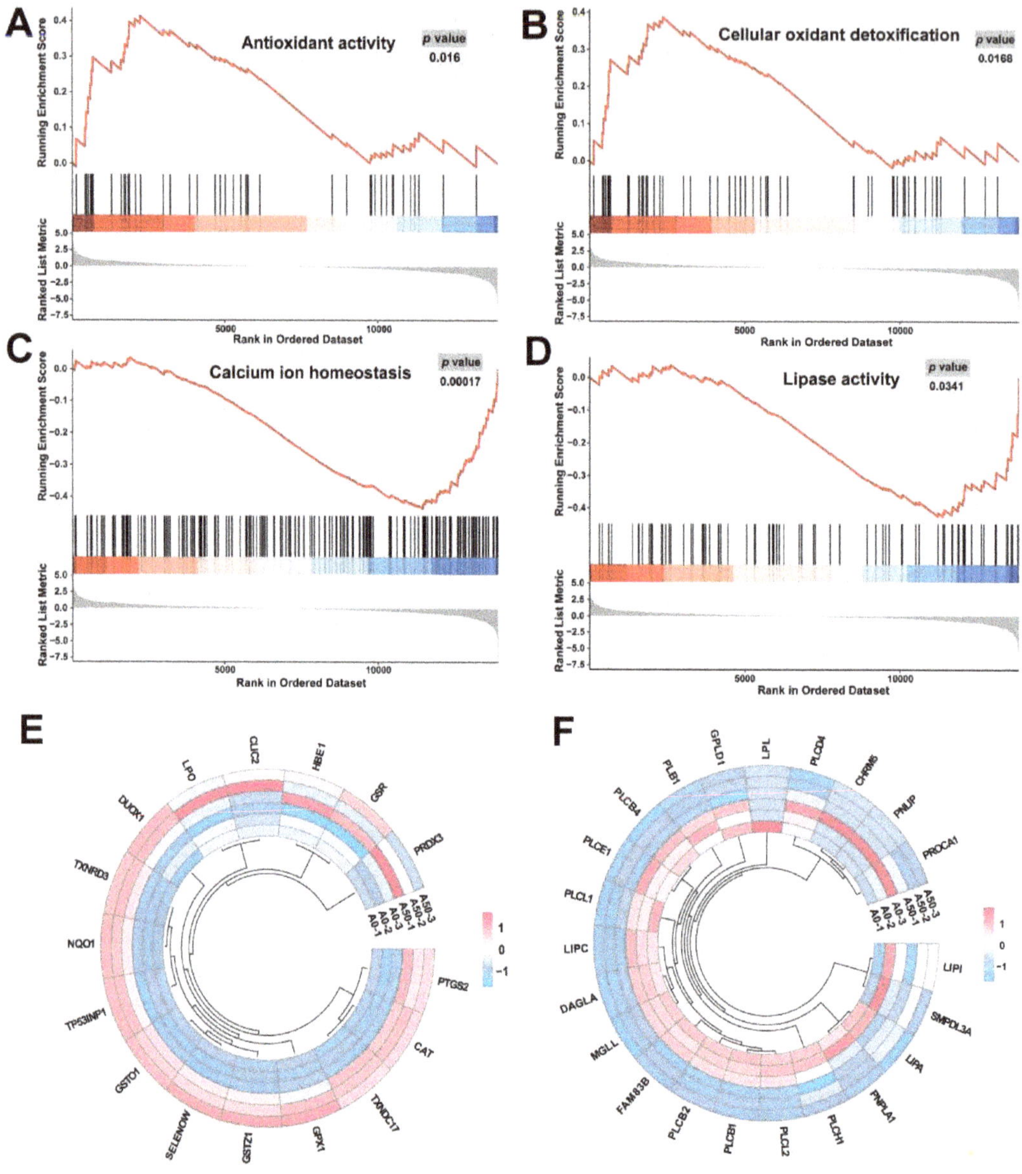

Figure 3. GSEA demonstrated that genes in AFB1 group were enriched in antioxidant activity (**A**), cellular oxidant detoxification (**B**), calcium ion homeostasis (**C**), and lipase activity (**D**). The Circos heatmaps revealed the association of the antioxidant activity (**E**) and lipase activity (**F**) related genes in AFB1 and control groups.

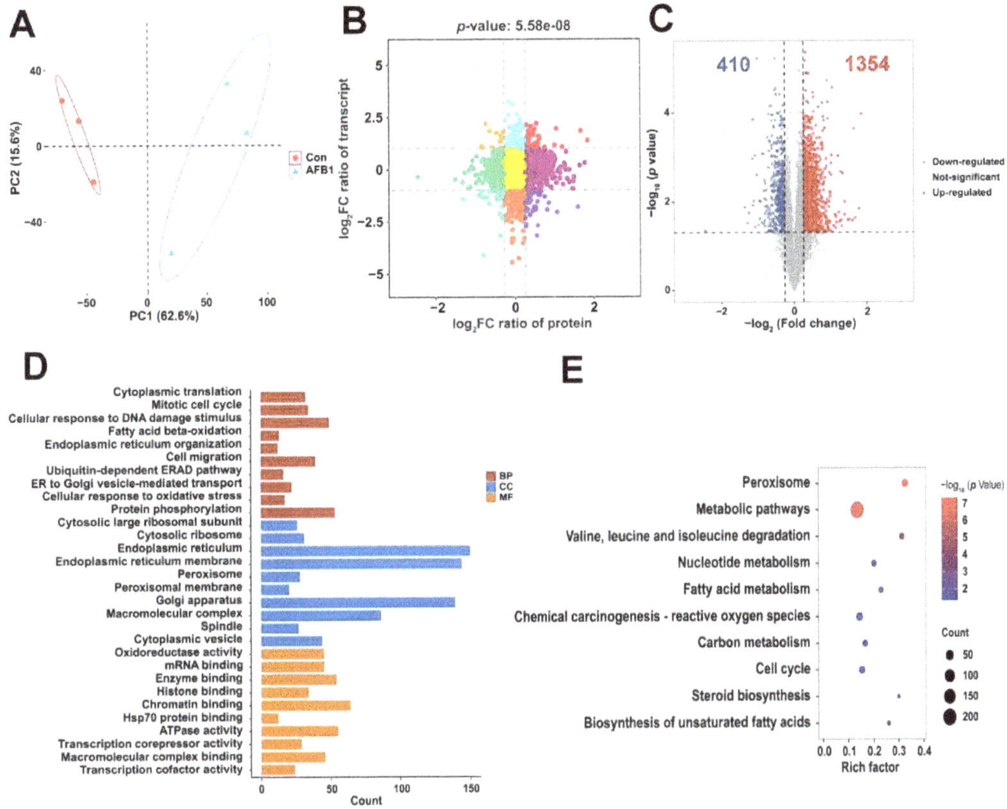

Figure 4. (**A**) The PCA of the protein expression between the AFB1 and control groups. (**B**) The nine-quadrant diagram depicting the correlation between protein expression and mRNA expression. (**C**) Volcano plot showcasing protein variations between AFB1 and the control group. Red represents up-regulated alterations, while blue represents down-regulated alterations. Analysis of enrichment of (**D**) GO terms ($p < 0.05$) and (**E**) KEGG pathway ($p < 0.05$) for DEPs.

2.4. Analysis of the Metabolomics Data between the Group Treated with AFB1 and the Control Group

The PCA score scatter plots for all samples are presented in Figure 6A, revealing a marked distinction between the AFB1-treated and control groups. Further analysis of the data was carried out by means of orthogonal partial least squares discriminant analysis (OPLS-DA) (Figure 6B). The model-fitting parameter (R^2X) and the model-discrimination parameter (R^2Y) of the untargeted metabolomics approaches were, respectively, 0.663 and 0.998, suggesting that the model had a suitable fit (Figure 6C). The Q^2 values in the original model were extremely close to 1, indicating the model's reliability. In this study, 2410 metabolites were identified. The metabolites that have significant differences according to VIP values are shown in the OPLS-DA S-plot. (Figure 6D). The metabolite set enrichment analysis (MSEA) pathway analysis revealed that the metabolites were significantly enriched in pantothenate and CoA biosynthesis, glutathione metabolism, valine, leucine and isoleucine degradation, and fatty acid degradation (Figure 6E).

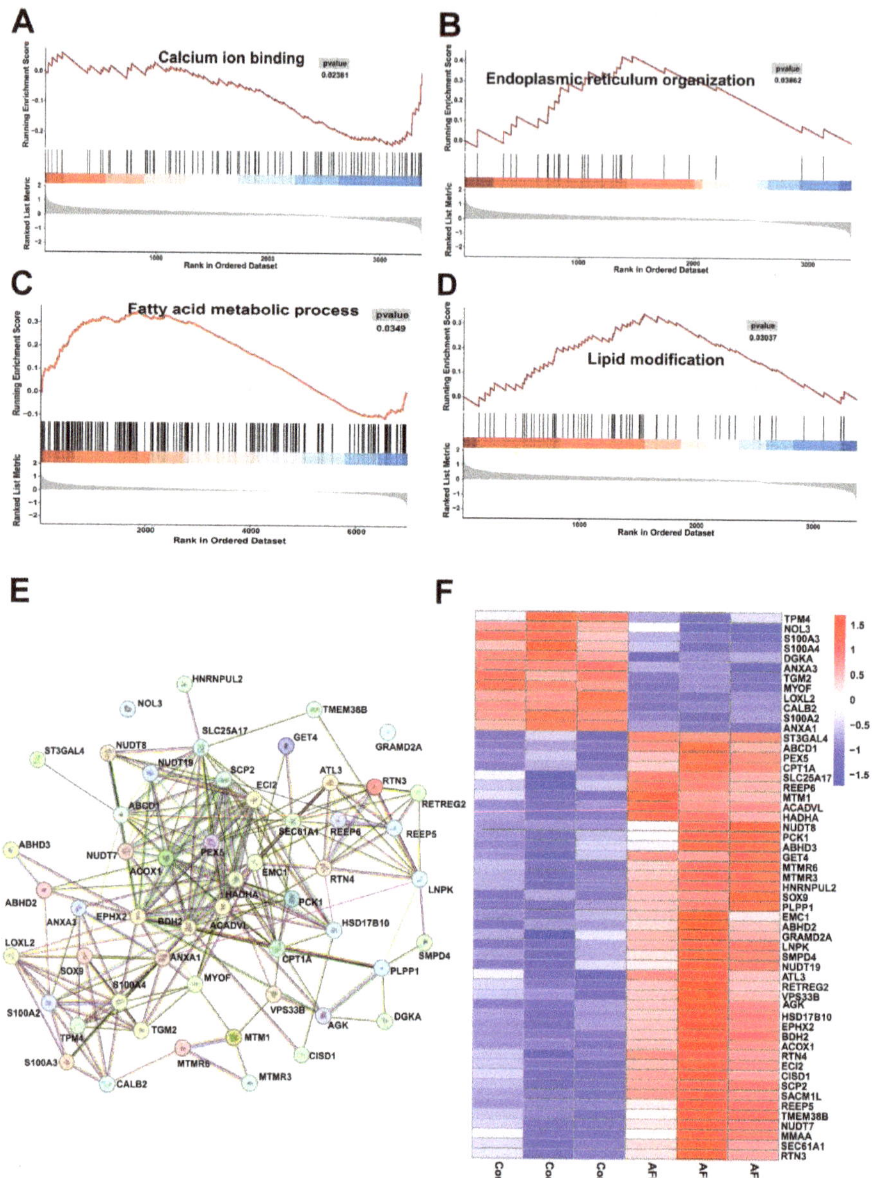

Figure 5. Comparison of the AFB1 group with the control group using GSEA. Calcium ion binding (**A**), endoplasmic reticulum organization (**B**), fatty acid metabolic process (**C**), and lipid modification (**D**). (**E**) The interaction among the genes related to calcium ion binding, endoplasmic reticulum organization, fatty acid metabolic process, and lipid modification. (**F**) The heatmap clearly demonstrates the significant disparity in the DEPs of calcium ion binding, endoplasmic reticulum organization, fatty acid metabolic process, and lipid modification.

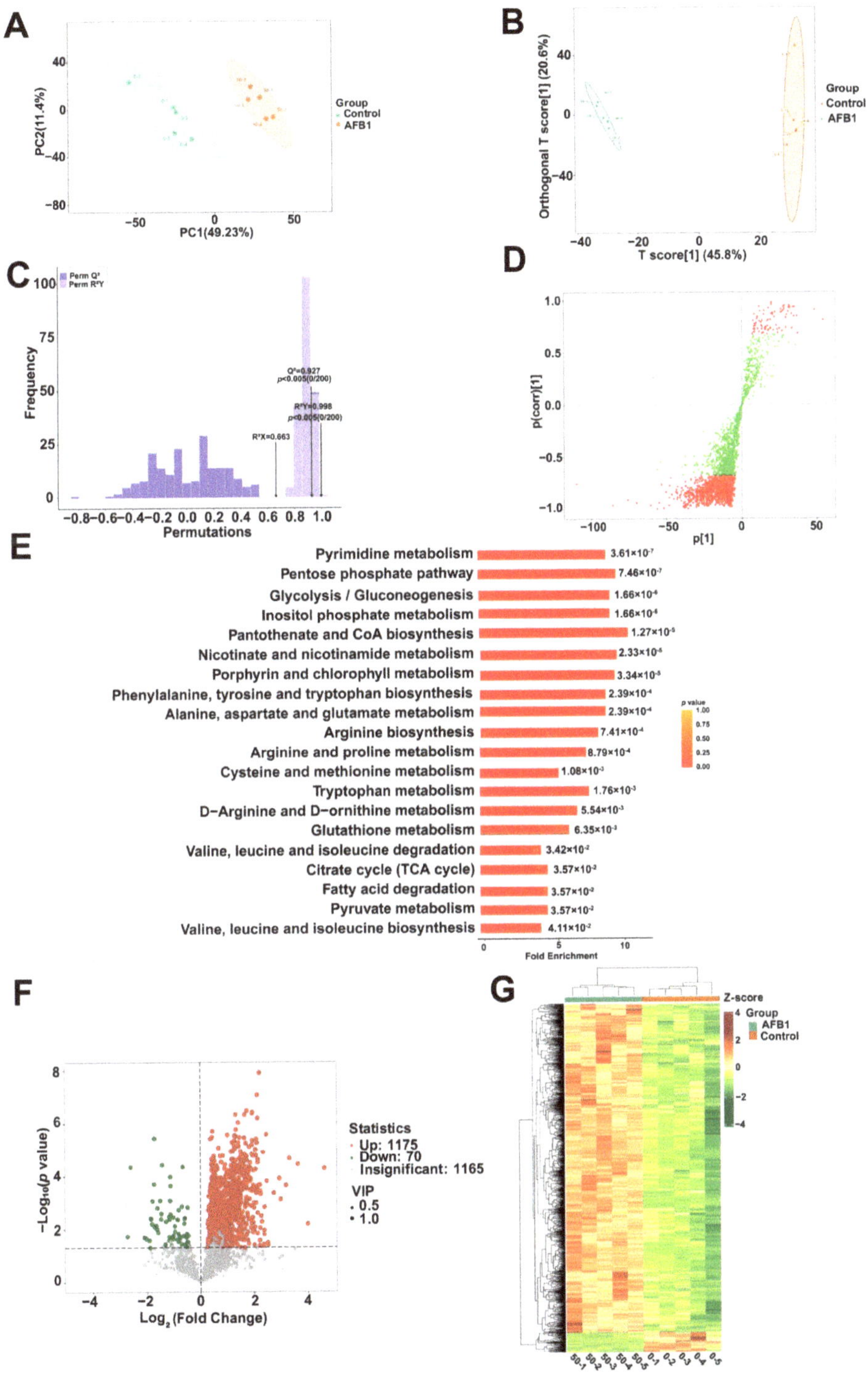

Figure 6. (**A**) The two-dimensional images resulting from the PCA analysis. (**B**) The OPLS-DA score plots indicate significant isolation and discrimination between the AFB1 and control groups. (**C**) The

OPLS-DA model underwent testing with 200 random permutations. (**D**) OPLS-DA S-plot of the AFB1 and control groups. (**E**) MSEA enrichment analysis of DEMs between AFB1 and control groups. (**F**) The volcano plot shows DEMs between control groups and AFB1-treated group. (**G**) Heatmap of metabolomics between AFB1 and control groups. The DEMs with VIP scores > 1 and p value < 0.05 are displayed in the heatmap.

Differential expression metabolites (DEMs) were defined as metabolites with VIP > 1 and $p < 0.05$. Subsequently, it was found that 70 metabolites were down-regulated, while 1175 metabolites were up-regulated (Figure 6F). As depicted in the volcano plot, the red and green dots represent up-regulated and down-regulated DEMs in SW480 cells, respectively. A heat map was constructed (Figure 6G) for the DEMs, which showed significant differences between the AFB1-treated and control groups. The DEMs detected were annotated using the KEGG database (Figure 7A). The KEGG classification results and enrichment analysis suggest that the DEMs revealed are related to cellular processes, human diseases, organismal systems, and environmental information processing. The enriched KEGG pathways include biosynthesis of unsaturated fatty acids, chemical carcinogenesis–DNA adducts, chemical carcinogenesis–reactive oxygen species, nicotinate and nicotinamide metabolism, porphyrin metabolism, lipid and atherosclerosis, and glutathione metabolism. Selection of DEMs in the chemical carcinogenesis–reactive oxygen species and nicotinate and nicotinamide metabolism pathways were made for cluster analysis (Figure 7B–C). A bar chart was constructed relying on the magnitude of the log2Fold change values for the DEMs (Figure 7D). In the AFB1-treated group, the most remarkable up-regulation compared to the control group was witnessed in His–Asp–Glu, 7′,8′-Dihydro-8′-hydroxyreticulataxanthin, and the S-adenosylmethionine metabolite. In the OPLS-DA model, the top 20 DEMs having the most significant VIP values were chosen to construct a scatter plot (Figure 7E) and violin plot (Figure 7F) to show the differences and changes in metabolites more lucidly and visually. The names of the 20 metabolites are listed in Table S1 corresponding to the IDs.

2.5. The Potential Target for AFB1 Toxicity

It can be seen from Table 1 and Figure 8A–D that the interactions between AFB1 and related proteins were predicted. Specifically, AFB1 bound CAT through the formation of three hydrogen bonds at Arg72 and Arg112, as well as ten hydrophobic contacts with Val73, Tyr358, Val74, Arg354, Val146, His75, Phe132, Gly131, Ser114, and Ala133 (Figure 8A). AFB1 bound ACSL1 by forming one hydrogen bond at Ser581 and twelve hydrophobic contacts with Met620, His683, Val743, Met499, Thr650, Val675, Ala622, Leu767, His580, Ser497, Leu766, and Gly498 (Figure 8B). Additionally, AFB1 bound SCP2 through the formation of two hydrogen bonds at So4122 and Lys115, along with seven hydrophobic contacts with Leu114, Val82, Gln111, Val83, Phe79, Ile15, and Val11 (Figure 8C). Finally, AFB1 bound SEC61A1 by forming seven hydrophobic contacts with TIle292, Ser287, Gln456, Ser180, Gly176, Ser177, and Asn288 (Figure 8D). The protein SCP2 plays a crucial role in facilitating the redistribution of cholesterol from the plasma membrane to various intracellular locations, and promoting efficient cholesterol transport from the plasma membrane to the endoplasmic reticulum [23]. In Figure 8E–H, the time evolution of the root mean square deviation (RMSD), radius of gyration (Rg), total binding free energy, and hydrogen bond number are shown. The binding of the AFB1-SCP2 complex remained stable throughout the 50 ns dynamic simulation, with RMSD and root mean square fluctuation (RMSF) values (Figure 8I) fluctuating by less than 5Å. The hydrogen bond occupancy rate of each residue is depicted in Figure 8J, with three sites exhibiting an occupancy rate exceeding 10%. The occupancy rate of the hydrogen bond involving isoleucine in complex 82 exhibited the highest value, reaching 50%.

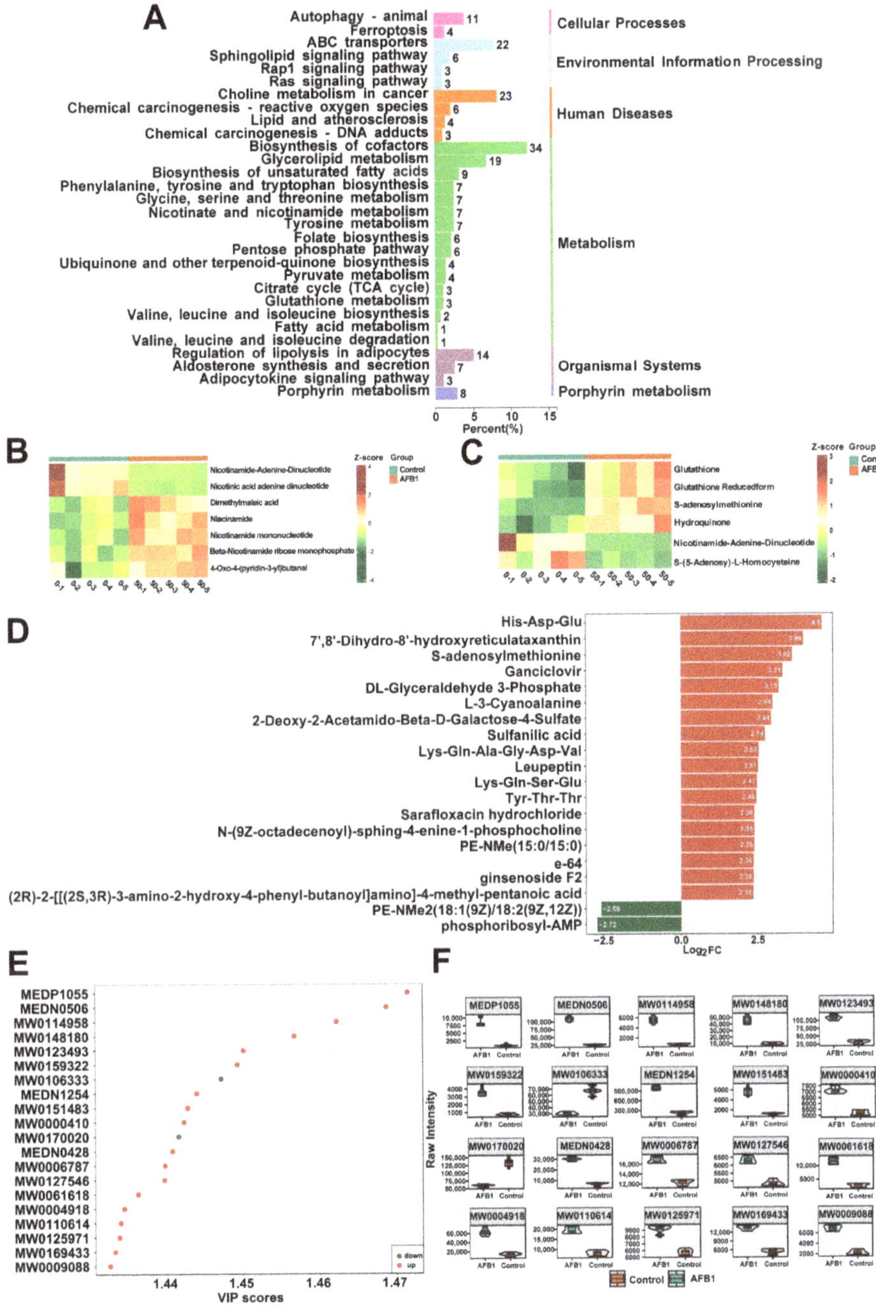

Figure 7. (**A**) The KEGG pathway analysis of DEMs between AFB1-treated and control groups. Heatmap of DEMs of nicotinic acid and nicotinamide metabolism (**B**) and chemical carcinogenesis–reactive oxygen species (**C**). (**D**) The bar plot shows representative metabolites that were significantly up-regulated or down-regulated in the top 20 DMEs. (**E**) The scatter plot of the top 20 DMEs for VIP values. (**F**) Metabolite violin diagram based on the top 20 VIP values.

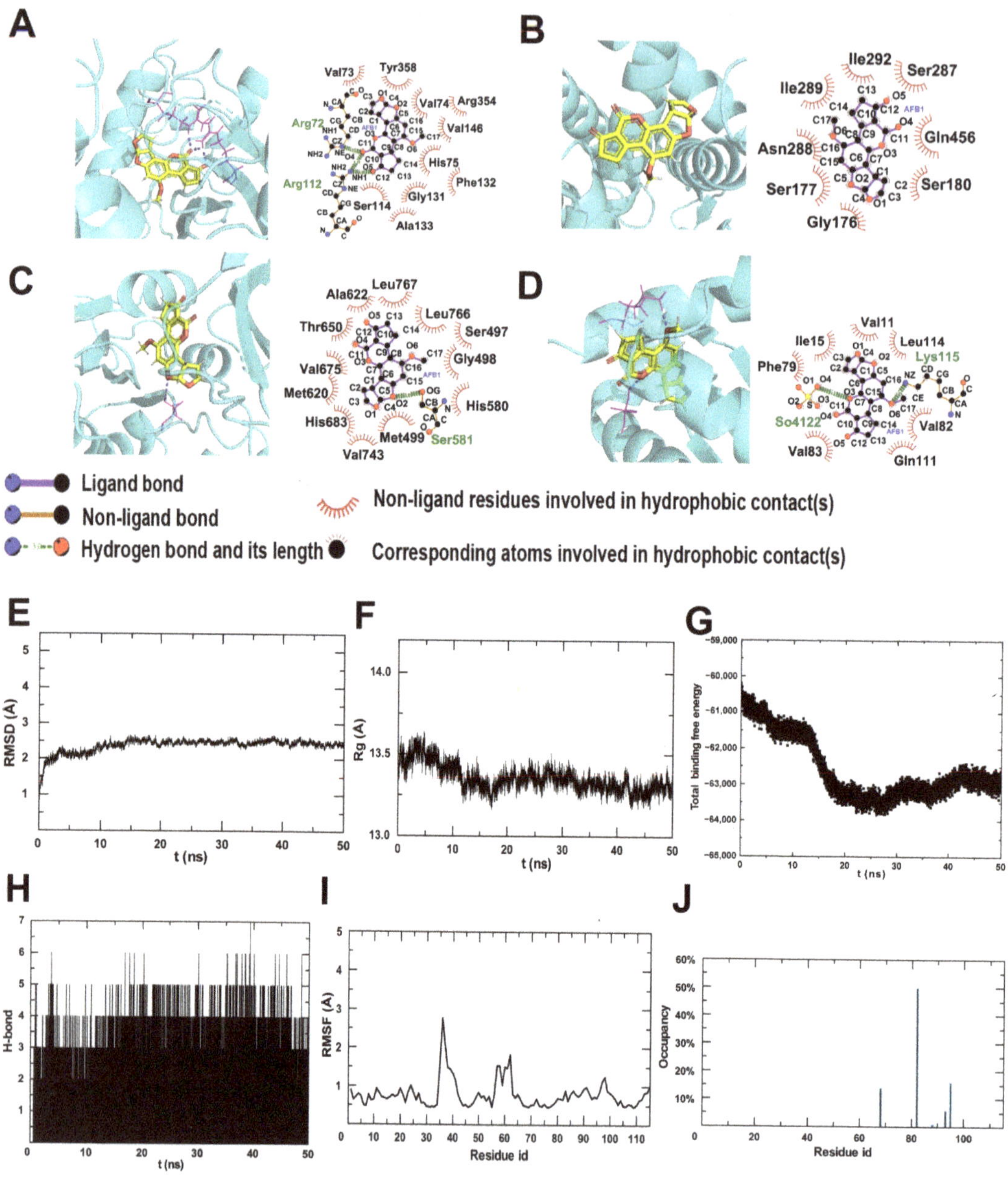

Figure 8. The interactions between AFB1 and related proteins of CAT (**A**), SEC61A1 (**B**), ACSL1 (**C**), and SCP2 (**D**). (**E**) Time evolution of RMSD, Rg (**F**), total binding free energy (**G**), and hydrogen bond number (**H**). (**I**) RMSF of each residue from SCP2. (**J**) H-bond occupancy rate of each residue.

Table 1. Interactions between AFB1 and oxidation and lipid-related proteins.

Protein	PDB ID	Total Score	Crash	Polar	H-Bond Number	Residues Involved in H-Bond Formation	Hydrophobic Contact Number	Residues Involved in Hydrophobic Contacts
CAT	8HID	7.7967	−0.2167	0	3	Arg72, Arg112 (2 H-bonds)	10	Val73, Tyr358, Val74, Arg354, Val146, His75, Phe132, Gly131, Ser114, Ala133
SEC61A1	8DNV	5.6242	−0.7089	2.9825	0	–	7	Ile292, Ser287, Gln456, Ser180, Gly176, Ser177, Asn288
ACSL1	2JFK	5.5354	−0.9486	1.9938	1	Ser581	12	Met620, His683, Val743, Met499, Thr650, Val675, Ala622, Leu767, His580, Ser497, Leu766, Gly498
SCP2	1IKT	5.1034	−0.5708	1.8757	2	So4122, Lys115,	7	Leu114, Val82, Gln111, Val83, Phe79, Ile15, Val11
LPIN1	5WGB	4.8557	−0.6691	3.1075	2	Tyr26, Arg34	5	Val33, Thr415, Glu417, Thr30, Glu416
NQO1	5EA2	4.24	−0.6254	1.0619	2	Lys113, Tyr126	4	Tyr67, Phe120, Glu117, Phe178

2.6. Oxidative Stress Induced by AFB1

Intracellular reactive oxygen species (ROS) levels in SW480 cells were assessed using the fluorescent probe DCFH-DA. A significant increase in ROS levels was observed when cells were exposed to 50 μM AFB1 compared to the control group (Figure 9A). The fluorescence intensity of DCFH-DA showed a significant increase from 100% in the control group to 304% in the group treated with 50 μM AFB1. Overall, AFB1 stimulation triggered the production of intracellular ROS in SW480 cells.

2.7. The Antioxidant Enzyme Activity

Figure 9B–C demonstrate a significant increase in CAT and GST levels in AFB1-treated SW480 cells compared to untreated cells. The quantitative values of CAT activity were 0.58 ± 0.07, 1.60 ± 0.12, 2.22 ± 0.09, and 3.24 ± 0.1 U/mg prot in the 0, 5, 10, and 50 μM AFB1 groups, respectively. The quantitative values of GST activity were 0.122 ± 0.002, 0.165 ± 0.011, 0.157 ± 0.005, and 0.156 ± 0.003 U/mg prot, respectively. These results imply an imbalance in the cellular oxidative and antioxidant levels, leading to the occurrence of oxidative stress.

2.8. The Exposure to AFB1 Leads to Excessive Accumulation of Intracellular Ca^{2+}

The representative images depicting intracellular Ca^{2+} levels are presented in Figure 9D, illustrating the dose-dependent impact of AFB1 on Ca^{2+} concentrations in SW480 cells. The p values for the 0, 5, 10, and 50 μM AFB1 groups were found to be 0.004, 0.001, and 0.000001, respectively. The green fluorescence results revealed the occurrence of cellular Ca^{2+} overload.

2.9. Fatty Acid Localization

BODIPY 500/510 C1, C12 is a BODIPY fluorescent-labeled saturated fatty acid capable of traversing the cellular membrane to specifically stain polar lipids within the cell. When the probe is incorporated into living cells, an excitation dimer can be formed with a maximum excitation wavelength of 500 nm and a maximum emission wavelength of 510 nm. As shown in Figure 9E, the green fluorescence indicated the condensation of lipid droplets after AFB1 treatment, and the control group was labeled with a normal cell membrane. The result indicates that the transport and metabolism of fatty acids in SW480 cells exhibit impairments.

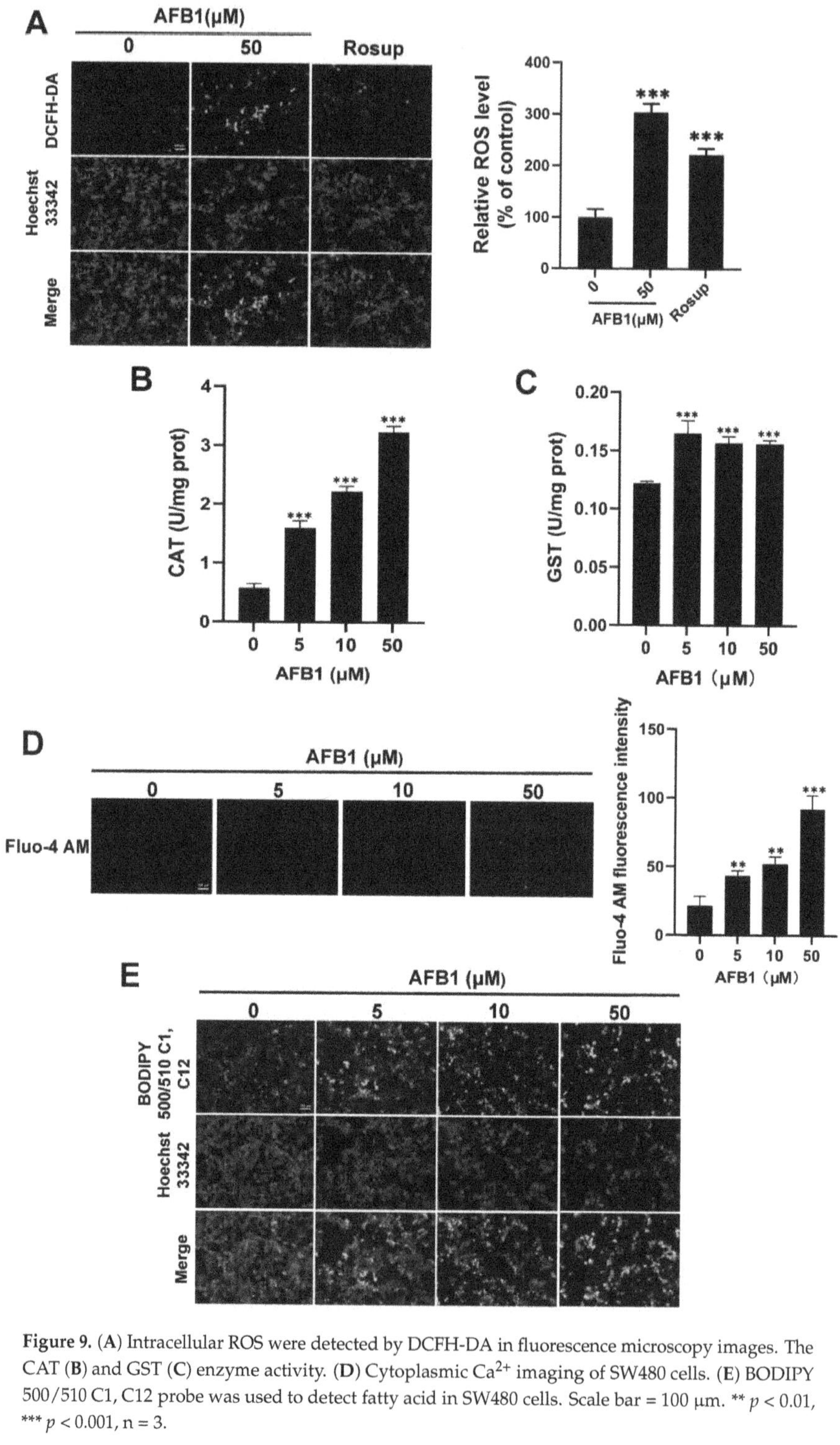

Figure 9. (**A**) Intracellular ROS were detected by DCFH-DA in fluorescence microscopy images. The CAT (**B**) and GST (**C**) enzyme activity. (**D**) Cytoplasmic Ca^{2+} imaging of SW480 cells. (**E**) BODIPY 500/510 C1, C12 probe was used to detect fatty acid in SW480 cells. Scale bar = 100 μm. ** $p < 0.01$, *** $p < 0.001$, n = 3.

2.10. AFB1-Induced Mitochondrial Dysfunction

In comparison with the control group, the red–green fluorescence ratio in the groups treated with 0, 5, 10, and 50 μM of AFB1 exhibited a downward trend, with values of 2.14 ± 0.18, 1.52 ± 0.09, and 1.07 ± 0.09, respectively (Figure 10A). This demonstrated that treating with AFB1 caused a decrease in the MMP of SW480 cells. The assessment of mitochondrial function was conducted using the mPTP. The mPTP is a non-specific channel composed of the inner and outer mitochondrial membranes which serves as the structural basis for mitochondrial function. Under physiological conditions, the mPTP toggles between the open and closed states with the uptake and release of Ca^{2+} by mitochondria. However, in the event of oxidative damage or cell death, the mPTP would remain continuously open.

The Calcein AM probe can passively enter cells and accumulate in mitochondria, where the nearly nonfluorescent Calcein AM is hydrolyzed by esterase to remove the acetyl methyl ester, resulting in the formation of the polar fluorescent dye calcein and the emission of intense green fluorescence. Additionally, when the opening degree of the mPTP increases, Co^{2+} can enter the mitochondria and form a complex with calcein, causing the quenching of the green fluorescence signal. Therefore, we can determine the opening degree of the mPTP based on the intensity of the green fluorescence. Ionomycin is a Ca^{2+} ionophore that induces the influx of extracellular Ca^{2+} into the mitochondrial matrix, leading to the opening of the mPTP, and is commonly used as a positive control. In Figure 10B, the Calcein AM and Calcein AM + Ionomycin + $Cocl_2$ groups are the negative and positive controls, respectively. The fluorescence intensities of the calcein were measured as 95.51 ± 3.84, 73.64 ± 3.60, 65.89 ± 4.69, and 34.23 ± 1.47 in the 0, 5, 10, and 50 μM AFB1 groups ($p < 0.001$), respectively. The significant decrease in calcein fluorescence intensity indicated that AFB1 over-activated the opening of the mPTP, causing mitochondrial damage in the SW480 cells. After AFB1 treatment, the fluorescence intensity results of MitoTracker Red CMXRos indicated that the mitochondrial activity of the SW480 cells decreased (Figure 10C).

2.11. mRNA and Protein Expression Involved in Oxidative Stress and Lipid Signaling Pathways

The toxicity effect of AFB1 on the intestine was assessed by examining 19 genes associated with oxidative stress and lipid signaling pathways. Compared to the control group, the *GRP78, PERK, IRE1α, XBP1, SEC61A1, CAT, GPX1, GSS, NQO1, COX2, HO-1* genes related in the oxidative stress pathway were up-regulated (Figure 11A). In the lipid synthesis and fatty acid metabolism pathways, except for the up-regulated expression of *ACSBG1, ACSL1, LDLR*, and *SCP2* genes, the *ACSL5, LPIN1, CYP4F3, CYP4F12* genes were down-regulated (Figure 11B). Figure 11C–D show that the results of western blotting analysis indicate a considerable enhancement in the expression levels of SEC61A1, COX2, HO-1, NQO1, PERK, SCP2, and ACSL1 in the AFB1 group as opposed to the control group. Furthermore, the expression levels of LPIN1 and LDLR also dropped in a concentration-dependent manner in response to AFB1.

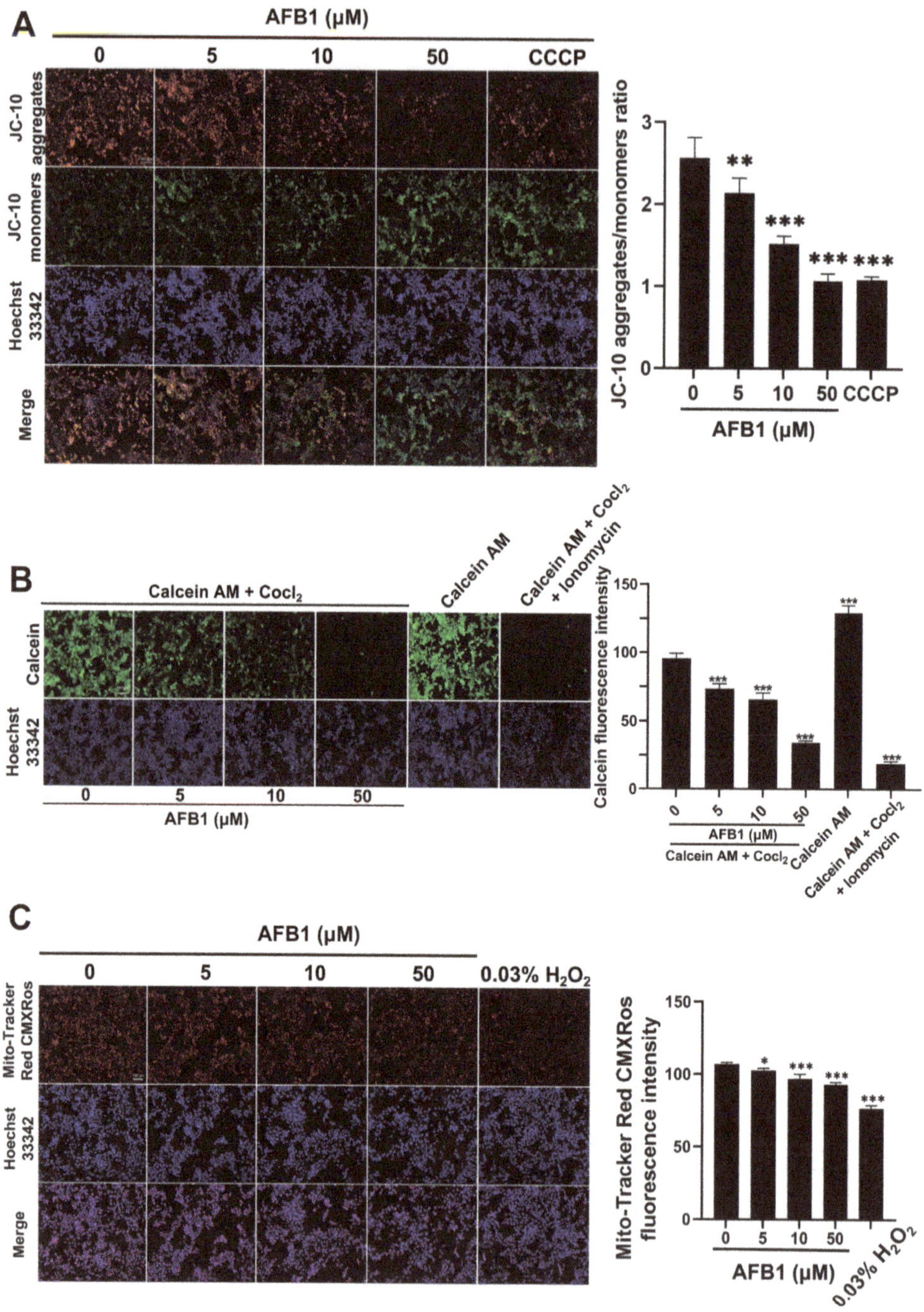

Figure 10. (**A**) JC-10 probe was used to detect the MMP in SW480 cells. (**B**) Representative fluorescence image and quantitative results of mPTP detection in SW480 cells using calcein AM probe after 72 h of AFB1 treatment. (**C**) The MitoTracker Red CMXRos probe was used to detect mitochondrial activity in SW480 cells. Scale bar = 100 μm. * $p < 0.05$, ** $p < 0.01$, *** $p < 0.001$, n = 3.

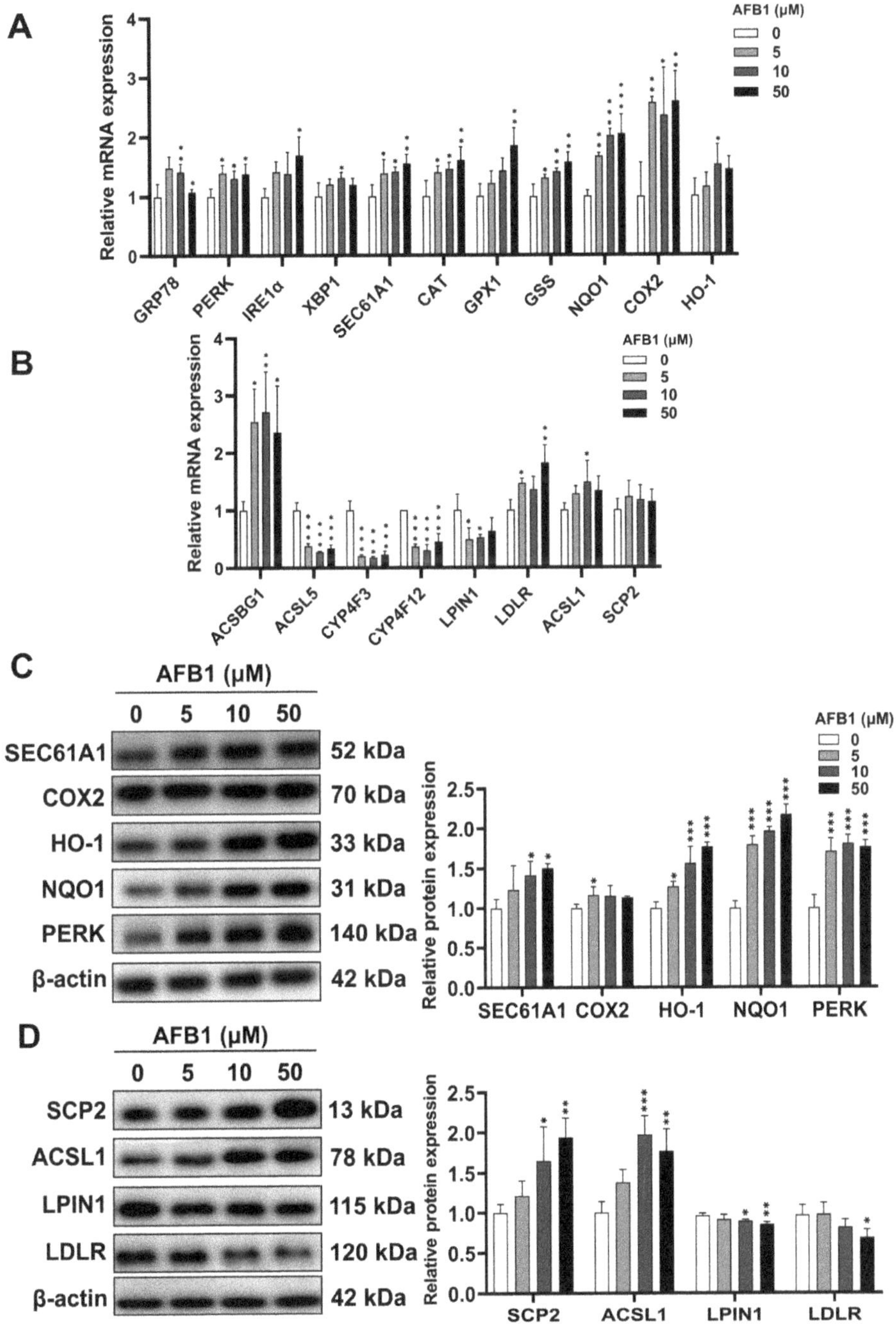

Figure 11. The levels of mRNA were assessed via RT-qPCR analysis pertaining to oxidative stress (**A**) and lipid metabolism (**B**). (**C**) The protein expression levels of SEC61A1, COX2, HO-1, NQO1, and PERK in 0, 5, 10, and 50 μM AFB1-treated SW480 cells. (**D**) The protein expression levels of SCP2, ACSL1, LPIN1 and LDLR in 0, 5, 10, and 50 μM AFB1-treated SW480 cells. * $p < 0.05$, ** $p < 0.01$, and *** $p < 0.001$, n = 3.

3. Discussion

The intestinal tract functions as a crucial barrier against toxins and detrimental substances. Although AFB1 has been demonstrated to possess hepatotoxic properties, it can also induce intestinal damage both in vivo and in vitro [24,25]. Histological examination revealed that the intestinal inflammation triggered by the pathogenic dosage of AFB1 was less than that in the liver [26]. Furthermore, colon injury was observed following low-dose AFB1 treatment prior to any discernible signs of liver damage.

With the development of high-throughput sequencing and microarray technologies, researchers were allowed to monitor the expression levels of thousands of genes across a genome simultaneously. Wu et al. [27] showed the m6A methylome and transcriptome profiles in HCT116 cells treated with AFB1, revealing that AFB1-induced HCT116 cytotoxicity is associated with perturbations in the levels of m6A methylation modifications and abnormal regulation of m6A regulators. Transcriptomic analysis was employed to elucidate the mechanism underlying afb1 toxicity in aquatic environments and organisms. In a study investigating the hepatotoxicity of AFB1 in developing juvenile zebrafish, transcriptome sequencing analysis revealed that AFB1 perturbed liver redox levels, thereby inducing oxidative stress, steatosis, and lipid droplet accumulation [28]. In our study, the 1037 DEGs in SW480 cells treated with AFB1 were enriched in pathways related to calcium, triglyceride lipase activity, long-chain fatty acid metabolic processes, and responses to oxidative stress. The genes *CAT* and *NQO1* were found to be enriched in response to oxidative stress, as evidenced by the significant increase in their levels observed in the IGV diagram. The DEGs identified through RT-qPCR were consistent with the sequencing results, thereby validating the findings of transcriptome analysis. These findings suggest an imbalance between oxidation and antioxidant levels within the cell, leading to oxidative stress occurrence. These findings provide a theoretical basis for understanding the mechanisms of AFB1 toxicity and present new strategies for the prevention, diagnosis, and treatment of AFB1 toxicity.

According to the genetic central dogma, genetic information encoded in the DNA sequence of a gene is transcribed into mRNA, which subsequently undergoes translation to synthesize proteins and execute diverse cellular functions. Proteomics is an effective tool for comprehensively understanding genome expression, and can complement the study of genome modification and translation. Multi-omics strategies, such as combining epitranscriptomics, proteomics, and metabolomics, help to reveal new mechanisms and identify new targets. In the previous study, through the multi-omics combined analysis of miRNA sequencing, transcriptomics, and proteomics, it was found that AFB1 and AFM1 exert an additive toxic effect on NCM460 cells via the p53 signaling pathway [29]. In our study, proteomic data showed that after the treatment of SW480 cells with AFB1, the regulation of terms, such as endoplasmic reticulum organization, fatty acid metabolic process, lipid modification, and calcium ion binding, were altered. Furthermore, the key proteins related to the terms have been discovered in previous studies to be associated with intestinal lipid metabolism. The protein SCP2, which is differentially expressed and enriched by the proteome in this study, has been demonstrated to facilitate the translocation of cytoplasmic lipid peroxidation (LPO) into mitochondria during RSL3-induced ferroptosis in chondrocytes. This process leads to mitochondrial membrane impairment and the release of ROS [30]. Among them, the up-regulation of SCP2 led to changes in pathways involved in lipid uptake and absorption in zebrafish of intestinal epithelial cells [31]. These indicated that the involvement of SCP2 in AFB1-induced cytotoxicity may represent a significant target worthy of investigation.

Metabolomics, a frequently utilized approach for uncovering metabolic alterations induced by external stimuli, has found extensive application in diverse research domains such as toxicology, drug development, and environmental science [32–34]. Metabolites are the products of gene–environment interactions and have a close association with disease phenotypes [35]. In the present study, metabolomics detected 2410 metabolites, and MSEA results indicated their association with glutathione metabolism and fatty acid degradation.

Glutathione is the most abundant soluble sulfhydryl antioxidant in cells, which plays an important role in antioxidant defense, protein function regulation, gene expression, and cell proliferation [36]. Fatty acids are structural components of cell membrane phospholipids, able to esterify with glycerol or sterol backbones to generate triacylglycerols or sterol esters, respectively, and then kept in lipid droplets [37,38]. Therefore, the significant changes in glutathione metabolism and fatty acid degradation caused by AFB1 treatment might have led to a redox imbalance and lipid metabolism disorder in SW480 cells, resulting in decreased cell viability and mitochondrial damage.

The validation in this study was limited to a single cell line SW480 cells, which are cancer cells with distinct metabolic and physiological states compared to normal cells. This limits the ability to further elucidate the potential mechanism of AFB1-induced cytotoxicity. Future research will include the establishment of animal and organoid models for rats and mice, as well as the isolation of human primary intestinal epithelial cells, in order to investigate the toxic mechanism of AFB1 in the intestinal tract.

4. Conclusions

This study demonstrated the toxic effects of AFB1 on SW480 cells. Multi-omics analysis revealed that oxidative stress and lipid metabolism may be involved in AFB1-mediated cytotoxicity. According to the experimental results, AFB1 can reduce cellular viability, increase ROS and calcium ions, decrease mitochondrial activity, lead to fatty acid metabolism disorders, and eventually lead to mitochondrial damage and lipid metabolism disorders. The aforementioned findings serve as a crucial foundation for elucidating the cytotoxicity of AFB1.

5. Materials and Methods

5.1. Chemical Reagent and Cell Culture

The chemical formula of AFB1 is $C_{17}H_{12}O_6$, and its chemical structure is depicted in Figure 1A. AFB1 was obtained from Macklin (Shanghai, China), with a purity level of 98.275%. The human SW480 cells utilized in this study were cultured in a constant temperature incubator at 37 °C using Dulbecco's Modified Eagle medium supplemented with 10% fetal bovine serum. The AFB1 compound was dissolved in DMSO, with subsequent experiments conducted using working concentrations of 5, 10, 50, and 100 µM.

5.2. Cell Viability Assay

The MTT cytotoxicity assay was utilized to evaluate the viability of SW480 cells that were exposed to AFB1 [39]. SW480 cells were seeded in 96-well plates and treated with 0, 5, 10, 50, and 100 µM AFB1 for 72 h. We prepared a 5 mg/mL MTT solution and added 10 µL of the solution along with 90 µL of the medium to each well at 37 °C with 5% CO_2. After a duration of 4 h, the OD at 490 nm was quantified for each well subsequent to the addition of DMSO.

5.3. RNA Extraction and Transcriptomics

The SW480 cells were seeded in 6 cm dishes with a density of 3×10^6 cells per dish. The RNA was extracted from SW480 cells cultured with 0 and 50 µM AFB1 for 72 h using the Trizol reagent (Invitrogen, Carlsbad, CA, USA) method. The mRNA sequencing was conducted by Seqhealth Technology Co., Ltd. (Wuhan, China) on triplicate RNA samples for each group. Briefly, the polyadenylated RNA enrichment was performed using VAHTS mRNA capture beads with a total of 10 µg of RNA, while the resulting mRNA fragments were predominantly distributed within the range of 100–200 nt. The sequencing library was constructed using the KC-Digital stranded mRNA library prep kit (seqhealth). Sequencing of the final libraries was carried out on the Novaseq 6000 (Illumina) with a read length of 200–500 bps. The sequencing mRNA data were qualified twice. HISAT2 was used for comparison against the human genome, and the resulting count number was applied to identify DEGs with the aid of the DeSeq2 package.

5.4. Tandem Mass Tag Proteomics Analysis

The SW480 cells were exposed to 0 and 50 μM AFB1 for 72 h, followed by the addition of a lysate buffer containing 100 mM triethylammonium bicarbonate and 1% SDS. The protein was reduced using Tris(2-carboxyethyl) phosphine hydrochloride and subsequently alkylated with iodoacetamide. This was followed by precipitation with six volumes of ice-cold acetone and subsequent blow-drying. After resuspending in a triethylammonium bicarbonate buffer, enzymolysis was performed using trypsin and incubated at 37 °C for 20 h. A set of TMTpro isobaric tags were utilized to tag the peptides, and the labeling reaction was carried out at ambient temperature for 60 min. The High-pH Reversed-Phase Peptide Fractionation Kit (Invitrogen) was employed to fractionate the labeled peptides into 15 fractions. The separated peptides were analyzed using the EASY-nLC 1200 UHPLC System and Orbitrap Exploris 480 MS for combined detection. The mass spectrometry files were processed using Proteome Discoverer version 3.0, and the statistical analysis of the expression matrix was conducted utilizing RStudio (2022.12.0-353).

5.5. Sample Preparation and Analysis for Metabolomics

SW480 cells treated with AFB1 for 72 h were washed with PBS. Into the cell sample, a 500 μL solution containing the internal standard was added and vortexed to suspend the sample. The sample was frozen rapidly by immersing it in liquid nitrogen for 5 min and then thawed in dry ice and ice for 5 min, respectively. Then, the solution was vortexed for 2 min to ensure thorough mixing, and this process was repeated three times. The centrifugation of the samples was performed at 12,000 r/min for 10 min at 4 °C. The supernatant was placed in −20 °C for 30 min and underwent centrifugation at 12,000 r/min for 3 min at 4 °C, followed by collecting the supernatant for LC-MS analysis. A 2.1 mm × 100 mm ACQUITY UPLC HSS T3 C18 column from Waters (Milford, MA, USA) was employed for the purpose of elution. After gradient elution, the column temperature was set to 40 °C, accompanied by an injection volume of 4 μL and a flow rate of 0.4 mL/min. Finally, the metabolomics data were processed and analyzed.

5.6. Molecular Docking

Molecular docking was completed using SYBYL-X 2.0 software, after the 3D conformations of proteins were obtained from the Protein Data Bank [40]. Water molecules and heteroatoms of protein conformation were removed, and hydrogen atoms were added for preprocessing using SYBYL-X 2.0. The 2D structure of AFB1 was retrieved PubChem and imported into Open Babel GUI 2.4.1 to generate the mol2 format. Semi-flexible docking was utilized for the docking of proteins with AFB1. The interaction between the protein and the molecule was regarded as stable when the score value exceeded 5.

5.7. Molecular Dynamics (MD)

The MD simulations and calculations were performed using the SPONGE 1.4 and XPONGE2 software packages [41]. The ff14SB united-atom force field3 was employed throughout the simulation study. By using the SPC/E water model in cubic boxes, the ligand–protein complex was solved, ensuring that the boxes were filled with water and that there was a minimum distance of 1.2 nm from each side of the cube. Prior to the formal dynamics simulation, preliminary NVT and NPT simulations were conducted for a duration of 100 ps. The system underwent a 50 ns dynamic simulation, with the temperature held at 300 K and the atmospheric pressure set to 1 bar.

5.8. ROS Measurement by DCFH-DA

To investigate the effects of AFB1 on oxidative stress in SW480 cells and measure ROS levels, the SW480 cells treated with 50 μM AFB1 for 72 h were incubated with 1 μM DCFH-DA probe (Beyotime, Shanghai, China) at a temperature of 37 °C for 40 min. After washing the stained cells with PBS, we observed them using a fluorescence microscope (Zeiss, Oberkochen, Germany) and fluorescence intensity was quantitatively detected using

ImageJ software (version 1.54d), and the mean gray value, defined as the Integrated Density divided by area, was taken as the detection result.

5.9. Assay of Catalase (CAT) and Glutathione S-Transferase (GST)

SW480 cells in 6-well plates were scraped and collected after a 72 h treatment with AFB1 at concentrations of 0, 5, 10, and 50 μM. The reagent provided in the kit was added to the cell precipitate, followed by placing the cells on ice for ultrasonication. The subsequent suspension was subjected to centrifugation at 8000× g and 4 °C for a duration of 10 min. The supernatant was utilized for the quantification of CAT and GST activities, with absorbance measurements performed at 405 nm and 340 nm as per the provided instructions, respectively.

5.10. Measurement of Cellular Calcium

The cellular calcium detection was performed in SW480 cells following a 72 h incubation with treatments of 0, 5, 10, and 50 μM AFB1. The calcium concentration was assessed by quantifying the fluorescence intensity of Fluo-4 AM dye (Beyotime) loaded in the cytoplasm. The cell samples were incubated at 37 °C for 40 min with a diluted Fluo-4 AM to a final concentration of 1 μM in PBS. The variations in green fluorescence at 494 and 516 nm excitation and emission wavelengths were captured by a fluorescence microscope (Zeiss).

5.11. Fatty Acid Green Fluorescence Probe

The BODIPY 500/510 C1, C12 fatty acid green fluorescence probe was employed in this study to investigate lipid accumulation. Cells treated with 0, 5, 10, 50 μM AFB1 for 72 h in 24-well plates were incubated with 5 μM BODIPY 500/510 C1, C12 for 15 min. After washing with PBS, the green fluorescence with an emission wavelength of 510 nm was observed by fluorescence microscope (Zeiss).

5.12. Detection of Mitochondrial Membrane Potential (MMP)

MMP was assessed by making use of the JC-10 probe, which was obtained from Solarbio (Beijing, China). Generally, MMP is maintained at a relatively high level. In the mitochondrial matrix, the JC-10 probe forms aggregates and emits red fluorescence. However, when mitochondria are damaged, JC-10 exists as monomers and results in green fluorescence. SW480 cells were treated to 0, 5, 10, and 50 μM of AFB1 for 72 h, followed by incubation with the JC-10 probe and Hoechst 33342 at 37 °C in the dark for 20 min and 10 min, respectively. Finally, fluorescent images are obtained with the help of a microscope (Zeiss).

5.13. Detection of Mitochondrial Permeability Transition Pore (mPTP)-Opening

To assess the impact of AFB1 on mitochondrial function in SW480 cells, we utilized the Calcein AM probe (Beyotime) to detect the opening of the mPTP. Briefly, SW480 cells were seeded into 24-well plates at 8×10^4 cells per well and treated with 0, 5, 10, and 50 μM AFB1 for 72 h. Subsequently, 250 μL of Calcein AM staining solution was added to each well and incubated at 37 °C for 40 min. Following that, the culture medium was added to replace the Calcein AM staining solution and further incubated for 20 min. Finally, the cell fluorescence was evaluated using a fluorescence microscope (Zeiss) and then the Image J software was employed for the quantitative analysis.

5.14. Assessment of Mitochondrial Activity

Mitochondrial activity was measured using the MitoTracker Red CMXRos probe (Beyotime), which specifically tags bioactive mitochondria and fluoresces in red. SW480 cells were treated with 0, 5, 10, and 50 μM of AFB1 for 72 h, followed by incubation with the MitoTracker Red CMXRos probe and Hoechst 33342 at 37 °C in the absence of light for 30 min and 10 min, respectively. Finally, a fluorescence microscope (Zeiss) was used to observe the fluorescence intensity of SW480 cells.

5.15. RT-qPCR

The RNA-extraction procedure is consistent with the previously described methodology. According to the directions of the Evo M-MLV RT Reaction Mix kit (Accurate Biology, Changsha, China), the total RNA was reverse-transcribed into a cDNA template. The primer sequences of the selected genes utilized in this investigation are presented in Table 2. RT-qPCR was implemented by means of the AtiaMX real-time PCR system from Agilent (Santa Clara, CA, USA). The PCR comprised an initial denaturation step at 95 °C for 30 s, followed by 40 cycles at 95 °C for 5 s, and 60 °C for 30 s. The dissociation curve is obtained through the dissolution procedure specified by the machine's default settings. By using the $2^{-\Delta\Delta CT}$ method [27], the relative mRNA level of each targeted gene was standardized to *β-actin*.

Table 2. Sequences of primers used in RT-qPCR.

Gene	Forward Primer (5′-3′)	Reverse Primer (5′-3′)
GRP78	CAGTTGTTACTGTACCAGCCTA (F)	CATTTAGGCCAGCAATAGTTCC (R)
PERK	CCAGTTTTGTACTCCAATTGCA (F)	CAGATACAGCTGGCCTCTATAC (R)
IRE1α	CGTGAGCGACAGAATAGAAAAG (F)	GCTTCTTATTTCTCATGGCTCG (R)
XBP1	CTTGTAGTTGAGAACCAGGAGT (F)	CCCAACAGGATATCAGACTCTG (R)
SEC61A1	AATTTGAAGGTGCTATCATCGC (F)	ACGTATAGAAGAGCTTGATGGG (R)
CAT	GAGCACAGCATCCAATATTCTG (F)	CTCATTCAGCACGTTCACATAG (R)
GPX1	GTTGCCTGGAACTTTGAGAAG (F)	CTCGATGTCAATGGTCTGGAAGG (R)
GSS	CTCTTTGACATCCACAAGCAAG (F)	AACATGTAGTCTGAGCGATTCA (R)
NQO1	CAACCACGAGCCCAGCCAATC (F)	ACTCCACCACCTCCCATCCTTTC (R)
COX2	TGTCAAAACCGAGGTGTATGTA (F)	AACGTTCCAAAATCCCTTGAAG (R)
HO-1	CCTCCCTGTACCACATCTATGT (F)	GCTCTTCTGGGAAGTAGACAG (R)
ACSBG1	AAACAAGATGGCCAATGTGTAC (F)	CGTGATATTGTCTTGACTCAGC (R)
ACSL5	ATATGATGCTGAGAACCTAGGC (F)	CATGCTCCACACATTTGAGAAA (R)
CYP4F3	AGAAGAAGGGGAGAGGAGGTTGTG (F)	GAGGCGGCAGCAGTTGTCATAG (R)
CYP4F12	AAGGCATTGTCAGATGAGGATA (F)	TTCAATCTCTTTAGGATCGCGG (R)
LPIN1	GAGCAGCAGAACTCTTCCTAAT (F)	CTTTTGCAATCTACCAGGCTAC (R)
LDLR	CTGTAGGGGTCTTTACGTGTTC (F)	GTTTTCCTCGTCAGATTTGTCC (R)
ACSL1	CTGTGTCATGGAGCTAAAATCG (F)	GTGTTTGCTTGTCCGAAAATTC (R)
SCP2	GGAGCTGAGAATTCAAGAGACT (F)	AGTCACCAAAAACATAGCCAAC (R)
β-ACTIN	GCACTCTTCCAGCCTTCCTTCC (F)	GCGGATGTCCACGTCACACTTC (R)

5.16. Western Blotting

Using RIPA buffer (Beyotime), SW480 cells were lysed to extract protein after being treated with 0, 5, 10, and 50 µM AFB1 for 72 h. Proteins were fractionated on 10% sodium dodecyl sulfate–polyacrylamide gel electrophoresis and subsequently transferred to a PVDF membrane. The bands were blocked with 5% non-fat milk powder for 1.5 h. Primary antibodies were added to the membranes and incubated overnight at 4 °C [42]. The antibodies utilized were SEC61A1 (ABclonal, Wuhan, China), COX2 (Abclonal), HO-1 (Proteintech, Wuhan, China), NQO1 (Proteintech), PERK (Proteintech), SCP2 (Proteintech), ACSL1 (Proteintech), LPIN1 (Proteintech), LDLR (Proteintech), and β-actin (Proteintech), which were used at a dilution of 1:1000, 1:1000, 1:3000, 1:4000, 1:1000, 1:2000, 1:3000, 1:5000, 1:2000, and 1:2500, respectively. Then, peroxidase-conjugated goat anti-rabbit IgG secondary antibodies (1:10,000, Proteintech) were incubated with the membranes for 90 min. Exposure was carried out via the Amersham Imager 680 from GE Healthcare Bio-Sciences AB (Uppsala, Sweden), followed by the utilization of the Image J software for the quantification of the target bands.

5.17. Statistical Analysis

All statistical data were analyzed by using SPSS 26 (IBM, New York, NY, USA). One-way ANOVA was utilized for group differences and presented as the mean ± standard deviation (SD). A p value less than 0.05 was considered statistically significant.

Supplementary Materials: The following supporting information can be downloaded at: https://www.mdpi.com/article/10.3390/toxins16090375/s1, Table S1: The names of the top 20 metabolites with VIP values corresponding to ID numbers.

Author Contributions: Conceptualization, A.Z.; methodology, A.Z., Z.W. and X.L.; software, M.C., X.G., J.Z. and X.L.; formal analysis, M.C., J.W., Y.Q. and Z.W.; investigation M.C., J.W. and Y.Q.; data curation, M.C., J.W. and Y.Q.; writing—original draft preparation, M.C., J.W., Y.Q., X.G., J.Z. and Y.L.; writing—review and editing, A.Z.; funding acquisition, A.Z. All authors have read and agreed to the published version of the manuscript.

Funding: This research was funded by the National Natural Science Foundation of China (82104520), the Natural Science Foundation of Fujian Province (2021J05045), the Fujian Medical University High-level Talent Research Startup Funding Project (XRCZX2020014), Open Research Fund of Key Laboratory of Environment and Health, Fujian Province University (GWGXZD-202303), Open Research Fund of Fujian Key Laboratory of Tumor Microbiology, Fujian Medical University (FMUTM-202301), and Fujian Research and Training Grants for Young and Middle-aged Leaders in Healthcare.

Institutional Review Board Statement: Not applicable.

Informed Consent Statement: Not applicable.

Data Availability Statement: Data are contained within the article and Supplementary Materials.

Conflicts of Interest: The authors declare no conflicts of interest.

References

1. Raysyan, A.; Eremin, S.A.; Beloglazova, N.V.; De Saeger, S.; Gravel, I.V. Immunochemical approaches for detection of aflatoxin B1 in herbal medicines. *Phytochem. Anal.* **2020**, *31*, 662–669. [CrossRef]
2. Sun, G.; Wang, S.; Hu, X.; Su, J.; Zhang, Y.; Xie, Y.; Zhang, H.; Tang, L.; Wang, J.S. Co-contamination of aflatoxin B1 and fumonisin B1 in food and human dietary exposure in three areas of China. *Food Addit. Contam. Part A Chem. Anal. Control Expo. Risk Assess.* **2011**, *28*, 461–470. [CrossRef] [PubMed]
3. Rushing, B.R.; Selim, M.I. Aflatoxin B1: A review on metabolism, toxicity, occurrence in food, occupational exposure, and detoxification methods. *Food Chem. Toxicol.* **2019**, *124*, 81–100. [CrossRef] [PubMed]
4. Cao, W.; Yu, P.; Yang, K.; Cao, D. Aflatoxin B1: Metabolism, toxicology, and its involvement in oxidative stress and cancer development. *Toxicol. Mech. Methods* **2022**, *32*, 395–419. [CrossRef]
5. Damiano, S.; Jarriyawattanachaikul, W.; Girolami, F.; Longobardi, C.; Nebbia, C.; Andretta, E.; Lauritano, C.; Dabbou, S.; Avantaggiato, G.; Schiavone, A.; et al. Curcumin supplementation protects broiler chickens against the renal oxidative stress induced by the dietary exposure to low levels of aflatoxin B1. *Front. Vet. Sci.* **2022**, *8*, 822227. [CrossRef]
6. Crick, F. Central dogma of molecular biology. *Nature* **1970**, *227*, 561–563. [CrossRef]
7. Lander, E.S.; Linton, L.M.; Birren, B.; Nusbaum, C.; Zody, M.C.; Baldwin, J. Initial sequencing and analysis of the human genome. *Nature* **2001**, *409*, 860–921. [PubMed]
8. Waterston, R.H.; Lindblad-Toh, K.; Birney, E.; Rogers, J.; Abril, J.F.; Agarwal, P. Initial sequencing and comparative analysis of the mouse genome. *Nature* **2002**, *420*, 520–562. [PubMed]
9. Gibbs, R.A.; Weinstock, G.M.; Metzker, M.L.; Muzny, D.M.; Sodergren, E.J.; Scherer, S. Genome sequence of the Brown Norway rat yields insights into mammalian evolution. *Nature* **2004**, *428*, 493–521.
10. Chimpanzee Sequencing and Analysis Consortium. Initial sequence of the chimpanzee genome and comparison with the human genome. *Nature* **2005**, *437*, 69–87. [CrossRef]
11. Wu, Q.; Gao, X.; Lin, Y.; Wu, C.; Zhang, J.; Chen, M.; Wen, J.; Wu, Y.; Tian, K.; Bao, W.; et al. Integrating epigenetics, proteomics, and metabolomics to reveal the involvement of Wnt/β-catenin signaling pathway in oridonin-induced reproductive toxicity. *Toxics* **2024**, *12*, 339. [CrossRef]
12. Budayeva, H.G.; Kirkpatrick, D.S. Monitoring protein communities and their responses to therapeutics. *Nat. Rev. Drug Discov.* **2020**, *19*, 414–426. [CrossRef] [PubMed]
13. Sun, T.; Li, Y.; Li, Y.; Li, H.; Gong, Y.; Wu, J.; Ning, Y.; Ding, C.; Xu, Y. Proteomic analysis of copper toxicity in human fungal pathogen *Cryptococcus neoformans*. *Front. Cell. Infect. Microbiol.* **2021**, *11*, 662404. [CrossRef] [PubMed]

14. Gao, Y.; Li, S.; Bao, X.; Luo, C.; Yang, H.; Wang, J.; Zhao, S.; Zheng, N. Transcriptional and proteomic analysis revealed a synergistic effect of aflatoxin M1 and ochratoxin A mycotoxins on the intestinal epithelial integrity of differentiated human Caco-2 cells. *J. Proteome Res.* **2018**, *17*, 3128–3142. [CrossRef] [PubMed]
15. Alseekh, S.; Aharoni, A.; Brotman, Y.; Contrepois, K.; D'Auria, J.; Ewald, J. Mass spectrometry-based metabolomics: A guide for annotation, quantification and best reporting practices. *Nat. Methods* **2021**, *18*, 747–756. [CrossRef]
16. Guo, H.H.; Feng, C.L.; Zhang, W.X.; Luo, Z.G.; Zhang, H.J.; Zhang, T.T.; Ma, C.; Zhan, Y.; Li, R.; Wu, S.; et al. Liver-target nanotechnology facilitates berberine to ameliorate cardio-metabolic diseases. *Nat. Commun.* **2019**, *10*, 1981. [CrossRef] [PubMed]
17. Wert, K.J.; Velez, G.; Kanchustambham, V.L.; Shankar, V.; Evans, L.P.; Sengillo, J.D.; Zare, R.N.; Bassuk, A.G.; Tsang, S.H.; Mahajan, V.B. Metabolite therapy guided by liquid biopsy proteomics delays retinal neurodegeneration. *EBioMedicine* **2020**, *52*, 102636. [CrossRef] [PubMed]
18. Michaudel, C.; Danne, C.; Agus, A.; Magniez, A.; Aucouturier, A.; Spatz, M.; Lefevre, A.; Kirchgesner, J.; Rolhion, N.; Wang, Y.; et al. Rewiring the altered tryptophan metabolism as a novel therapeutic strategy in inflammatory bowel diseases. *Gut* **2023**, *72*, 1296–1307. [CrossRef]
19. Zhou, J.; Tang, L.; Wang, J.S. Assessment of the adverse impacts of aflatoxin B1 on gut-microbiota dependent metabolism in F344 rats. *Chemosphere* **2019**, *217*, 618–628. [CrossRef]
20. Yang, X.; Li, X.; Gao, Y.; Wang, J.; Zheng, N. Integrated metabolomics and lipidomics analysis reveals lipid metabolic disorder in NCM460 cells caused by aflatoxin B1 and aflatoxin M1 alone and in combination. *Toxins* **2023**, *15*, 255. [CrossRef]
21. Harrison, J.C.; Carvajal, M.; Garner, R.C. Does aflatoxin exposure in the United Kingdom constitute a cancer risk? *Environ. Health Perspect.* **1993**, *99*, 99–105. [CrossRef] [PubMed]
22. Pu, J.; Yuan, Q.; Yan, H.; Tian, G.; Chen, D.; He, J.; Zheng, P.; Yu, J.; Mao, X.; Huang, Z.; et al. Effects of chronic exposure to low levels of dietary aflatoxin B1 on growth performance, apparent total tract digestibility and intestinal health in pigs. *Animals* **2021**, *11*, 336. [CrossRef] [PubMed]
23. Galano, M.; Venugopal, S.; Papadopoulos, V. Role of STAR and SCP2/SCPx in the transport of cholesterol and other lipids. *Int. J. Mol. Sci.* **2022**, *23*, 12115. [CrossRef] [PubMed]
24. Zhang, M.; Li, Q.; Wang, J.; Sun, J.; Xiang, Y.; Jin, X. Aflatoxin B1 disrupts the intestinal barrier integrity by reducing junction protein and promoting apoptosis in pigs and mice. *Ecotoxicol. Environ. Saf.* **2022**, *247*, 114250. [CrossRef]
25. Romero, A.; Ares, I.; Ramos, E.; Castellano, V.; Martínez, M.; Martínez-Larrañaga, M.R.; Anadón, A.; Martínez, M.A. Mycotoxins modify the barrier function of Caco-2 cells through differential gene expression of specific claudin isoforms: Protective effect of illite mineral clay. *Toxicology* **2016**, *353–354*, 21–33. [CrossRef]
26. Zhang, L.; Cheng, D.; Zhang, J.; Tang, H.; Li, F.; Peng, Y.; Duan, X.; Meng, E.; Zhang, C.; Zeng, T.; et al. Role of macrophage AHR/TLR4/STAT3 signaling axis in the colitis induced by non-canonical AHR ligand aflatoxin B1. *J. Hazard. Mater.* **2023**, *452*, 131262. [CrossRef] [PubMed]
27. Wu, Y.; Bao, W.; Ren, J.; Li, C.; Chen, M.; Zhang, D.; Zhu, A. Integrated profiles of transcriptome and mRNA m6A modification reveal the intestinal cytotoxicity of aflatoxin B1 on HCT116 cells. *Genes* **2022**, *14*, 79. [CrossRef]
28. Feng, C.; Bai, H.; Chang, X.; Wu, Z.; Dong, W.; Ma, Q.; Yang, J. Aflatoxin B1-induced early developmental hepatotoxicity in larvae zebrafish. *Chemosphere* **2023**, *340*, 139940. [CrossRef]
29. Gao, Y.N.; Yang, X.; Wang, J.Q.; Liu, H.M.; Zheng, N. Multi-omics reveal additive cytotoxicity effects of aflatoxin B1 and aflatoxin M1 toward intestinal NCM460 cells. *Toxins* **2022**, *14*, 368. [CrossRef]
30. Dai, T.; Xue, X.; Huang, J.; Yang, Z.; Xu, P.; Wang, M.; Xu, W.; Feng, Z.; Zhu, W.; Xu, Y.; et al. SCP2 mediates the transport of lipid hydroperoxides to mitochondria in chondrocyte ferroptosis. *Cell Death Discov.* **2023**, *9*, 234. [CrossRef]
31. Mu, X.; Qi, S.; Wang, H.; Yuan, L.; Wang, C.; Li, Y.; Qiu, J. Bisphenol analogues induced metabolic effects through eliciting intestinal cell heterogeneous response. *Environ. Int.* **2022**, *165*, 107287. [CrossRef] [PubMed]
32. Warth, B.; Raffeiner, P.; Granados, A.; Huan, T.; Fang, M.; Forsberg, E.M.; Benton, H.P.; Goetz, L.; Johnson, C.H.; Siuzdak, G. Metabolomics reveals that dietary xenoestrogens alter cellular metabolism induced by palbociclib/letrozole combination cancer therapy. *Cell Chem. Biol.* **2018**, *25*, 291–300. [CrossRef]
33. Wang, X.; Qiu, J.; Xu, Y.; Liao, G.; Jia, Q.; Pan, Y.; Wang, T.; Qian, Y. Integrated non-targeted lipidomics and metabolomics analyses for fluctuations of neonicotinoids imidacloprid and acetamiprid on Neuro-2a cells. *Environ. Pollut.* **2021**, *284*, 117327. [CrossRef] [PubMed]
34. Rombouts, C.; De Spiegeleer, M.; Van Meulebroek, L.; De Vos, W.H.; Vanhaecke, L. Validated comprehensive metabolomics and lipidomics analysis of colon tissue and cell lines. *Anal. Chim. Acta* **2019**, *1066*, 79–92. [CrossRef] [PubMed]
35. Surendran, P.; Stewart, I.D.; Au Yeung, V.P.W.; Pietzner, M.; Raffler, J.; Wörheide, M.A.; Li, C.; Smith, R.F.; Wittemans, L.B.L.; Bomba, L.; et al. Rare and common genetic determinants of metabolic individuality and their effects on human health. *Nat. Med.* **2022**, *28*, 2321–2332. [CrossRef]
36. Wang, L.; Ahn, Y.J.; Asmis, R. Sexual dimorphism in glutathione metabolism and glutathione-dependent responses. *Redox. Biol.* **2020**, *31*, 101410. [CrossRef] [PubMed]
37. Dyall, S.C.; Balas, L.; Bazan, N.G.; Brenna, J.T.; Chiang, N.; da Costa Souza, F.; Dalli, J.; Durand, T.; Galano, J.M.; Lein, P.J.; et al. Polyunsaturated fatty acids and fatty acid-derived lipid mediators: Recent advances in the understanding of their biosynthesis, structures, and functions. *Prog. Lipid Res.* **2022**, *86*, 101165. [CrossRef]

38. Currie, E.; Schulze, A.; Zechner, R.; Walther, T.C.; Farese, R.V., Jr. Cellular fatty acid metabolism and cancer. *Cell Metab.* **2013**, *18*, 153–161. [CrossRef] [PubMed]
39. Chen, X.; Hu, H.; Lin, X.; Chen, M.; Bao, W.; Wu, Y.; Li, C.; Gao, Y.; Hou, S.; Yang, Q.; et al. Euphorbia factor L1 inhibited transport channel and energy metabolism in human colon adenocarcinoma cell line Caco-2. *Biomed. Pharmacother.* **2023**, *169*, 115919. [CrossRef]
40. Wu, Y.; Chen, X.; Bao, W.; Hong, X.; Li, C.; Lu, J.; Zhang, D.; Zhu, A. Effect of humantenine on mRNA m6A modification and expression in human colon cancer cell line HCT116. *Genes* **2022**, *13*, 781. [CrossRef]
41. Wu, Z.; Zhang, J.; Wu, Y.; Chen, M.; Hu, H.; Gao, X.; Li, C.; Li, M.; Zhang, Y.; Lin, X.; et al. Gelsenicine disrupted the intestinal barrier of Caenorhabditis elegans. *Chem. Biol. Interact.* **2024**, *395*, 111036. [CrossRef] [PubMed]
42. Zou, L.; Bao, W.; Gao, Y.; Chen, M.; Wu, Y.; Wang, S.; Li, C.; Zhang, J.; Zhang, D.; Wang, Q.; et al. Integrated analysis of transcriptome and microRNA profile reveals the toxicity of *euphorbia* factors toward human colon adenocarcinoma cell line Caco-2. *Molecules* **2022**, *27*, 6931. [CrossRef] [PubMed]

Disclaimer/Publisher's Note: The statements, opinions and data contained in all publications are solely those of the individual author(s) and contributor(s) and not of MDPI and/or the editor(s). MDPI and/or the editor(s) disclaim responsibility for any injury to people or property resulting from any ideas, methods, instructions or products referred to in the content.

toxins

MDPI

Review

Aflatoxin B1: Challenges and Strategies for the Intestinal Microbiota and Intestinal Health of Monogastric Animals

Hyunjun Choi, Yesid Garavito-Duarte , Alexa R. Gormley and Sung Woo Kim *

Department of Animal Science, North Carolina State University, Raleigh, NC 27695, USA; hchoi25@ncsu.edu (H.C.); yrgaravi@ncsu.edu (Y.G.-D.); agormle@ncsu.edu (A.R.G.)
* Correspondence: sungwoo_kim@ncsu.edu

Abstract: The objective of this review is to investigate the impacts of aflatoxins, particularly aflatoxin B1 (AFB_1), on intestinal microbiota, intestinal health, and growth performance in monogastric animals, primarily chickens and pigs, as well as dietary interventions to mitigate these effects. Aflatoxin B1 contamination in feeds disrupts intestinal microbiota, induces immune responses and oxidative damage, increases antioxidant activity, and impairs jejunal cell viability, barrier function, and morphology in the small intestine. These changes compromise nutrient digestion and reduce growth performance in animals. The negative impact of AFB_1 on the % change in average daily gain (ΔADG) of chickens and pigs was estimated based on meta-analysis: $\Delta ADG\ (\%)_{chicken} = -0.13 \times AFB_1$ intake per body weight (ng/g·d) and $\Delta ADG\ (\%)_{pig} = -0.74 \times AFB_1$ intake per body weight (μg/kg·d), indicating that increasing AFB_1 contamination linearly reduces the growth of animals. To mitigate the harmful impacts of AFB_1, various dietary strategies have been effective. Mycotoxin-detoxifying agents include mycotoxin-adsorbing agents, such as clay and yeast cell wall compounds, binding to AFB_1 and mycotoxin-biotransforming agents, such as specific strains of *Bacillus subtilis* and mycotoxin-degrading enzyme, degrading AFB_1 into non-toxic metabolites such as aflatoxin D1. Multiple mycotoxin-detoxifying agents are often combined and used together to improve the intestinal health and growth of chickens and pigs fed AFB_1-contaminated feeds. In summary, AFB_1 negatively impacts intestinal microbiota, induces immune responses and oxidative stress, disrupts intestinal morphology, and impairs nutrient digestion in the small intestine, leading to reduced growth performance. Supplementing multi-component mycotoxin-detoxifying agents in feeds could effectively adsorb and degrade AFB_1 co-contaminated with other mycotoxins prior to its absorption in the small intestine, preventing its negative impacts on the intestinal health and growth performance of chickens and pigs.

Keywords: aflatoxin; chickens; growth performance; intestinal health; pigs

Key Contribution: Emphasis is placed on the significant negative impacts of aflatoxins on the intestine, particularly the small intestine, as the small intestine is the primary site of initial exposure and absorption when animals consume contaminated feed. The impacts of AFB_1 on the growth performance of chickens and pigs are evaluated, highlighting that increasing AFB_1 intake per body weight linearly reduces growth. The review also highlights the mode of action of various mycotoxin-detoxifying agents and other feed additives in managing AFB_1 to enhance intestinal health and growth performance.

1. Introduction

Mycotoxins are secondary metabolites produced by toxigenic fungi, and it is estimated that 25% of the world's cereals are contaminated with mycotoxins [1–3]. Mycotoxins play crucial roles in a range of toxic mechanisms, compromising various metabolic functions in both humans and domestic animals [4]. Monogastric animals, including chickens and pigs, are particularly susceptible to mycotoxins, unlike ruminants that harbor a diverse microbiota capable of metabolizing and detoxifying certain mycotoxins in the rumen [5–7]. Chickens and pigs are of high interest not only due to their vulnerability to mycotoxins but also their significant roles in global meat and egg production [8,9]. As two of the most widely farmed livestock species worldwide, the negative impacts of mycotoxin contamination on the health and growth performance of chickens and pigs can have substantial economic and food security implications [9,10]. Among the diverse types of mycotoxins, aflatoxins have gained high regulatory attention and scientific focus due to their toxicity and prevalence [1,11].

The name "aflatoxin" is derived from "A" for *Aspergillus*, "fla" for "*flavus*", and "toxin" for its toxic nature [12]. This indicates that aflatoxins are primarily produced by certain species of *Aspergillus* fungi, including *Aspergillus flavus* and *Aspergillus parasiticus.* There are four main types of aflatoxins, including aflatoxin B1 (AFB_1), aflatoxin B2, aflatoxin G1, and aflatoxin G2. In particular, AFB_1 is known as the most toxic mycotoxin to animals. A previous in vitro study reported that AFB_1 has a stronger inhibitory effect on cell growth compared to other aflatoxins [13]. Additionally, mixtures of aflatoxins further enhance the inhibitory effects on cell growth [13]. Due to the severe health risks specifically associated with AFB_1, AFB_1 is highly regulated, especially in feeds that are more likely to cause aflatoxicosis, a toxic condition resulting from aflatoxin ingestion by animals [14]. In terms of prevalence, AFB_1 is more commonly found than other aflatoxins in feedstuffs such as corn, corn co-products, peanut meal, and cottonseed meal [15]. These feedstuffs are often grown in warm and humid climates, which are conducive to fungal growth and aflatoxin B1 production, particularly during the pre-harvest and storage stages [15]. Lastly, *Aspergillus* fungi produce AFB_1 more than other aflatoxins, indicating that the contents of AFB_1 could be more pronounced in feeds [16]. As a result, AFB_1 is the primary concern among aflatoxins in feed safety management.

The gastrointestinal tract (GIT) is the first biological system to come into direct contact with aflatoxins upon ingestion, and it is here that the negative impacts of aflatoxin exposure begin to manifest [17]. However, despite the GIT being the initial site of exposure, much of the research on aflatoxicosis has focused on systemic effects, such as liver damage and immune suppression, and the specific impacts on intestinal health have been less thoroughly explored. Understanding the impacts of AFB_1 in the intestine is also crucial, particularly in the small intestine, as the small intestine is where mycotoxins would first affect the animal by altering intestinal microbiota and activating intestinal immune cells [18–20]. Additionally, AFB_1, like nutrients, are primarily absorbed in the small intestine, making their effects particularly significant in this region [21].

Aflatoxin B1 has been shown to alter the intestinal microbiota, reducing beneficial microbial populations such as *Lactobacillus* spp., and simultaneously promoting the growth of harmful bacterial species such as *Escherichia coli* in the small intestine of animals [19,22]. The negative modulation of luminal and mucosa-associated microbiota in the small intestine also has a great influence on the intestinal health of animals [23]. These changes can further compromise intestinal health, weakening the immune defenses that are critical for protecting the animal from infections and diseases [24]. Aflatoxin B1 can also compromise intestinal integrity by decreasing the relative gene expression of tight junction proteins, which weakens these connections and leads to increased intestinal permeability [25,26].

This allows pathogens and toxins to enter the bloodstream more easily, resulting in systemic inflammation and immune challenges. Moreover, AFB_1 can damage enterocytes, impairing the ability of the intestine to absorb nutrients effectively [26,27]. This not only leads to reduced growth performance but also exacerbates malnutrition in animals. This understanding is crucial for developing dietary interventions to mitigate the negative impacts of aflatoxins in feeds for monogastric animals. Therefore, the objective of this review is to investigate the impacts of AFB_1 on the intestine and growth performance after absorption, and to comparatively evaluate various dietary interventions that could effectively prevent these negative impacts on intestinal health and growth performance in monogastric animals.

2. Influence of AFB_1 on Intestinal Microbiota in Monogastric Animals

2.1. Intestinal Microbiota

The intestinal microbiota has gained increased recognition for its critical role in maintaining intestinal health and supporting growth in animals [28,29]. Beneficial bacteria in the intestine help prevent colonization by opportunistic pathogens or ammonia-producing bacteria, thereby reducing the risk of intestinal inflammation [30–32]. Aflatoxin B1 adversely affects the intestinal microbiota [33–35]. In pigs, AFB_1 exposure has been associated with an increase in *Escherichia coli* populations within the colonic digesta [26] (Table 1). These adverse impacts on the intestinal microbiota are likely due to the reduced production of short-chain fatty acids, a consequence of diminished populations of fiber-degrading bacteria [36]. Additionally, AFB_1 has been shown to promote pathogenic bacterial infections, including *Escherichia coli*, *Salmonella*, and *Klebsiella*, in the ileal digesta of chickens [22]. This rise in infections is likely compounded by immune suppression caused by AFB_1, which makes the intestine more susceptible to pathogenic invasion. Both the luminal and mucosa-associated microbiota (also called 'mucosal microbiota') in the small intestine play a critical role in preserving intestinal barrier functions, thereby preventing translocation of harmful pathogens and toxins to enterocytes of animals [33,37]. The mucosa-associated microbiota, however, directly interact with the mucus layer and intestinal cells [38–40], indicating the changes in the mucosa-associated microbiota could provide substantial evidence of the negative impacts of AFB_1 on epithelium cells in the small intestine of animals. The difference between the luminal and mucosa-associated microbiota is distinct [41]. Therefore, conclusions drawn from findings examining only the luminal microbiota could lead to misunderstandings about interactions between the intestinal microbiota and the intestine. Additionally, AFB_1 increases the relative abundance of *Staphylococcus* and *Escherichia-Shigella* and decreases *Lactobacillus*, *Burkholderia caballeronia paraburkholderia*, *Romboutsia*, and *Corynebacterium* in the jejunal tissue of chickens [19]. Similarly, AFB_1 increases the prevalence of *Staphylococcus xylosus* in the jejunal tissue of chickens [42]. In nursery pigs, AFB_1 contamination in feeds also negatively affects the mucosal microbiota in the jejunum, reducing *Lactobacillus kitasatonis* [43]. Thus, changes in mucosa-associated microbiota by AFB_1 are an important parameter to assess the influence of aflatoxins on the intestinal health of animals.

2.2. Crosstalk Between the Mucosa-Associated Microbiota and the Intestinal Immune System

Crosstalk between the mucosa-associated microbiota and the intestinal immune system is critical for maintaining mucosal homeostasis. The small intestine of animals contains receptors on immune cells, such as dendritic cells and microfold cells in the epithelium, which recognize intestinal antigens primarily through toll-like receptors (TLRs) and nucleotide-binding oligomerization domain-like receptors (NODs). These pattern recognition receptors modulate cytokine production to maintain the immune function and intestinal health of animals [23,44]. An increase in cytokine production triggered by AFB_1 leads to heightened pro-inflammatory cytokine production and oxidative damage within the jejunum of animals [26,45]. A previous study reported that 40 μg/kg of AFB_1 in feeds increased the relative abundance of pathogenic bacteria, such as *Staphylococcus*, whereas it decreased the relative abundance of beneficial bacteria, such as *Lactobacillus*, in the jejunal tissue [19]. Additionally, AFB_1 increased the relative mRNA abundance of TLR2, NOD1, inducible nitric oxide synthase (iNOS), and nuclear factor kappa-light-chain-enhancer of activated B cells (NF-κB) as well as increased the relative mRNA abundance of interleukin-6 (IL-6), interleukin-8 (IL-8), and tumor necrosis factor-alpha (TNF-α) and increased cysteine-aspartic acid protease-3 (CASPASE-3) protein expression in the jejunum of chickens [19]. An in vitro study using 200 μg/kg of AFB_1 similarly increased the relative mRNA abundance of NF-κB and iNOS, elevated pro-inflammatory cytokines including IL-6, IL-8, and TNF-α, and decreased intestinal epithelial cell viability in chicken embryo primary intestinal epithelium [20]. These findings suggest that AFB_1 is strongly correlated with changes in the mucosa-associated microbiota and epithelial cell receptors, which can induce immune responses and affect the viability of small intestinal cells in animals. However, another study reported that 600 μg/kg of AFB_1 reduced the mRNA expression of TLR2-2, TLR4, and TLR7 in the jejunum [46], indicating that excessively high AFB_1 contamination can impair the innate immunity of the small intestine by suppressing these receptors. Therefore, AFB_1 negatively modulates the mucosa-associated microbiota, induces microbial-sensing receptors, and directly impairs the intestinal immune system in animals.

Table 1. Intestinal microbiota, immune response and oxidative damage products, and intestinal barrier function in chickens and pigs fed diets contaminated with aflatoxin B1 (AFB_1). Changes were indicated using ↑ (increase) and ↓ (decrease).

Age or IBW [1]	Experimental Period (d)	AFB_1 (μg/kg)	Result [2,3]	Reference
Intestinal microbiota				
Chicken (d)				
1	42	42	Relative abundance (RA): ↑ *Staphylococcus xylosus* (jejunal tissue)	[42]
1	21	40	Cell count: ↑ *Escherichia coli*, ↑ *Clostridium perfringens*, and ↑ Gram-negative bacteria (ileal digesta)	[47]
1	42	600	Cell count: ↑ *Coliforms* (cecal digesta)	[48]
1	42	1000	↓ acetate, ↓ propionate, and ↓ butyrate (feces)	[49]
7	14	394 to 1574	Cell count: ↑ *Escherichia coli*, ↑ *Salmonella*, and ↑ *Klebsiella* (ileal digesta)	[22]
22	21	40	RA: ↑ *Staphylococcus*, ↑ *Escherichia-shigella*, ↓ *Lactobacillus*, ↓ *Burkholderia-caballeronia-paraburkholderia*, ↓ *Romboutsia*, and ↓ *Corynebacterium* (jejunal tissue)	[19]
Pig (kg)				
6	48	180	RA: ↓ *Lactobacillus kitasatonis* (jejunal mucosa)	[43]
38	102	102	Cell count: ↑ *Escherichia coli* (colonic digesta)	[26]

Table 1. *Cont.*

Age or IBW [1]	Experimental Period (d)	AFB_1 (μg/kg)	Result [2,3]	Reference
Immune response and oxidative damage products				
Chicken (d)				
1	7 to 21	300	↓ IgA^+ number and ↓ sIgA, ↓ IgA, and ↓ IgG (ileum)	[50]
1	7 to 21	600	mRNA expression: ↑ TNF-α, ↑ CASPASE-3, ↑ CASPASE-8, and ↑ CASPASE-10 (jejunum)	[51]
1	7 to 21	600	mRNA expression: ↓ TLR2-2, ↓ TLR4, and ↓ TLR7 (small intestine)	[46]
1	21	600	↓ IgA^+ number and mRNA expression: ↓ IgA, ↓ pIgR, ↓ IgM, and ↓ IgG (small intestine)	[52]
1	21	100	↑ diamine oxidase and ↑ endotoxin (small intestine)	[53]
1	21	2000	↓ IgA (small intestine)	[54]
1	42	1000	↓ sIgA, ↑ IL-1β, and ↑ TNF-α (ileum)	[49]
22	21	40	mRNA expression: ↑ TLR2, ↑ NOD1, ↑NF-κB, ↑ iNOS, ↑ IL-6, ↑ IL-8, and ↑ TNF-α (jejunum)	[19]
Pig (kg)				
6	48	180	↑ IgA and ↑ protein carbonyl (jejunal mucosa)	[43]
56	35	180	↑ IgG (duodenal and jejunal mucosa)	
7	31	180	↑ protein carbonyl and ↑ TNF-α (jejunal mucosa)	[45]
7	30	500	↓ nitric oxide (small intestine)	[55]
9	30	320	↓ IFN-γ, ↓ IL-1β, ↓ TNF-α, ↓ IL-6, ↓ CAT, ↓ GPx, and ↓ SOD (duodenum), ↑ IFN-γ, ↑ IL-1β, and ↑ IL-6 (colon), and ↑ TBARS (duodenum and colon)	[56]
38	102	102	mRNA expression: ↑ TNF-α and ↑ IL-1β and ↑ TGF-β and ↓ SOD (jejunal mucosa)	[26]
Intestinal barrier function				
Chicken (d)				
1	7 to 21	600	↓ goblet cells (small intestine)	[46]
1	21	40	mRNA expression: ↓ claudin-1, ↓ sIgA, and ↓ pIgR (jejunum)	[47]
1	42	1000	mRNA expression: ↓ occludin, ↓ claudin-1, and ↓ zonula occludens-1 (ileal mucosa)	[49]
1	42	200	mRNA expression: ↓ claudin-3, ↓ occludin, and ↑ claudin-2 (jejunum)	[57]
22	21	40	mRNA expression: ↑ CASPASE-3 (jejunum)	[19]

Table 1. *Cont.*

Age or IBW [1]	Experimental Period (d)	AFB_1 (μg/kg)	Result [2,3]	Reference
Pig (kg)				
38	102	102	mRNA expression: ↓ zonula occludens-1 (jejunal mucosa)	[26]
Not available [4]	12 h	10 to 50	mRNA expression: ↓ CASPASE-3, ↑ zonula occludens-1, and ↑ occludin at 10 μg/kg of AFB_1, mRNA expression: ↓ mucin 2 at 20 μg/kg of AFB, and mRNA expression: ↑ Bcl-2 and ↑ratio of Bax to Bcl-2 at 30 to 50 μg/kg of AFB_1 (jejunal cell culture)	[24]
Not available [4]	48 h	40	mRNA expression: ↓ Bcl-2, ↓ zonula occludens-1, ↑ BaX, ↑IL-6, and ↑ CASPASE-3 (jejunal cell culture)	[25]

[1] IBW = initial body weight. [2] Result description based on the comparison between diets contaminated with AFB_1 and the control diet. [3] IgA = immunoglobulin A; pIgR = polymeric immunoglobulin receptor; IgM = immunoglobulin M; IgG = immunoglobulin G; TLR2-2 = toll-like receptor 2 type-2; TLR4 = toll-like receptor 4; TLR7 = toll-like receptor 7; sIgA = secretory immunoglobulin A; IgG = immunoglobulin G; IL-1β = interleukin-1β; TNF-α = tumor necrosis factor-alpha; CASPASE-3 = cysteine-aspartic acid protease-3; CASPASE-8 = cysteine-aspartic acid protease-8; CASPASE-10 = cysteine-aspartic acid protease-10; IFN-γ = interferon gamma; IL-6 = interleukin-6; SOD = superoxide dismutase; CAT = catalase; GPx = glutathione peroxidases; TBARS = thiobarbituric acid reactive substances; TGF-β = transforming growth factor-beta; Bcl-2 = B-cell lymphoma protein 2; Bax = Bcl-2-associated X protein. [4] In vitro, porcine jejunal epithelial cells were used.

3. Influence of AFB_1 on the Intestinal Health Parameters of Chicken and Pig

3.1. Immune Response and Oxidative Damage Products in the Small Intestine

The intestinal mucosa acts as a physical barrier between the body and the external environment [17]. Aflatoxin B1 enters the body through the intestinal mucosa, primarily in the small intestine due to it being a site of high absorption [21], meaning the enterocytes are highly exposed to these external compounds [58]. Aflatoxin B1 is a major contaminant in feedstuffs, and enterocyte exposure to AFB_1 has negative impacts, impairing innate immunity in the small intestine of animals [18]. The impacts of AFB_1 on the jejunal immune system can be seen via changes in the expression of immunoglobulins (Igs), representing innate immune status. In the small intestine of chickens, AFB_1 decreases IgA^+ cell numbers and reduces the expression of immunoglobulin-related genes, including IgA, pIgR, IgM, and IgG, in the duodenum, jejunum, and ileum, thereby impacting immune function [52]. Furthermore, AFB_1 exposure reduces antibody production from certain immune cells, including IgA, IgG, and IgM [47] and also reduces secretory IgA levels in the jejunum of chickens [54].

Aflatoxin B1 also induces the production of pro-inflammatory cytokines and oxidative damage products in the jejunum. A previous study reported that aflatoxin exposure at 1000 μg/kg AFB_1 for 42 d increased the levels of IL-1β and TNF-α in the small intestine of chickens [49]. In contrast, aflatoxin exposure at 500 μg/kg for 30 d decreased interferon-gamma (IFN-γ), interleukin-1β (IL-1β), TNF-α, and IL-6 levels in the small intestine of nursery pigs, whereas in the colon and liver, inflammatory cytokines were increased [55]. A possible explanation for the reduction in pro-inflammatory cytokines in the jejunum in pigs is likely due to high AFB_1 exposure, which induces epithelial cells apoptosis and subsequently impairs the function of the immune system [24,46]. Aflatoxin B1 exposure also increases diamine oxidase and endotoxin levels in the intestine of chickens, which act as important biomarkers of intestinal health [53]. Similarly, this immune suppression caused by AFB_1 also increases oxidative damage in the intestine of pigs, as evidenced by increased protein carbonyl levels in the jejunum [45] and elevated thiobarbituric acid-reactive substances (TBARSs) levels in intestinal tissues of pigs [56]. Additionally, AFB_1 exposure affects antioxidant enzymes in the intestine, such as superoxide dismutase (SOD), catalase (CAT), and glutathione peroxidase (GPx) [56]. Aflatoxin B1 exposure increases SOD levels in the jejunal mucosa of pigs [26]. Thus, AFB_1 exposure induces intestinal inflammation, disrupts immune responses, increases oxidative damage products, and impairs antioxidant activity in the small intestine, all of which can negatively impact intestinal health parameters related to the growth of animals.

3.2. *Tight Junction Protein, Intestinal Morphology, Tissue Repair, and Nutrient Digestion*

Aflatoxin B1 negatively impacts the viability of intestinal cells in animals [24] by inhibiting intestinal development through mechanisms such as G_2/M cell cycle arrest in epithelial cells [59], reducing the number of goblet cells [46], and increasing apoptosis rates, as indicated by a rise in TUNEL-positive cells in the jejunum [51]. A possible mechanism for these impacts on intestinal cells is that AFB_1 induces oxidative stress by generating reactive oxygen species (ROS), which damage cellular components, including lipid, protein, and DNA [60,61]. This oxidative damage disrupts cellular integrity and triggers apoptosis, or programmed cell death, resulting in the loss of vital epithelial cells in the intestinal lining [62,63]. Aflatoxin B1 also compromises intestinal permeability by reducing the relative gene expression of tight junction proteins, such as claudins, occludin, and zonula occludens, increasing lactate dehydrogenase activity in enterocytes, and decreasing trans-epithelial electrical resistance (TEER) in the small intestine [24,64]. The TEER directly correlates to tight junction integrity in the intestinal epithelium, indicating that AFB_1 can more easily enter circulation when TEER is reduced [64]. Aflatoxin B1 also induces apoptosis in intestinal cells by disrupting the balance of key apoptotic markers, such as B-cell lymphoma protein 2 (Bcl-2), Bcl-2-associatetd X protein (Bax), and cysteine-aspartic acid protease-3 (CASPASE-3) [24]. Aflatoxin B1 increases Bax, a protein that promotes cell death by making the mitochondrial membrane more permeable and releasing apoptotic signals in jejunal cells of pigs, whereas AFB_1 reduces Bcl-2, a protein that protects cells from dying, thereby weakening the cellular defense against apoptosis [25]. As a result, AFB_1 increases in the Bax/Bcl-2 ratio, inducing apoptosis of intestinal cells [65]. Aflatoxin B1 also activates CASPASE-3, a crucial executioner enzyme in the apoptotic pathway, which cleaves essential cellular components, leading to the controlled cell breakdown and death of enterocytes in animals [24,65]. In addition, AFB_1 disrupts the cell cycle in intestinal epithelial cells, causing G_2/M phase arrest [59], which limits normal proliferation and regeneration. Increased intestinal permeability further exacerbates the negative impacts of AFB_1, as it facilitates entry into the body. This damage to intestinal integrity leads to impaired intestinal morphology and reduced nutrient utilization, and ultimately, a decline in animal growth performance [66]. Consequently, the adverse impacts of AFB_1 are pronounced under conditions of increased intestinal susceptibility [21].

Aflatoxin B1 impairs intestinal morphology, reducing villus height and increasing crypt depth, which reduces the villus height-to-crypt depth ratio (VH:CD), an indicator of impaired intestinal structure and nutrient absorption function (Table 2). This impaired morphology reduces nutrient digestion and absorption, resulting in malnutrition and consequently compromised growth performance (Tables 3 and 4). Studies have shown that AFB_1 exposure decreases both apparent the ileal digestibility (AID) and total tract digestibility (ATTD) of nutrients in chickens and pigs [26,27,45,67]. The small and large intestines have distinct physiological roles, with most nutrient absorption occurring in the small intestine, whereas water and other metabolites are absorbed in the large intestine of animals [68]. Aflatoxin B1 is primarily absorbed in the small intestine, similarly to nutrients, meaning their impacts could be more pronounced in the small intestine, whereas AFB_1 residues in the large intestine mainly result from bile secretion after absorption in the small intestine [21]; therefore, the negative impacts of aflatoxins are typically more concentrated in the small intestine of animals. Aflatoxin B1, in particular, is the most toxic aflatoxin due to its high absorption rate, which is influenced by its small size and lipid solubility, allowing it to readily cross cell membranes in the small intestine [69,70]. Altogether, the negative impacts of AFB_1 are major contributors to compromised intestinal morphology, tight junction integrity, and nutrient digestion and absorption. Collectively, aflatoxins disrupt the luminal and mucosal microbiota, particularly in the small intestine, trigger immune responses and oxidative damage, compromise intestinal morphology, and impair nutrient utilization, ultimately affecting animal growth.

Table 2. Intestinal morphology and nutrient digestion in chickens and pigs fed diets contaminated with aflatoxin B1 (AFB_1). Changes were indicated using ↑ (increase) and ↓ (decrease).

Age or IBW [1]	Experimental Period (d)	AFB_1 (μg/kg)	Result [2,3]	Reference
Intestinal morphology				
Chicken (d)				
1	7 to 21	600	↓ villus height, ↓ villus width, ↓ VH:CD, and ↑ crypt depth (small intestine)	[46]
1	7 to 21	600	↓ villus height, ↓ VH:CD, ↑ crypt depth, and ↑ G_2/M cell cycle arrest (jejunum)	[59]
1	7 to 21	600	↓ villus height, ↓ VH:CD, ↓ number of absorptive cells, ↑ number of TUNEL-positive cells, and ↑ apoptotic rate in cells (jejunum)	[51]
1	20	1500	↑ lactulose/rhamnose ratio (jejunum)	[71]
1	21	60 to 120	↑ crypt depth at 60 μg/kg of AFB_1 (jejunum) and ↓ villus height at 120 μg/kg of AFB_1 (ileum)	[72]
1	28	0.5 mL of 20 μg/d	↑ epithelial thickness, ↑ enterocyte proliferation, ↑ epithelial plasma cell infiltration, and ↑ goblet cell proliferation (small intestine)	[73]
1	42	42	↓ villus height, ↓ VH:CD, and ↑ crypt depth (jejunum)	[42]
1	35	1000	↓ villus height, ↓villus width, ↓ VH:CD, and ↑ crypt depth (jejunum)	[74]
7	14	394 to 1574	↓ villus height, ↓ VH:CD, ↑ crypt depth, ↑ goblet cell count, and ↑ lamina propria lymphoid follicles diameter (jejunum)	[22]
11	19	250	↓ villus height, ↓ villus width, ↓ VH:CD, and ↑ crypt depth (jejunum)	[27]
Pig (kg)				
6	48	180	↓ villus height (jejunum)	[43]
7	30	500	↓ fold size and ↑ villus height (small intestine)	[55]
Not available [4]	48 h	40	↓ rate of intestinal cell viability, ↑ the % of necrotic cell, ↑ late apoptotic cell, and ↑ early apoptotic cell (jejunal cell culture)	[25]

Table 2. *Cont.*

Age or IBW [1]	Experimental Period (d)	AFB_1 (μg/kg)	Result [2,3]	Reference
Not available [4]	12 h	30	↓ fluorescent intensity of Bcl-2 and ↑ fluorescent intensity of the ratio of BaX to Bcl-2 (jejunal cell culture)	[24]
		40 to 60	↓ cell viability (jejunal cell culture)	
Nutrient digestion				
Chicken (d)				
1	20	1500	↓ AID of GE, CP, Asp, Thr, Pro, Gly, Ala, Cys, Val, Ile, Leu, Tyr, Phe, and His	[71]
1	21	70 to 750	↑ ATTD of EE at 750 μg/kg of AFB_1	[17]
1	273	2500 to 3910	↓ Retention of DM, CP, and EE	[67]
11	19	250	↓ ATTD of GE, CP, and EE	[27]
Pig (kg)				
7	31	180	↓ AID of CP	[45]
38	102	102	↓ ATTD of DM, GE, and EE	[26]

[1] IBW = initial body weight. [2] Result description based on the comparison between diets contaminated with aflatoxins and control diet. [3] VH:CD = villus height-to-crypt depth ratio; AID = apparent ileal digestibility; ATTD = apparent total tract digestibility; DM = dry matter; GE = gross energy; CP = crude protein; EE = ether extract; Bcl-2 = B-cell lymphoma protein 2; Bax = Bcl-2-associated X protein. [4] In vitro, porcine jejunal epithelial cells were used.

3.3. Impacts of AFB_1 After Absorption on Overall Health and Growth Performance

After absorption, AFB_1 is partially metabolized in hepatic cells into an active form that binds to hepatic macromolecules, a process believed to contribute to its toxicity and carcinogenicity [75,76]. Upon entering the liver, AFB_1 undergoes metabolic activation, primarily by cytochrome P450 enzymes, particularly CYP1A2 and CYP3A4 [57]. This bioactivation converts AFB_1 into the highly reactive AFB_1-8,9-epoxide, which can form adducts with DNA, proteins, and other macromolecules, leading to hepatotoxicity and carcinogenicity [77,78]. The binding of AFB_1-8,9-epoxide to hepatic macromolecules disrupts normal cellular functions, adversely affecting immune function and enzyme activities [79,80], and increasing hepatotoxicity [81]. This hepatotoxicity often results in liver enlargement (hepatomegaly) as the liver attempts to counteract the damage [27,82]. Liver weights were increased with high AFB_1 doses, as observed in experiments with doses ranging from 100 to 2500 μg/kg in feeds [27,82,83]. When the AFB_1 detoxification capacity of the liver is overwhelmed, systemic impacts such as immunosuppression, metabolic dysfunction, and reduced growth performance in animals occur, as documented in literature reviews [84,85]. Additionally, AFB_1 exposure compromises cell membrane integrity, leading to increased lipid peroxidation and elevated levels of malondialdehyde (MDA), a marker of oxidative stress. Studies have reported increased MDA levels in the livers of chickens exposed to AFB_1, indicating oxidative damage and cell death [83,86–88]. The remaining AFB_1 is converted into less toxic forms, which are then transported from hepatic cells into the blood and bile, along with un-metabolized parent compounds, and eventually are excreted in feces and urine [21,89].

To determine the negative impacts of AFB_1 intake per BW on the % change in the average daily gain (ADG) of animals, a meta-analysis was conducted (Figure 1). The reason for using AFB_1 intake per BW as the independent variable for the meta-analysis is because this approach can accurately determine the impacts of AFB_1 on growth, as the trials involved animals with varying feed consumptions and body weights. The change in ADG relative to the control diet group was calculated as follows:

$$\Delta ADG\ (\%) = (ADG\ of\ AFB_1\ treatment\ group - ADG\ of\ control\ group)/ADG\ of\ control\ group \times 100$$

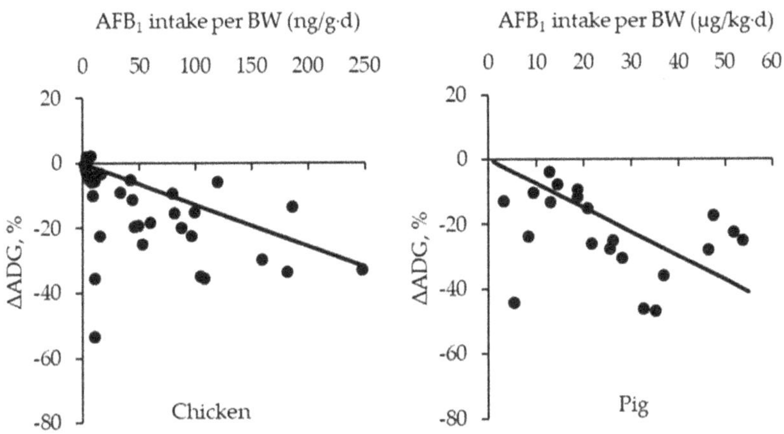

Figure 1. Change in average daily gain (ΔADG) of chickens and pigs fed diets contaminated with aflatoxin B1 (AFB_1). The meta-analysis was conducted by Proc REG using the data from 27 peer-reviewed papers on chickens and 13 peer-reviewed papers on pigs to determine the impacts of AFB_1 intake per body weight (BW) on ΔADG of animals. The equations were as follows: for chickens fed diets with AFB_1 (solid line, •): $\Delta ADG\ (\%)_{chicken} = -0.13 \times AFB_1$ intake per BW (ng/g·d) with standard error of slope = 0.02, $r^2 = 0.48$, and $p < 0.01$; and for pigs fed diets with AFB_1 (solid line, •): $\Delta ADG\ (\%)_{pig} = -0.74 \times AFB_1$ intake per BW (μg/kg·d), with standard error of the slope = 0.11, $r^2 = 0.70$, and $p < 0.01$. The AFB_1 intake per BW ranged from 1.2 to 350.9 ng/g·d in chickens and 3.0 to 53.7 μg/kg·d in pigs.

A literature search was conducted using the databases of PubMed and Google Scholar with keywords including aflatoxins, intestinal health, growth performance, and chickens and pigs, followed by screening after reading each article. The initial body weight (IBW) and experimental period ranges for chickens were 40 to 850 g and 7 to 62 d, respectively (Table 3), whereas the IBW and experimental period ranges for pigs were 7 to 50 kg and 21 to 90 d, respectively (Table 4). The AFB_1 intake per body weight (BW) ranged from 1.2 to 350.9 ng/g·d in chickens and 3.0 to 53.7 μg/kg·d in pigs. Data on AFB_1 co-contaminated with other mycotoxins were excluded to determine the impact of AFB_1 on the ADG of animals. The AFB_1 intake per BW of animals in each research article was calculated by multiplying the dietary AFB_1 content with the overall average daily feed intake of each treatment divided by the mean BW of animals. In the meta-analysis, the impact of AFB_1 intake per BW of animals on % change in body weight gain was evaluated using a linear regression with the Proc REG procedure in SAS (SAS Inst. Inc., Cary, NC, USA). The independent variable was the AFB_1 per BW (ng/kg·d for chicken and μg/kg·d for pig), and the dependent variable was the calculated % change in ADG (ΔADG, %) relative to the control group. To ensure the regression model passed through the origin, the NOINT option was applied, forcing the intercept of the regression line to be zero. This approach assumes that in the absence of AFB_1 intake, there would be no impact on ADG (ΔADG = 0%). Based on the meta-analysis, the equations for chicken and pig in diets containing AFB_1 were as follows (Figure 1):

For chicken: $\Delta ADG\ (\%)_{chicken} = -0.13 \times AFB_1$ intake per BW (ng/g·d)

For pig: $\Delta ADG\ (\%)_{pig} = -0.74 \times AFB_1$ intake per BW (μg/kg·d).

These equations indicate that for every 1 μg/kg·d of AFB_1 intake per BW, ADG decreases by 0.13% in chickens and by 0.74% in pigs. For example, AFB_1-contaminated diets (300 μg/kg) were fed to nursery pigs (weighing 7 to 25 kg, averaging 16 kg) with an ADFI of 0.50 kg, estimating a 6.9% reduction in ADG compared to the ADG of pigs fed non-contaminated diets.

Studies report that AFB_1 absorption is higher in early life stages compared to older life stage in rats [21]. This increased absorption is likely due to an immature digestive system, differences in lipid composition of the epithelial cell membrane in the small intestine at different ages, and age-related changes in liver metabolic activity that may also have impacts of AFB_1 on animal growth [90,91]. Additionally, in pigs, the weaning period could leave them more susceptible to AFB_1 exposure due to the greater likelihood of opportunistic pathogenic bacterial colonization in the intestine, triggered by dietary challenges and reduced feed intake. Considering the collective findings on the negative impacts of AFB_1 on intestinal microbiota, intestinal health, overall health, and growth performance of animals, reducing AFB_1 absorption could be a key factor in mitigating its adverse impacts on animals.

Table 3. Growth performance of chickens fed diets contaminated with aflatoxin B1 (AFB_1) and AFB_1 co-contaminated with other mycotoxins [1,2].

IBW (g) [3]	Experimental Period (d)	Dietary AFB_1 (μg/kg) [4]	AFB_1 Intake per BW (ng/g·d) [5]	Contamination Type	Growth Performance (% Change)			Reference
					ADG	ADFI	G:F	
40	44	25	2.0	Natural	−3.17 **	−2.96 **	−0.22	[88]
40	21	40	5.1	Natural	−7.39 **	−5.32 **	−2.19 **	[92]
40	35	1000	98.7	Culture	−14.99 **	−4.34	−11.14 **	[74]
33	35	1000	119.7	Not available	−5.69 **	4.65	−9.88 **	[93]
35	62	15 30 45 60	1.2 2.3 3.5 4.7	Culture	−0.32 −1.92 −2.88 −4.79	−0.13 −1.29 −1.03 −2.83	−0.19 −0.64 −1.87 ** −2.02 **	[94]
35	21	100	15.2	Pure AFB_1 with LPS (1.7×10^6 EU/bird)	−22.44 **	−13.80 **	−10.02 **	[53]
40	28	1000	104.3	Culture	−34.74 **	−28.93 **	−8.17 **	[83]
40	42	500	52.6	Culture	−25.00 **	−7.62 **	−18.81 **	[95]
40	42	100	10.7	Culture	−35.56 **	−26.50 **	−12.31 **	[96]
40	35	1000	108.3	Culture	−35.33 **	−26.29 **	−12.54 **	[97]
40	21	2500	350.9	Culture	−16.77 **	−14.77 **	−2.35	[98]
40	49	500 1000 2000	41.4 87.6 181.6	Culture	−5.28 −19.90 ** −33.73 **	−0.31 −9.83 ** −21.45 **	−4.98 −11.17 ** −15.64 **	[99]
40	42	100	8.6	Pure AFB_1	−9.78 **	−5.69	−4.34 **	[100]
40	23	750 1500	79.6 158.9	Culture	−9.42 ** −29.70 **	−9.67 ** −28.25 **	0.27 −2.02 **	[101]

Table 3. *Cont.*

IBW (g) [3]	Experimental Period (d)	Dietary AFB_1 (μg/kg) [4]	AFB_1 Intake per BW (ng/g·d) [5]	Contamination Type	Growth Performance (% Change)			Reference
					ADG	ADFI	G:F	
40	49	129 385 895	10.2 32.7 96.8	Pure AFB_1	−5.53 ** −8.91 ** −22.48 **	5.41 9.67 ** 19.57 **	−10.37 ** −16.95 ** −35.17 **	[102]
40	42	50 100	4.1 7.9	Pure AFB_1	−4.05 ** −5.74	−9.67 ** −14.77	6.23 ** 10.60	[103]
40	42	500	45.3	Culture	−19.44	−16.17	−3.91	[104]
40	42	200	15.7	Pure AFB_1	−3.29 **	−1.46	−1.82 **	[57]
40	20	1500	185.4	Culture	−13.51 **	−8.20 **	−6.88 **	[71]
42	37	40	3.2	Culture	1.96	3.95 **	−1.91 **	[105]
43	42	500	43.4	Culture	−11.29 **	−1.88	−9.6	[106]
43	42	600	48.7	Culture	−19.23 **	−9.85 **	−10.40 **	[48]
45 833	21 21	60 120	6.6 11.0	Pure AFB_1	2.07 −4.00	2.02 −1.43	0.05 −2.61	[72]
47	42	1000	80.8	Pure AFB_1	−15.47 **	−6.94	−9.16 **	[82]
48	21	2000	269.6		−25.12 **	3.33	−27.54 **	[54]
838	7	100	10.4	Pure AFB_1	−53.33 **	−4.76 **	−51.00 **	[87]
140	14	394 1574	60.0 247.5	Culture	−18.34 ** −32.87 **	−7.17 ** −14.88 **	−12.04 ** −21.14 **	[22]
				Mycotoxin [6], μg/kg				
45	42	25	1.7	DON: 1000, ZEA: 90, and OTA: 90	−4.10 **	−4.85 **	0.78	[107]
		50	3.3	DON: 1000, ZEA: 90, and OTA: 475	−6.15 **	−8.13 **	2.16	
50	21	42	4.9	DON: 86 and ZEA: 50 μg/kg	−9.11 **	−15.71 **	7.83 **	[42]
	42	42	4.3	DON: 86 and ZEA: 50 μg/kg	−8.80	3.67	−12.02	
40	49	500 1000 2000	38.1 82.0 194.1	OTA: 1000 OTA: 2000 OTA: 4000	−20.22 ** −35.41 ** −51.00 **	−21.83 ** −30.69 ** −35.98 **	2.06 −6.81 −23.47 **	[99]
162	34	330	30.2	ZEA: 4, AFB_2: 80, AFG_1: 30, and AFG_2: 7	−4.52 **	−2.73	−1.84 **	[108]

4.1. Mycotoxin-Adsorbing Agent

Mycotoxin-adsorbing agents are extensively supplemented in monogastric animal feeds to mitigate the negative impacts of aflatoxins (Table 5) [31,113,114]. The mode of action for mycotoxin-adsorbing agents relies on the characteristics of the agents, which bind the toxins and excrete them along with undigested waste [131,132]. Inorganic compounds, such as clay and minerals, are naturally occurring or modified aluminosilicates. Aluminosilicates include bentonites, montmorillonites, kaolinites, and zeolite, characterized by their layered structure and high cation-exchange capacity [133,134]. These inorganic compounds adsorb toxins through ionic and hydrophobic interactions, preventing their absorption in the GIT [131], thereby reducing its toxicity and improving intestinal health and growth performance in animals [47,56]. In vitro trials evaluating the efficiency of clay compounds in adsorbing AFB_1 reported high adsorption rates, ranging from 92 to 99% [131,132,135], indicating that clay compounds have a high affinity for AFB_1. Charcoal can also be used to adsorb multiple mycotoxins as a mycotoxin-adsorbing agent; however, activated charcoal also adsorbs essential nutrients in feed [136]. Notably, the clay compounds show less interference in the utilization of vitamins or micro-minerals in the GIT of animals, making clay compounds commonly utilized to prevent the impacts of AFB_1 on intestinal health and growth of animals. For other mycotoxins, such as deoxynivalenol (DON) and zearalenone (ZEA), the adsorption efficacy of clay compounds is relatively low and highly variable among different adsorbing agents when compared to the adsorption efficiency for AFB_1 [131]. The reason for the high adsorption of AFB_1 by clay may be due to the hydrophobic characteristics of AFB_1, causing strong ionic and electrostatic interactions with clay [131]. Aflatoxin B1 also has a large, planar, and rigid aromatic structure [137], allowing it to fit closely with the flat, layered surfaces of clays like montmorillonite [131].

In the jejunum of chickens, the inorganic compound, hydrated sodium calcium aluminosilicate (HSCA), decreased *Escherichia coli* and gram-negative bacteria in the ileum and increased the relative mRNA expression of claudin-1, secretory IgA, and pIgR [47]. Also, the inorganic compound, 0.40% bentonite, increased intestinal morphology and increased the nutrient digestion of chickens [27]. The beneficial effects of clay compounds also improved the growth performance of animals (Table 6). Thus, clay compounds effectively adsorb AFB_1 in the digesta of animals, thereby mitigating its negative impacts on intestinal health, which are highly related to growth performance. The clay compound, however, could reduce feed intake if the clay compound was supplemented to chicken feeds at a rate of 1% [96], potentially indicating that high doses of clay compounds could reduce feed palatability.

Yeast cell wall compounds, another type of mycotoxin-adsorbing agent, can efficiently adsorb AFB_1 due to the manno-oligosaccharides and β-glucan structures present in their composition [138,139]. Specifically, β-glucan in the yeast cell wall has high adsorbability for AFB_1 in the digesta of pigs, with AFB_1 detoxification by *Saccharomyces cerevisiae* strains reaching 65% after 24 h of incubation [140]. The yeast cell wall also has a prebiotic effect on intestinal health and growth by positively modulating immune function through jejunal dectin-1 activation [141,142], which could mitigate the negative impacts of AFB_1 on the intestinal health of animals [139]. A previous study reported that yeast cell wall supplementation in feed contaminated with 180 μg/kg of AFB_1 decreased jejunal IgA and protein carbonyl, increased jejunal villus height and crypt cell proliferation, and improved AID of DM and CP in nursery pigs [43]. Manno-oligosaccharide supplementation in feeds, a compound found in high concentrations in yeast cell walls, can mitigate the negative impacts of AFB_1 on the intestinal health and growth of animals, decreasing the *Escherichia coli*, *Salmonella*, *Klebsiella*, and total gram-negative bacteria counts in the ileum and improving intestinal morphology including villus height, crypt depth, and VH:CD [22]. Additionally, the detoxification mechanism of *Lactobacillus*, *Bifidobacterium*, and *Enterococcus* generally involves binding AFB_1, which helps reduce the negative impacts of AFB_1 on the intestinal health of animals [143,144].

Table 5. Intestinal health of chickens and pigs fed diets contaminated with aflatoxin B1 (AFB_1) including mycotoxin-detoxifying agents or other feed additives. Changes were indicated using ↑ (increase) and ↓ (decrease).

Age or IBW [1]	Experimental Period (d)	AFB_1 (μg/kg)	Type	Level (%)	Result [2]	Reference
Mycotoxin-adsorbing agent						
Chicken (d)						
1	21	40	Clay (HSCA [3])	0.30	Cell count: ↓ *Escherichia coli* and ↓ Gram-negative bacteria (ileal digesta) and mRNA expression: ↑ claudin-1, ↑ sIgA, and ↑ pIgR (jejunum)	[47]
11	19	250	Clay (bentonite)	0.40	↑ villus height and ↑ villus surface area (ileum) and ↑ ATTD of CP and EE and AME	[27]
			Mineral (zeolite)	0.40	↑ ATTD of CP	
7	21	394 to 1574	Manno-oligosaccharides	0.10 to 0.20	Cell count: ↓ *Escherichia coli*, ↓ *Salmonella*, *Klebsiella*, and ↓ Gram-negative bacteria (ileal digesta) and ↑ villus height, ↑ VH:CD, ↓ crypt depth, and ↓ goblet cell counts (jejunum)	[22]
	35	394 to 1574	Manno-oligosaccharides	0.10 to 0.20	Cell count: ↓ *Escherichia coli*, ↓ *Salmonella*, and ↓ *Klebsiella* (ileal digesta) and ↑ villus height, ↑ crypt depth, ↑ VH:CD, and ↓ goblet cell counts (jejunum)	
1	21	2000	Cellulosic polymer	0.30	↓ relative weight of intestine	[54]
1	42	1000	*Lactobacillus salivarius*	10^8 CFU/kg	↑ acetate, ↑ propionate, and ↑ butyrate (feces) and ↓ IL-1β and ↓ TNF-alpha (ileum)	[49]

Table 5. *Cont.*

Age or IBW [1]	Experimental Period (d)	AFB_1 (μg/kg)	Type	Level (%)	Result [2]	Reference
1	21	40	*Lactobacillus acidophilus*, *Lactobacillus plantarum*, and *Enterococcus faecium*	3×10^{10} CFU/kg	Cell count: ↓ *Clostridium perfringens*, ↓ *Escherichia coli*, and ↓ Gram-negative bacteria (ileal digesta) and mRNA expression: ↑ claudin-1, ↑ sIgA, and ↑ pIgR (jejunum) and ↓ visceral lesion score (small intestine)	[47]
Pig (kg)						
6	48	180	Yeast cell wall	0.20	↓ IgA and ↓ protein carbonyl (jejunal mucosa) and ↑ AID of DM and CP	[43]
Multi-component mycotoxin-detoxifying agent						
Chicken (d)						
1	21 to 42	200	Clay (bentonite) + yeast cell wall	0.20	mRNA abundance: ↑ claudin-1 on d 21 and mRNA abundance: ↑ claudin-2 and ↑ occludin on d 42 (jejunum)	[57]
1	42	600	Adsorbing gent [clay (bentonite), activated charcoal, *Lactobacillus* sp., and *Bifidobacterium* sp.] + biotransforming agent (*Bacillus* sp.)	0.10	Cell count: ↓ *Coliforms* (cecal digesta)	[48]
1	21 to 42	42	Adsorbing gent [4] [clay (montmorillonite), *Lactobacillus casei*, and *Enterococcus faecalis*] + biotransforming agent [5] (*Bacillus subtilis*, *Candida utilis*, and mycotoxin-degrading enzyme)	0.10	↑ ATTD of CP	[42]
				0.15	↓ crypt depth, ↑ VH:CD, and ↑ relative jejunum weight on d 42 (jejunum) and ↑ ATTD of CP	
Pig (kg)						
7	30	500	Clay + yeast cell wall	0.10	↓ villus height (small intestine)	[55]

Table 5. *Cont.*

Age or IBW [1]	Experimental Period (d)	AFB_1 (μg/kg)	Type	Level (%)	Result [2]	Reference
Not available [6]	48 h	40	Adsorbing agent (*Lactobacillus casein* and *Candida utilis*) + biotransforming agent (*Aspergillus oryzae, Bacillus subtilis*, and mycotoxin-degrading enzyme)]	5.00	mRNA abundance: ↓ IL-6, ↑ occludin, ↑ ZO-1, ↓ Bax, ↓ CASPASE-3, and ↑ Bcl-2 and ↑ rate of cell viability, ↓ % of necrotic cell, ↓ early apoptotic cell, and ↓ viable cell rate (jejunal cell culture)	[25]
Other feed additive						
Chicken (d)						
1	34	250	Phytobiotics	0.03 to 0.05	↑ villus height, ↑ VH:CD, and ↓ crypt depth (jejunum)	[145]
1	42	600	Milk thistle (*Silybum marianum*)	1.00	Cell count: ↓ *Coliforms* (cecal digesta)	[48]
1	35	1000	Phytobiotics	0.05	No difference in the number of bacteria (cecal digesta) and ↑ villus height, ↑ VH:CD, and ↓ crypt depth (jejunum)	[74]
1	7 to 21	600	Selenium	0.4 mg/kg	↑ villus height, ↑ VH:CD, ↑ number of absorptive cells, ↓ crypt depth, and ↓ TUNEL-positive cells (jejunum)	[51]
1	14 to 21	300	Sodium selenite	0.4 mg/kg	↑ IgA^+ number and ↑ IgG (ileum)	[50]
1	21	2000	Cellulosic polymer + curcumin	0.50	↑ IgA and ↓ relative weight of intestine (small intestine)	[54]
1	35	1000	Toxin binders (adsorbing agents + mycotoxin-degrading enzyme + plant extract)	0.05	No difference on the number of bacteria (cecal digesta) and ↑ villus height, ↑ VH:CD, and ↓ crypt depth (jejunum)	[74]

Table 5. *Cont.*

Age or IBW [1]	Experimental Period (d)	AFB_1 (μg/kg)	Type	Level (%)	Result [2]	Reference
Pig (kg)						
9	30	320	Grape seed waste	8.00	↑ SOD, ↑ antioxidant capacity, ↑ thiobartituric acid reactive substances, ↑ IL-6, and ↑ IL-8 (duodenum) and ↑ CAT, ↑ GPx, ↓ IFN-γ, ↓ IL-1β, ↓ TNF-α, and ↓ IL-6 (colon)	[56]

[1] IBW = initial body weight. [2] Comparison of the effects of mycotoxin mitigation agents in diets contaminated with AFB_1, compared to the effects of diets contaminated with AFB_1 without mycotoxin mitigation agents, in chickens and pigs; VH:CD = villus height-to-crypt depth ratio; SOD = superoxide dismutase; CAT = catalase; GPx = glutathione peroxidase; IL-6 = interleukin-6; IL-8 = interleukin-8; IFN-γ = interferon-gamma, IL-1β = interleukin-1β, TNF-α = tumor necrosis factor α; AID = apparent ileal digestibility; AME = apparent metabolizable energy; ATTD = apparent total tract digestibility; CP = crude protein. [3] HSCA = hydrated sodium calcium aluminosilicate. [4] Adsorbing agent = *Lactobacillus casei* (1.0×10^8 CFU/g) and *Enterococcus faecalis* (1.0×10^{10} CFU/g) were included. [5] Biotransforming agent = *Bacillus subtilis* (1.0×10^8 CFU/g) and *Candida utilis* (1.0×10^8 CFU/g) were included. [6] In vitro, porcine jejunal epithelial cells were used.

Table 6. Growth performance of chickens and pigs fed diets contaminated with aflatoxin B1 (AFB_1) including mycotoxin-adsorbing agent.

Age or IBW [1]	Experimental Period (d)	AFB_1 (μg/kg)	Type	Level (%)	Growth Performance [2] (% Change)						Reference
					ADG		ADFI		G:F		
					vs. AFB_1 [3]	vs. Control [4]	vs. AFB_1	vs. Control	vs. AFB_1	vs. Control	
Chicken (d)											
1	21	40	Clay (HSCA [4])	0.30	4.47 **	−3.26 **	3.66 **	−1.85 **	0.78 **	−1.43 **	[92]
1	20	250	Clay (bentonite)	0.40	17.05 **	−6.36	6.88	−0.33	9.52 **	−6.05	[27]
			Clay (zeolite)	0.40	18.41 **	−5.27	10.26 **	2.83	7.40 **	−7.87 **	
1	42	100	Clay (sodium bentonite)	0.50	29.12 **	−16.70 **	16.21 **	−14.77 **	11.10 **	−2.26 **	[96]
			Clay (sodium bentonite)	1.00	4.85 **	−32.35 **	−1.91 **	−28.06 **	6.89 **	−5.96 **	
			Silicate (sorbatox)	0.50	4.08 **	−32.85 **	−2.42 **	−28.43 **	6.65 **	−6.17 **	
			Silicate (klinofeed)	0.20	3.12 **	−33.47 **	0.54 **	−26.27 **	2.56 **	−9.77 **	
1	42	500	Clay (smectite)	0.20	14.29 **	−14.29 **	0.00	−7.62 **	14.29 **	−7.22 **	[95]
1	42	1000	Diatomaceous earth extracted from a quarry	0.10	2.15 **	−13.62 **	2.29	−4.79	−0.14	−9.27 **	[82]
				0.20	18.84 **	0.50	11.87 **	4.13	6.23	−3.48	
				0.50	14.29	−3.35 **	6.77	−0.62	7.04 **	−2.74	
1	35	1000	Yeast cell wall	0.20	3.45 **	−31.82 **	3.45 **	−24.05 **	0.00	−10.23 **	[97]
7	14	500	Mannanoligosaccharides	0.10	4.35 **	2.33 **	1.98 **	−12.73 **	−5.38 **	−7.77 **	[22]
				0.20	10.87 **	6.98 **	3.64 **	−7.27 **	−1.08 **	−6.26 **	
		2000	Mannanoligosaccharides	0.10	13.16 **	3.70 **	9.12 **	−21.82 **	−9.68 **	−13.44 **	
				0.20	18.42 **	6.17 **	11.54 **	−18.18 **	−7.53 **	−11.52 **	
1	21	2000	Cellulosic polymer	0.30	25.15 **	−6.19 **	−3.23 **	0.12	21.56 **	−2.13 **	[54]
1	42	500	Mineral (toxin binder)	0.10	7.55 **	−5.00 **	1.98 **	0.00	5.46 **	−5.00 **	[106]
Pig (kg)											
9	35	922	Clay (sodium bentonite)	1.00	20.15 **	−6.73	21.59 **	−2.73	−1.18	−4.12	[113]
9	42	150	Clay (montmorillonite)	0.02	5.84	−5.67	4.22	−3.06	1.9	−2.69	[124]
9	42	500	Clay (sodium bentonite)	0.50	37.46 **	−0.41	9.43	2.96	25.62 **	−3.27	[146]
			Clay (montmorillonite) + zeolite	0.50	18.03 **	−14.49	−4.83	−10.45	24.01 **	−4.51	
			Clay (calcium bentonite)	0.50	22.54 **	−11.22	−5.16	−10.77	29.21 **	−0.51	
			Clay (attapulgite)	0.50	−0.56	−27.96	−8.98	−14.36	9.25	−15.88	
9	42	373	Clay (maifanite)	1.00	12.24	1.85	8.16	−2.75	3.77	4.73	[115]
10	21	500	Clay (HSCA [5])	0.50	36.20 **	−1.28	39.14 **	−1.47	−2.11	0.19	[116]
11	28	840	Clay (HSCA)	0.50	71.43 **	−7.69	74.63 **	3.54	−1.83	−10.85	[114]

Table 6. *Cont.*

Age or IBW [1]	Experimental Period (d)	AFB_1 (μg/kg)	Type	Level (%)	Growth Performance [2] (% Change)						Reference
					ADG		ADFI		G:F		
					vs. AFB_1 [3]	vs. Control [4]	vs. AFB_1	vs. Control	vs. AFB_1	vs. Control	
11	28	800	Clay (calcium bentonite)	0.50	29.17 **	−3.13	13.68	0.76	13.63 **	−3.85	[113]
			Clay (HSCA)	0.50	20.83 **	−9.38	8.55	−3.79	11.32	−5.81	
			Clay mineral (palygorskite)	0.50	10.42	−17.19	0.00	−11.36	10.42	−6.57	
			Clay mineral (sepiolite)	0.50	25.00 **	−6.25	0.85	−10.61	23.94 **	4.87	
11	41	200	Clay (bentonite)	0.40 0.50	26.19 ** 23.81 **	−3.64 −5.45 **	7.41 11.11	−3.33 0.00	17.49 ** 11.43	−0.31 −5.45	[118]
20	21	1100	Clay (HSCA)	0.50	166.67 **	−18.99	38.82	−19.73	92.09 **	0.92	[77]
30	90	110	Clay (montmorillonite)	0.30	5.56	−8.06	5.44	−1.27	0.11	−6.88	[122]
			Clay (montmorillonite nanocomposite)	0.30	12.96 **	−1.61	7.48	0.64	5.10 **	−2.24	
			Clay (montmorillonite nanocomposite)	0.30	12.96 **	−1.61	7.54	0.73	5.04 **	−2.33	
6	35	20	Yeast cell wall	0.20	5.42 **	7.8	6.10 **	7.98	−0.64	−0.17	[123]

[1] IBW = initial body weight. [2] Asterisk marks (**) represent statistical tendency ($p < 0.10$) and significant difference ($p < 0.05$), respectively. [3] The percentage increase or decrease in the average daily gain (ADG), average daily feed intake (ADFI), and gain-to-feed ratio (G:F) was determined in AFB_1 contamination with mitigation agents relative to the aflatoxin contamination group. [4] The percentage increase or decrease in the ADG, ADFI, and G:F was determined in aflatoxin contamination with mitigation agents relative to the control group. [5] HSCA = hydrated sodium calcium aluminosilicate.

4.2. Mycotoxin-Biotransforming Agent

To mitigate the negative impacts of AFB_1, biotransforming agents such as mycotoxin-degrading enzymes and specific bacteria that secrete enzymes that degrade mycotoxins into non-toxic metabolites can be supplemented in feeds to counteract its negative impacts on the intestinal health and growth of animals [42,47]. Selected strains of *Bacillus* spp. have generally demonstrated a high ability to degrade AFB_1 into non-toxic metabolites [147]. A possible explanation is that these bacteria produce various enzymes, such as laccase and lactonase, which target AFB_1 and convert it to aflatoxin Q1 [148] and aflatoxin D1 [149], respectively, both of which are less toxic metabolites [150,151]. *Bacillus cereus* and *Bacillus subtilis* are known to produce lactonase, which cleaves the lactone ring of AFB_1 by approximately 40% and 50%, respectively, after 72 h of incubation in an in vitro study [152]. Previous studies also reported that *Bacillus TUBF1 B.* [153] and *Bacillus subtilis UTBSP1* [154] achieved over 80% AFB_1 degradation after 72 h of incubation in in vitro studies. Additionally, supplementation with *Bacillus subtiltis* showed additional beneficial effects on the intestinal health of animals by modulating the intestinal microbiota, reducing immune responses [155], and improving the intestinal morphology of animals [156]. A previous study reported that *Nocardia corynebacteroides* showed 70% AFB_1 degradation after 44 h of incubation in an in vitro study [157]. Similarly, another study reported that *Nocardia corynebacteroides* degraded AFB_1 [158], decreased lesion scores in the duodenum of chickens, reduced residual AFB_1 in the liver, and improved growth performance compared to chickens fed an AFB_1-contaminated diet without *Nocardia corynebacteroides* [159]. However, the specific less toxic metabolite produced by *Nocardia corynebacteroides* has not been investigated, and the degradation rate of AFB_1 by *Nocardia corynebacteroides* is influenced by heat and proteinase, which affect enzyme efficiency [158].

4.3. Multi-Component Mycotoxin-Detoxifying Agent

To maximize the mitigation of the negative impacts of AFB_1 on the intestinal health and growth performance of monogastric animals, multi-component mycotoxin-detoxifying agents can be used in chicken and pig feeds (Table 7) [57,96,160]. Naturally contaminated feedstuffs often contain multiple mycotoxins; a global survey of over 70,000 samples found that 64% of sampled feeds and feedstuffs were contaminated with more than one mycotoxin [1]. Multi-component mycotoxin-detoxifying agents may be more effective in mitigating the impacts of AFB_1 and other mycotoxins than single-component mycotoxin-detoxifying agents, by combining adsorption and detoxification properties and positive effects on intestinal health, offering broader coverage when mycotoxin contamination is unspecified.

Table 7. Growth performance of chickens and pigs fed diets contaminated with aflatoxin B1 (AFB_1) including a multi-component mycotoxin-detoxifying agent.

Age or IBW [1]	Experimental Period (d)	AFB_1 (μg/kg)	Type	Level (%)	Growth Performance [2] (% Change)						Reference
					ADG		ADFI		G:F		
					vs. AFB_1 [3]	vs. Control [4]	vs. AFB_1	vs. Control	vs. AFB_1	vs. Control	
Chicken (d)											
1	42	200	Clay (bentonite) + yeast cell wall	0.20	0.17	−3.09	0.90	−0.57	−0.72	−2.53	[57]
1	35	1000	Clay (bentonite and HSCA) + oligomannose	0.05	7.77 **	−8.37 **	4.65	0.10	2.97	−8.46 **	[74]
1	42	42	Adsorbing agent [5] [Clay (montmorillonite), *Lactobacillus casei*, and *Enterococcus faecalis*)] + biotransforming agent [6] (*Bacillus subtilis*, *Candida utilis*, and mycotoxin-degrading enzyme)	0.50	5.35	−8.47	−8.47	−3.92	15.10	1.26	[42]
				0.10	8.42	−12.65	24.44 **	−1.12	24.44 **	9.19	
				0.15	9.75 **	−11.16	24.44 **	0.10	24.44 **	8.69	
1	42	600	Adsorbing agent [clay (bentonite), activated charcoal, *Lactobacillus* sp., and *Bifidobacterium* sp.] + biotransforming agent (*Bacillus* sp.)	0.10	18.75 **	−5.00 **	8.14 **	−2.11	9.81 **	−2.96 **	[48]
1	37	40	Adsorbing gent (*Lactobacillus acidophilus*) + biotransforming agent (*Bacillus subtitlis*)	0.005	3.45 **	7.14 **	−5.68 **	−1.19 **	9.68 **	8.43 **	[105]
1	42	500	Adsorbing gent (*Streptococcus salivarius sp. Thermophilus*, *Lactobacillus* spp. [7], *Bifidobacterium bifidum*, *Enterococcus faecium*, and *Candida pintolopesii*) + biotransforming agent (*Aspergillus oryzae*)	0.50	5.66 **	−6.67 **	−2.97 **	−4.85 **	8.89 **	−1.90 **	[106]
Pig (kg)											
7	30	500	Yeast cell wall + calcium carbonate	0.10	15.15 **	−7.32	−3.77	−22.73	19.67	19.94	[55]
9	42	150	Clay (sodium bentonite and sepiolite) + dried yeast	0.011	4.10	−6.86	0.53	−6.50	4.32	−0.39	[124]
			Clay (sodium bentonite and sepiolite) + yeast culture	0.015	2.12	−8.98	0.66	−6.37	1.80	−2.79	
6	32	217	Clay (bentonite) + yeast cell wall	0.20	6.76	−11.76	6.76	−11.76	−1.04	1.33	[66]
				0.40	19.93 *	−0.88	19.93 *	−0.88	1.86	4.30	

[1] IBW = initial body weight. [2] Asterisk marks (*, **) represent statistical tendency ($p < 0.10$) and significant difference ($p < 0.05$), respectively. [3] The percentage increase or decrease in the average daily gain (ADG), average daily feed intake (ADFI), and gain-to-feed ratio (G:F) was determined in aflatoxin contamination with mitigation agents relative to the aflatoxin contamination group. [4] The percentage increase or decrease in the ADG, ADFI, and G:F was determined in aflatoxin contamination with mitigation agents relative to the control group. [5] Adsorbing agent = *Lactobacillus casei* (1.0×10^8 CFU/g) and *Enterococcus faecalis* (1.0×10^{10} CFU/g) were included. [6] Biotransforming agent = *Bacillus subtilis* (1.0×10^8 CFU/g) and *Candida utilis* (1.0×10^8 CFU/g) were included. [7] *Lactobacillus* spp. included *Lactobacillus delbrueckii* subsp. *bulgaricus*, *Lactobacillus acidophilus*, *Lactobacillus plantarum*, and *Lactobacillus rhamnosus*.

The specificity of mycotoxin-adsorbing agents further supports the use of multi-component mycotoxin-detoxifying agents. Clay compounds, common and effective mycotoxin-adsorbing agents for AFB_1, have a low affinity for mycotoxins such as DON; therefore, the use of clay in combination with additional mycotoxin-detoxifying agents could be advantageous for use in situations of AFB_1 co-contaminated with other mycotoxins [42]. The addition of β-glucans from yeast and algae may be more effective than minerals and clay to mitigate the impacts of mycotoxins other than AFB_1 [161]. A previous in vitro study reported that yeast cell walls had a higher affinity for adsorption to DON when compared to bentonite, cellulose, and activated charcoal, exemplifying that no single mitigation product will be appropriate in every situation [132]. Additionally, the combined use of mycotoxin-degrading enzyme supplementation with bacteria including adsorbing agents (*Lactobacillus casei* and *Enterococcus faecalis*) and biotransforming agents (*Bacillus subtilis* and *Candida utilis*) showed beneficial effects on intestinal health and growth, thereby mitigating the negative impacts of AFB_1 on animals [42]. Overall, multi-component mycotoxin-detoxifying agents may provide enhanced detoxification efficiency in managing AFB_1 co-contamination with other mycotoxins.

4.4. Other Feed Additives That Mitigate Mycotoxin Impacts

Select non-nutritive feed additives, such as antioxidants, have been shown to help chickens and pigs mitigate the impacts of mycotoxins on growth performance by improving intestinal barrier functions and anti-oxidative capacity (Table 8) [72,129,130].

Table 8. Growth performance of chickens and pigs fed diets contaminated with aflatoxin B1 (AFB_1) including other feed additives that mitigate mycotoxin impacts.

Age or IBW [1]	Experimental Period (d)	AFB_1 (μg/kg)	Type	Level (%)	Growth Performance [2] (% Change)						Reference
					ADG		ADFI		G:F		
					vs. AFB_1 [3]	vs. Control [4]	vs. AFB_1	vs. Control	vs. AFB_1	vs. Control	
Chicken (d)											
1	35	1000	Clay (bentonite and HSCA) + oligomannose + phytobiotics	0.05	7.88 **	−8.27 **	2.74	−1.73	5.01 **	−6.65 **	[74]
1	42	100	Clay (sodium bentonite) + gention violet	0.50	11.60 **	−27.99 **	3.50 **	−24.10 **	7.83 **	−5.14 **	[96]
				1.00	10.61 **	−28.63 **	4.91 **	−23.06 **	5.44 **	−7.24 **	
			Clay (sodium bentonite) + acetic acid	0.50	13.02 **	−27.08 **	3.65 **	−23.99 **	9.05 **	−4.07 **	
				1.00	8.14 **	−30.23 **	−2.12 **	−28.22 **	10.48 **	−2.81 **	
1	35	38	Nano-composite magnetic graphene oxide + chitosan	0.25	6.56 **	−2.36	7.86 **	5.24	−1.20	−0.28	[109]
				0.50	8.44 **	−0.65	2.11	−0.36	6.19 **	−0.28	
1	35	1000	Phytobiotics	0.05	7.13 **	−8.90 **	1.52	−2.89	5.53 **	−6.19 **	[74]
1	44	122	Sporoderm-broken spores of *Ganoderma lucidum*	0.02	6.55 **	3.54 **	4.41 **	0.92	2.05 *	1.82 **	[88]
1	42	600	Milk thistle (*Silybum marianum*)	1.00	12.50 **	−10.00 **	10.47 **	0.00	1.84 **	−10.00 **	[48]
1	28	1000	Grapeseed extract	0.025	19.43 **	−21.57 **	12.50 **	2.65 **	6.67 **	−2.47 **	[83]
				0.050	18.44 **	−22.22 **	16.07 **	3.54 **	3.38	−5.47 **	
1	21	60	Tannic acid	0.025	2.63 **	5.41 **	6.25 **	8.51 **	1.98 **	−7.77 **	[72]
				0.050	2.63 **	5.41 **	4.17 **	6.38 **	3.64 **	−6.26 **	
1	21	100	Lycopene	0.020	6.52 **	−3.92 **	3.45 **	−2.17 **	2.97 **	−1.79 **	[100]
1	42	500	Licorice extract	0.30	7.55 **	−5.00 **	0.00	−1.94 **	7.55 **	−3.12 **	[106]
				0.60	5.66 **	−6.67 **	−0.99 **	−2.91 **	6.72 **	−3.87 **	
			Poultry litter biochar	0.50	3.77 **	−8.33 **	0.99 **	−0.97 **	2.76 **	−7.43 **	
1	42	50	Calcium propionate	0.25	4.12	−0.09	9.87	−0.76	−5.24	0.67	[103]
				0.50	4.51	0.28	10.40	−0.28	−5.33	0.57	
		100	Calcium propionate	0.25	5.79	−0.28	15.67	−1.42	−8.54	1.15	
				0.50	6.29	0.19	16.89	−0.38	−9.07	0.57	
1	21	2000	Curcumin	0.20	19.80 **	−10.21 **	1.33	4.83 **	−0.35	−19.77 **	[54]
1	21	2000	Cellulosic polymer + curcumin	0.50	23.22 **	−7.64 **	−2.12 **	1.27	24.20 **	0.00	[54]
1	42	600	Algae (*Spirulina platensis*)	1.00	12.50 **	−10.00 **	3.49 **	−6.32 **	8.71 **	−3.93 **	[48]
Pig (kg)											
7	40	500	Vegetable biocholine	0.08	6.55	−9.62	0.56	−11.11 **	5.95	1.68 **	[112]
9	30	320	Grape seed waste	8.00	51.05 **	−11.84	-	-	-	-	[56]
11	28	420	Selenium	0.006	2.17	−9.62	5.26	−11.50	−2.93	2.13	[114]
		840	Selenium	0.006	10.71 **	−40.38	1.49	−39.82	9.09	−0.93	
		840	Folic acid	0.020	32.14 **	−28.85	23.88	−26.55	6.67	−3.13	

[1] IBW = initial body weight. [2] Asterisk marks (*, **) represent statistical tendency ($p < 0.10$) and significant difference ($p < 0.05$), respectively. [3] The percentage increase or decrease in the average daily gain (ADG), average daily feed intake (ADFI), and gain-to-feed ratio (G:F) was determined in aflatoxin contamination with mitigation agents relative to the aflatoxin contamination group. [4] The percentage increase or decrease in the ADG, ADFI, and G:F was determined in aflatoxin contamination with mitigation agents relative to the control group.

Phytobiotics offer an additional solution to counteract the adverse impacts of AFB_1 on monogastric animals by promoting intestinal health and growth performance through their anti-inflammatory, antimicrobial, and anti-oxidative properties [72,83,106]. The phenolic compounds in phytobiotics mitigate oxidative damage to epithelial cells, support tissue repair, and enhance mucus secretion, improving overall intestinal health [74]. Tannic acid, a polyphenol known for its ability to bind polar organic compounds and prevent lipid oxidation, enhances enzyme activity and reduces oxidative stress in the liver [129]. Similarly, grape seed extract, which is rich in proanthocyanidins with strong antioxidant and anti-inflammatory effects, counters the adverse impacts of AFB_1 by removing ROS, inhibiting lipid peroxidation in epithelial cells, and reducing pro-inflammatory cytokine production and expression in the small intestine [162–164].

Selenium, curcumin, and lycopene mitigate oxidative stress caused by AFB_1 by protecting epithelial cells in the intestine and liver through distinct mechanisms [54,165,166]. Selenium, a critical component of selenoprotein enzymes such as glutathione peroxidase, supports antioxidant defenses by modulating the cellular dysfunction–apoptosis switch, decreasing pro-apoptotic protein concentrations, and increasing anti-apoptotic proteins in the liver [130,167]. Curcumin, a phenolic compound derived from the rhizome of *Curcuma longa*, exhibits powerful radical scavenging abilities due to the number and position of its hydroxyl groups [168]. Curcumin counteracts AFB_1-induced oxidative stress by reducing protein carbonylation, lipid peroxidation, and mitochondrial permeability transition and activating antioxidant-related genes such as CAT, SOD, and glutathione S-transferase, and detoxification genes. Additionally, curcumin inhibits pyroptosis signaling in the liver [169], which may prevent intestinal cell membrane damage [54]. A previous study demonstrated that combining curcumin and a cellulosic polymer was more effective than using either alone, improving immune and intestinal parameters in chickens fed AFB_1-contaminated diets [54]. This could be attributed to the immunomodulatory properties of curcumin [170] and the adsorptive properties of the cellulosic polymer [171] working synergistically. Lycopene, a carotenoid, reduces oxidative mitochondrial damage through the activation of NrF2 signaling pathways, stimulation of mitochondrial antioxidant capacity, and maintenance of mitochondrial biogenesis [172]. Lycopene also decreases inflammatory cytokines such as IFN-γ and IL-1β and reduces lipid oxidation by increasing the antioxidant enzyme activities and their mRNA expression in the broiler jejunum [173]. Together, these compounds could play a significant role in mitigating the impacts of AFB_1 exposure on intestinal health, primarily by enhancing antioxidant activity and maintaining cell cycle function.

5. Conclusions

The GIT is the primary biological system exposed to aflatoxins upon ingestion, where their detrimental impacts first manifest in monogastric animals, including chickens and pigs. Aflatoxins, particularly AFB_1, disrupt the intestinal microbiota, induce immune responses and the production of oxidative damage products, thereby negatively affecting intestinal morphology and reducing nutrient digestion. After AFB_1 absorption in the small intestine of the animals, the AFB_1 causes systemic negative impacts on animals, especially in the liver. These combined impacts of AFB_1 reduce growth performance in animals. Based on the meta-analysis, for every 1 μg/kg·d of AFB_1 intake per BW, average daily gain decreases by 0.13% in chickens and by 0.74% in pigs, indicating that increasing AFB_1 intake linearly reduces animal growth.

Mycotoxin-adsorbing agents, such as clay and yeast cell walls, effectively bind AFB_1 in the digesta, reducing its adverse impacts on intestinal health and growth performance. Additionally, biotransforming agents further support intestinal integrity, functionality, and growth by degrading AFB_1 into less toxic metabolites in the digesta. Multi-component detoxifying agents targeting AFB_1 could offer enhanced efficacy by adsorption and degradation together in the digesta of animals, which prevent negative impacts on intestinal health and growth performance. Naturally contaminated feedstuffs are often co-contaminated with AFB_1 and other mycotoxins. Therefore, multi-component mycotoxin-detoxifying agents offer different binding and degrading efficiencies for mycotoxins, that could mitigate the negative impacts on intestinal health and growth performance in animals, in the context of feeds with naturally occurring AFB_1, co-contaminated with other mycotoxins.

Author Contributions: Conceptualization, H.C. and S.W.K.; methodology, S.W.K.; formal analysis, H.C. and S.W.K.; investigation, H.C., Y.G.-D., A.R.G. and S.W.K.; data curation, H.C., Y.G.-D., A.R.G. and S.W.K.; writing—original draft preparation, H.C., Y.G.-D. and A.R.G.; writing—review and editing, H.C., Y.G.-D., A.R.G. and S.W.K.; supervision, S.W.K. All authors have read and agreed to the published version of the manuscript.

Funding: North Carolina Agricultural Foundation (660101 and 662825, Raleigh, NC, USA) and USDA-NIFA (Hatch 02893, Washington DC, USA).

Institutional Review Board Statement: Not applicable.

Informed Consent Statement: Not applicable.

Data Availability Statement: The original contributions presented in this study are included in the article. Further inquiries can be directed to the corresponding author(s).

Conflicts of Interest: The authors declare no conflicts of interest.

References

1. Gruber-Dorninger, C.; Jenkins, T.; Schatzmayr, G. Global mycotoxin occurrence in feed: A ten-year survey. *Toxins* **2019**, *11*, 375. [CrossRef] [PubMed]
2. Marin, S.; Ramos, A.J.; Cano-Sancho, G.; Sanchis, V. Mycotoxins: Occurrence, toxicology, and exposure assessment. *Food Chem. Toxicol.* **2013**, *60*, 218–237. [CrossRef]
3. Streit, E.; Schwab, C.; Sulyok, M.; Naehrer, K.; Krska, R.; Schatzmayr, G. Multi-mycotoxin screening reveals the occurrence of 139 different secondary metabolites in feed and feed ingredients. *Toxins* **2013**, *5*, 504–523. [CrossRef] [PubMed]
4. Schelstraete, W.; Devreese, M.; Croubels, S. Comparative toxicokinetics of fusarium mycotoxins in pigs and humans. *Food Chem. Toxicol.* **2020**, *137*, 111140. [CrossRef] [PubMed]
5. Holanda, D.M.; Kim, S.W. Mycotoxin occurrence, toxicity, and detoxifying agents in pig production with an emphasis on deoxynivalenol. *Toxins* **2021**, *13*, 171. [CrossRef] [PubMed]
6. Pierron, A.; Alassane-Kpembi, I.; Oswald, I.P. Impact of mycotoxin on immune response and consequences for pig health. *Anim. Nutr.* **2016**, *2*, 63–68. [CrossRef]
7. Haque, M.A.; Wang, Y.; Shen, Z.; Li, X.; Saleemi, M.K.; He, C. Mycotoxin contamination and control strategy in human, domestic animal and poultry: A review. *Microb. Pathog.* **2020**, *142*, 104095. [CrossRef] [PubMed]
8. Kim, S.W.; Gormley, A.; Jang, K.B.; Duarte, M.E. Current status of global pig production: An overview and research trends. *Anim. Biosci.* **2023**, *37*, 719–729. [CrossRef] [PubMed]
9. Bryden, W.L. Mycotoxin contamination of the feed supply chain: Implications for animal productivity and feed security. *Anim. Feed Sci. Technol.* **2012**, *173*, 134–158. [CrossRef]
10. Magnoli, A.P.; Poloni, V.L.; Cavaglieri, L. Impact of mycotoxin contamination in the animal feed industry. *Curr. Opin. Food Sci.* **2019**, *29*, 99–108. [CrossRef]
11. Patriarca, A.; Pinto, V.F. Prevalence of mycotoxins in foods and decontamination. *Curr. Opin. Food Sci.* **2017**, *14*, 50–60. [CrossRef]
12. Abbas, H.K.; Wilkinson, J.R.; Zablotowicz, R.M.; Accinelli, C.; Abel, C.A.; Bruns, H.A.; Weaver, M.A. Ecology of *Aspergillus flavus*, regulation of aflatoxin production, and management strategies to reduce aflatoxin contamination of corn. *Toxin reviews* **2009**, *28*, 142–153. [CrossRef]
13. Braicu, C.; Berindan-Neagoe, I.; Chedea, V.S.; Balacescu, L.; Brie, I.; Soritau, O.; Socaciu, C.; Irimie, A. Individual and combined cytotoxic effects of the major four aflatoxins in different in vitro stabilized systems. *J. Food Biochem.* **2010**, *34*, 1079–1090. [CrossRef]
14. Lizárraga-Paulín, E.G.; Moreno-Martínez, E.; Miranda-Castro, S.P. Aflatoxins and their impact on human and animal health: An emerging problem. In *Aflatoxin-Biochemistry and Molecular Biology*; Guevara-González, R.G., Ed.; InTech: Rijeka, Croatia, 2011; pp. 255–282.
15. Bervis, N.; Lorán, S.; Juan, T.; Carramiñana, J.J.; Herrera, A.; Ariño, A.; Herrera, M. Field monitoring of aflatoxins in feed and milk of high-yielding dairy cows under two feeding systems. *Toxins* **2021**, *13*, 201. [CrossRef] [PubMed]
16. Schrenk, D.; Bignami, M.; Bodin, L.; Chipman, J.K.; del Mazo, J.; Grasl-Kraupp, B.; Hogstrand, C.; Hoogenboom, L.; Leblanc, J.C. Risk assessment of aflatoxins in food. *EFSA J.* **2020**, *18*, e06040. [PubMed]
17. Yunus, A.W.; Razzazi-Fazeli, E.; Bohm, J. Aflatoxin B1 in affecting broiler's performance, immunity, and gastrointestinal tract: A review of history and contemporary issues. *Toxins* **2011**, *3*, 566–590. [CrossRef] [PubMed]
18. Holanda, D.M.; Kim, Y.I.; Parnsen, W.; Kim, S.W. Phytobiotics with adsorbent to mitigate toxicity of multiple mycotoxins on health and growth of pigs. *Toxins* **2021**, *13*, 442. [CrossRef] [PubMed]
19. Guo, H.; Wang, P.; Liu, C.; Zhou, T.; Chang, J.; Yin, Q.; Wang, L.; Jin, S.; Zhu, Q.; Lu, F. Effects of compound mycotoxin detoxifier on alleviating aflatoxin B1-induced inflammatory responses in intestine, liver and kidney of broilers. *Toxins* **2022**, *14*, 665. [CrossRef]
20. Guo, H.-W.; Chang, J.; Wang, P.; Yin, Q.-Q.; Liu, C.-Q.; Xu, X.-X.; Dang, X.-W.; Hu, X.-F.; Wang, Q.-L. Effects of compound probiotics and aflatoxin-degradation enzyme on alleviating aflatoxin-induced cytotoxicity in chicken embryo primary intestinal epithelium, liver and kidney cells. *AMB Express* **2021**, *11*, 35. [CrossRef]
21. Kumagai, S. Intestinal absorption and excretion of aflatoxin in rats. *Toxicol. Appl. Pharmacol.* **1989**, *97*, 88–97. [CrossRef] [PubMed]

22. Jahanian, E.; Mahdavi, A.H.; Asgary, S.; Jahanian, R. Effect of dietary supplementation of mannanoligosaccharides on growth performance, ileal microbial counts, and jejunal morphology in broiler chicks exposed to aflatoxins. *Livest. Sci.* **2016**, *190*, 123–130. [CrossRef]
23. Duarte, M.E.; Kim, S.W. Intestinal microbiota and its interaction to intestinal health in nursery pigs. *Anim. Nutr.* **2022**, *8*, 169–184. [CrossRef]
24. Zhang, M.; Li, Q.; Wang, J.; Sun, J.; Xiang, Y.; Jin, X. Aflatoxin B1 disrupts the intestinal barrier integrity by reducing junction protein and promoting apoptosis in pigs and mice. *Ecotoxicol. Environ. Saf.* **2022**, *247*, 114250. [CrossRef] [PubMed]
25. Huang, W.; Chang, J.; Wang, P.; Liu, C.; Yin, Q.; Song, A.; Gao, T.; Dang, X.; Lu, F. Effect of compound probiotics and mycotoxin degradation enzymes on alleviating cytotoxicity of swine jejunal epithelial cells induced by aflatoxin B1 and zearalenone. *Toxins* **2019**, *11*, 12. [CrossRef] [PubMed]
26. Pu, J.; Yuan, Q.; Yan, H.; Tian, G.; Chen, D.; He, J.; Zheng, P.; Yu, J.; Mao, X.; Huang, Z. Effects of chronic exposure to low levels of dietary aflatoxin B1 on growth performance, apparent total tract digestibility and intestinal health in pigs. *Animals* **2021**, *11*, 336. [CrossRef]
27. Alharthi, A.S.; Al Sulaiman, A.R.; Aljumaah, R.S.; Alabdullatif, A.A.; Ferronato, G.; Alqhtani, A.H.; Al-Garadi, M.A.; Al-Sornokh, H.; Abudabos, A.M. The efficacy of bentonite and zeolite in reducing aflatoxin B1 toxicity on production performance and intestinal and hepatic health of broiler chickens. *Ital. J. Anim. Sci.* **2022**, *21*, 1181–1189. [CrossRef]
28. Jang, K.B.; Purvis, J.M.; Kim, S.W. Supplemental effects of dietary lysophospholipids in lactation diets on sow performance, milk composition, gut health, and gut-associated microbiome of offspring. *J. Anim. Sci.* **2020**, *98*, skaa227. [CrossRef] [PubMed]
29. Mahmud, M.R.; Jian, C.; Uddin, M.K.; Huhtinen, M.; Salonen, A.; Peltoniemi, O.; Venhoranta, H.; Oliviero, C. Impact of intestinal microbiota on growth performance of suckling and weaned piglets. *Microbiol. Spectr.* **2023**, *11*, e0374422. [CrossRef] [PubMed]
30. Duarte, M.E.; Deng, Z.; Kim, S.W. Effects of dietary *Lactobacillus postbiotics* and bacitracin on the modulation of mucosa-associated microbiota and pattern recognition receptors affecting immunocompetence of jejunal mucosa in pigs challenged with enterotoxigenic F18$^+$ *Escherichia coli*. *J. Anim. Sci. Biotechnol.* **2024**, *15*, 139. [CrossRef] [PubMed]
31. Liu, M.; Liu, W.; Zhang, W.; Yao, J.; Mo, X. Ultrasound-assisted extraction of boulardii yeast cell wall polysaccharides: Characterization and its biological functions on early-weaned lambs. *Food Sci. Nutr.* **2021**, *9*, 3617–3630. [CrossRef] [PubMed]
32. Liu, S.; Hu, J.; Li, L.; Xing, S.; Yang, Y.; Liao, X. Sodium butyrate reduces ammonia production in the cecum of laying hens by regulating ammonia-producing bacteria. *Poult. Sci.* **2023**, *102*, 102241. [CrossRef] [PubMed]
33. Wang, W.; Wang, Y.; Yang, J.; Wagner, K.M.; Hwang, S.H.; Cheng, J.; Singh, N.; Edwards, P.; Morisseau, C.; Zhang, G.; et al. Aflatoxin B1 exposure disrupts the intestinal immune function via a soluble epoxide hydrolase-mediated manner. *Ecotoxicol. Environ. Saf.* **2023**, *249*, 114417. [CrossRef] [PubMed]
34. Poloni, V.; Magnoli, A.; Fochesato, A.; Cristofolini, A.; Caverzan, M.; Merkis, C.; Montenegro, M.; Cavaglieri, L. A *Saccharomyces cerevisiae* RC016-based feed additive reduces liver toxicity, residual aflatoxin B1 levels and positively influences intestinal morphology in broiler chickens fed chronic aflatoxin B1-contaminated diets. *Anim. Nutr.* **2020**, *6*, 31–38. [CrossRef]
35. Gao, Y.; Li, S.; Wang, J.; Luo, C.; Zhao, S.; Zheng, N. Modulation of intestinal epithelial permeability in differentiated Caco-2 cells exposed to aflatoxin M1 and ochratoxin a individually or collectively. *Toxins* **2018**, *10*, 13. [CrossRef] [PubMed]
36. Zhao, J.; Bai, Y.; Tao, S.; Zhang, G.; Wang, J.; Liu, L.; Zhang, S. Fiber-rich foods affected gut bacterial community and short-chain fatty acids production in pig model. *J. Funct. Foods* **2019**, *57*, 266–274. [CrossRef]
37. Zheng, D.; Liwinski, T.; Elinav, E. Interaction between microbiota and immunity in health and disease. *Cell Res.* **2020**, *30*, 492–506. [CrossRef]
38. Soderholm, A.T.; Pedicord, V.A. Intestinal epithelial cells: At the interface of the microbiota and mucosal immunity. *Immunology* **2019**, *158*, 267–280. [CrossRef]
39. Martín, R.; Chamignon, C.; Mhedbi-Hajri, N.; Chain, F.; Derrien, M.; Escribano-Vázquez, U.; Garault, P.; Cotillard, A.; Pham, H.P.; Chervaux, C.; et al. The potential probiotic *Lactobacillus rhamnosus* CNCM I-3690 strain protects the intestinal barrier by stimulating both mucus production and cytoprotective response. *Sci. Rep.* **2019**, *9*, 5398. [CrossRef] [PubMed]
40. Graziani, F.; Pujol, A.; Nicoletti, C.; Dou, S.; Maresca, M.; Giardina, T.; Fons, M.; Perrier, J. *Ruminococcus gnavus* E1 modulates mucin expression and intestinal glycosylation. *J. Appl. Microbiol.* **2016**, *120*, 1403–1417. [CrossRef] [PubMed]
41. Adhikari, B.; Kim, S.W.; Kwon, Y.M. Characterization of microbiota associated with digesta and mucosa in different regions of gastrointestinal tract of nursery pigs. *Int. J. Mol. Sci.* **2019**, *20*, 1630. [CrossRef]
42. Guo, H.; Wang, P.; Liu, C.; Chang, J.; Yin, Q.; Wang, L.; Jin, S.; Zhu, Q.; Lu, F. Compound mycotoxin detoxifier alleviating aflatoxin B1 toxic effects on broiler growth performance, organ damage and gut microbiota. *Poult. Sci.* **2023**, *102*, 102434. [CrossRef]
43. Kim, S.W.; Holanda, D.M.; Gao, X.; Park, I.; Yiannikouris, A. Efficacy of a yeast cell wall extract to mitigate the effect of naturally co-occurring mycotoxins contaminating feed ingredients fed to young pigs: Impact on gut health, microbiome, and growth. *Toxins* **2019**, *11*, 633. [CrossRef] [PubMed]

44. Warger, T.; Osterloh, P.; Rechtsteiner, G.; Fassbender, M.; Heib, V.; Schmid, B.; Schmitt, E.; Schild, H.r.; Radsak, M.P. Synergistic activation of dendritic cells by combined Toll-like receptor ligation induces superior CTL responses in vivo. *Blood* **2006**, *108*, 544–550. [CrossRef] [PubMed]
45. Holanda, D.M.; Kim, S.W. Impacts of weaning weights and mycotoxin challenges on jejunal mucosa-associated microbiota, intestinal and systemic health, and growth performance of nursery pigs. *J. Anim. Sci. Biotechnol.* **2022**, *13*, 43. [CrossRef] [PubMed]
46. Wang, F.; Zuo, Z.; Chen, K.; Gao, C.; Yang, Z.; Zhao, S.; Li, J.; Song, H.; Peng, X.; Fang, J.; et al. Histopathological injuries, ultrastructural changes, and depressed tlr expression in the small intestine of broiler chickens with aflatoxin B1. *Toxins* **2018**, *10*. [CrossRef] [PubMed]
47. Liu, N.; Wang, J.Q.; Liu, Z.Y.; Wang, Y.C.; Wang, J.P. Comparison of probiotics and clay detoxifier on the growth performance and enterotoxic markers of broilers fed diets contaminated with aflatoxin B1. *J. Appl. Poult. Res.* **2018**, *27*, 341–348. [CrossRef]
48. Feshanghchi, M.; Baghban-Kanani, P.; Kashefi-Motlagh, B.; Adib, F.; Azimi-Youvalari, S.; Hosseintabar-Ghasemabad, B.; Slozhenkina, M.; Gorlov, I.; Zangeronimo, M.G.; Swelum, A.A. Milk thistle (*Silybum marianum*), marine algae (*Spirulina platensis*) and toxin binder powders in the diets of broiler chickens exposed to aflatoxin-B1: Growth performance, humoral immune response and cecal microbiota. *Agriculture* **2022**, *12*, 805. [CrossRef]
49. Chen, X.; Ishfaq, M.; Wang, J. Effects of *Lactobacillus salivarius* supplementation on the growth performance, liver function, meat quality, immune responses and *Salmonella pullorum* infection resistance of broilers challenged with aflatoxin B1. *Poult. Sci.* **2022**, *101*, 101651. [CrossRef] [PubMed]
50. He, Y.; Fang, J.; Peng, X.; Cui, H.; Zuo, Z.; Deng, J.; Chen, Z.; Geng, Y.; Lai, W.; Shu, G.; et al. Effects of sodium selenite on aflatoxin B1-induced decrease of ileal IgA^+ cell numbers and immunoglobulin contents in broilers. *Biol. Trace Elem. Res.* **2014**, *160*, 49–55. [CrossRef]
51. Fang, J.; Zheng, Z.; Yang, Z.; Peng, X.; Zuo, Z.; Cui, H.; Ouyang, P.; Shu, G.; Chen, Z.; Huang, C. Ameliorative effects of selenium on the excess apoptosis of the jejunum caused by AFB1 through death receptor and endoplasmic reticulum pathways. *Toxicol. Res.* **2018**, *7*, 1108–1119. [CrossRef]
52. Jiang, M.; Fang, J.; Peng, X.; Cui, H.; Yu, Z. Effect of aflatoxin B_1 on IgA^+ cell number and immunoglobulin mrna expression in the intestine of broilers. *Immunopharmacol. Immunotoxicol.* **2015**, *37*, 450–457. [CrossRef] [PubMed]
53. Gao, S.; Zhang, L.; Zhu, D.; Huang, J.; Yang, J.; Jiang, J.; Wu, H.; Lv, G. Effects of glucose oxidase and *Bacillus subtilis* on growth performance and serum biochemical indicexs of broilers exposed to aflatoxin B1 and endotoxin. *Anim. Feed Sci. Technol.* **2022**, *286*, 115186. [CrossRef]
54. Solis-Cruz, B.; Hernandez-Patlan, D.; Petrone, V.M.; Pontin, K.P.; Latorre, J.D.; Beyssac, E.; Hernandez-Velasco, X.; Merino-Guzman, R.; Owens, C.; Hargis, B.M.; et al. Evaluation of cellulosic polymers and curcumin to reduce aflatoxin B1 toxic effects on performance, biochemical, and immunological parameters of broiler chickens. *Toxins* **2019**, *11*, 121. [CrossRef]
55. Tarasconi, L.; Dazuk, V.; Molosse, V.L.; Cécere, B.G.O.; Deolindo, G.L.; Mendes, R.E.; Gloria, E.M.; Ternus, E.M.; Galli, G.M.; Paiano, D.; et al. Nursery pigs fed with feed contaminated by aflatoxin B1 (*Aspergillus flavus*) and anti-mycotoxin blend: Pathogenesis and negative impact on animal health and weight gain. *Microb. Pathog.* **2024**, *186*, 106474. [CrossRef]
56. Taranu, I.; Marin, D.E.; Palade, M.; Pistol, G.C.; Chedea, V.S.; Gras, M.A.; Rotar, C. Assessment of the efficacy of a grape seed waste in counteracting the changes induced by aflatoxin B1 contaminated diet on performance, plasma, liver and intestinal tissues of pigs after weaning. *Toxicon* **2019**, *162*, 24–31. [CrossRef] [PubMed]
57. Lai, Y.; Sun, M.; He, Y.; Lei, J.; Han, Y.; Wu, Y.; Bai, D.; Guo, Y.; Zhang, B. Mycotoxins binder supplementation alleviates aflatoxin B1 toxic effects on the immune response and intestinal barrier function in broilers. *Poult. Sci.* **2022**, *101*, 101683. [CrossRef] [PubMed]
58. Chelakkot, C.; Ghim, J.; Ryu, S.H. Mechanisms regulating intestinal barrier integrity and its pathological implications. *Exp. Mol. Med.* **2018**, *50*, 1–9. [CrossRef]
59. Yin, H.; Jiang, M.; Peng, X.; Cui, H.; Zhou, Y.; He, M.; Zuo, Z.; Ouyang, P.; Fan, J.; Fang, J. The molecular mechanism of G_2M cell cycle arrest induced by AFB1 in the jejunum. *Oncotarget* **2016**, *7*, 35592–35606. [CrossRef] [PubMed]
60. Ma, J.; Liu, Y.; Guo, Y.; Ma, Q.; Ji, C.; Zhao, L. Transcriptional profiling of aflatoxin B1-induced oxidative stress and inflammatory response in macrophages. *Toxins* **2021**, *13*, 401. [CrossRef]
61. Yilmaz, S.; Kaya, E.; Kiscam, M.A. The effect on oxidative stress of aflatoxin and protective effect of lycopene on aflatoxin damage. In *Aflatoxin-Control, Analysis, Detection and Health Risks*; Lukman Bola, A.U., Ed.; IntechOpen: London, UK, 2017; pp. 67–90.
62. Yin, J.; Wu, M.; Li, Y.; Ren, W.; Xiao, H.; Chen, S.; Li, C.; Tan, B.; Ni, H.; Xiong, X.; et al. Toxicity assessment of hydrogen peroxide on Toll-like receptor system, apoptosis, and mitochondrial respiration in piglets and IPEC-J2 cells. *Oncotarget* **2017**, *8*, 3124–3131. [CrossRef]
63. Miller, M.A.; Zachary, J.F. Mechanisms and morphology of cellular injury, adaptation, and death. In *Pathologic Basis of Veterinary Disease*, 6th ed.; Zachary, J.F., Ed.; Mosby: St. Louis, MO, USA, 2017; pp. 2–43.
64. Gao, Y.; Bao, X.; Meng, L.; Liu, H.; Wang, J.; Zheng, N. Aflatoxin B1 and aflatoxin M1 induce compromised intestinal integrity through clathrin-mediated endocytosis. *Toxins* **2021**, *13*, 184. [CrossRef]

65. Jebali, R.; Salah-Abbès, J.B.; Abbès, S.; Hassan, A.M.; Abdel-Aziem, S.H.; El-Nekeety, A.A.; Oueslati, R.; Abdel-Wahhab, M.A. Lactobacillus plantarum alleviate aflatoxins (B1 and M1) induced disturbances in the intestinal genes expression and DNA fragmentation in mice. *Toxicon* **2018**, *146*, 13–23. [CrossRef] [PubMed]
66. Deng, Z.; Jang, K.B.; Jalukar, S.; Du, X.; Kim, S.W. Efficacy of feed additive containing bentonite and enzymatically hydrolyzed yeast on intestinal health and growth of newly weaned pigs under chronic dietary challenges of fumonisin and aflatoxin. *Toxins* **2023**, *15*, 433. [CrossRef]
67. Pandey, I.; Chauhan, S.S. Studies on production performance and toxin residues in tissues and eggs of layer chickens fed on diets with various concentrations of aflatoxin AFB1. *Br. Poult. Sci.* **2007**, *48*, 713–723. [CrossRef] [PubMed]
68. Kiela, P.R.; Ghishan, F.K. Physiology of intestinal absorption and secretion. *Best Pract. Res. Clin. Gastroenterol.* **2016**, *30*, 145–159. [CrossRef]
69. Wang, Z.; Li, X.; Wang, T.; Liao, G.; Gu, J.; Hou, R.; Qiu, J. Lipidomic profiling study on neurobehavior toxicity in zebrafish treated with aflatoxin B1. *Sci. Total Environ.* **2023**, *898*, 165553. [CrossRef] [PubMed]
70. Hernández-Ramírez, J.O.; Merino-Guzmán, R.; Téllez-Isaías, G.; Vázquez-Durán, A.; Méndez-Albores, A. Mitigation of AFB1-related toxic damage to the intestinal epithelium in broiler chickens consumed a yeast cell wall fraction. *Front. Vet. Sci.* **2021**, *8*, 677965. [CrossRef] [PubMed]
71. Chen, X.; Naehrer, K.; Applegate, T.J. Interactive effects of dietary protein concentration and aflatoxin B1 on performance, nutrient digestibility, and gut health in broiler chicks. *Poult. Sci.* **2016**, *95*, 1312–1325. [CrossRef]
72. Zhang, Z.F.; Xi, Y.; Wang, S.T.; Zheng, L.Y.; Qi, Y.; Guo, S.S.; Ding, B.Y. Effects of chinese gallnut tannic acid on growth performance, blood parameters, antioxidative status, intestinal histomorphology, and cecal microbial shedding in broilers challenged with aflatoxin B1. *J. Anim. Sci.* **2022**, *100*, skac099. [CrossRef]
73. Domingues, J.M.; Schreiner Spiassi, B.; Sanches, A.D.; Belote, B.L.; Santin, E.; Wagner, R. The use of histological parameters to assess intestinal and liver health on broilers challenged isolatedly and simultaneously with cyclopiazonic acid and aflatoxin B1. *Acta. Toxicol. Argent.* **2021**, *29*, 67–76.
74. Tavangar, P.; Gharahveysi, S.; Rezaeipour, V.; Irani, M. Efficacy of phytobiotic and toxin binder feed additives individually or in combination on the growth performance, blood biochemical parameters, intestinal morphology, and microbial population in broiler chickens exposed to aflatoxin B1. *Trop. Anim Health Prod.* **2021**, *53*, 335. [CrossRef] [PubMed]
75. Xu, F.; Li, Y.; Cao, Z.; Zhang, J.; Huang, W. AFB1-induced mice liver injury involves mitochondrial dysfunction mediated by mitochondrial biogenesis inhibition. *Ecotoxicol. Environ. Saf.* **2021**, *216*, 112213. [CrossRef]
76. Liu, X.; Kumar Mishra, S.; Wang, T.; Xu, Z.; Zhao, X.; Wang, Y.; Yin, H.; Fan, X.; Zeng, B.; Yang, M. AFB1 induced transcriptional regulation related to apoptosis and lipid metabolism in liver of chicken. *Toxins* **2020**, *12*, 290. [CrossRef] [PubMed]
77. Di Gregorio, M.C.; Jager, A.V.; Souto, P.C.M.C.; Costa, A.A.; Rottinghaus, G.E.; Passarelli, D.; Budiño, F.E.L.; Corassin, C.H.; Oliveira, C.A.F. Determination of serum aflatoxin B1-lysine to evaluate the efficacy of an aflatoxin-adsorbing feed additive in pigs fed an aflatoxin B1-contaminated diet. *Mycotoxin Res.* **2017**, *33*, 93–102. [CrossRef] [PubMed]
78. Macé, K.; Aguilar, F.; Wang, J.S.; Vautravers, P.; Gómez-Lechón, M.; Gonzalez, F.J.; Groopman, J.; Harris, C.C.; Pfeifer, A.M. Aflatoxin B1-induced DNA adduct formation and P53 mutations in CYP450-expressing human liver cell lines. *Carcinogenesis* **1997**, *18*, 1291–1297. [CrossRef] [PubMed]
79. Ye, L.; Chen, H.; Tsim, K.W.K.; Shen, X.; Li, X.; Li, X.; Lei, H.; Liu, Y. Aflatoxin B1 induces inflammatory liver injury via gut microbiota in mice. *J. Agric. Food Chem.* **2023**, *71*, 10787–10797. [CrossRef] [PubMed]
80. Singh, K.B.; Maurya, B.K.; Trigun, S.K. Activation of oxidative stress and inflammatory factors could account for histopathological progression of aflatoxin-B1 induced hepatocarcinogenesis in rat. *Mol. Cell. Biochem.* **2015**, *401*, 185–196. [CrossRef]
81. Vipin, A.V.; Rao, R.; Kurrey, N.K.; KA, A.A.; Venkateswaran, G. Protective effects of phenolics rich extract of ginger against aflatoxin B1-induced oxidative stress and hepatotoxicity. *Biomed. Pharmacother.* **2017**, *91*, 415–424.
82. Denli, M.; Blandon, J.C.; Guynot, M.E.; Salado, S.; Perez, J.F. Effects of dietary afladetox on performance, serum biochemistry, histopathological changes, and aflatoxin residues in broilers exposed to aflatoxin B(1). *Poult. Sci.* **2009**, *88*, 1444–1451. [CrossRef]
83. Ali Rajput, S.; Sun, L.; Zhang, N.; Mohamed Khalil, M.; Gao, X.; Ling, Z.; Zhu, L.; Khan, F.A.; Zhang, J.; Qi, D. Ameliorative effects of grape seed proanthocyanidin extract on growth performance, immune function, antioxidant capacity, biochemical constituents, liver histopathology and aflatoxin residues in broilers exposed to aflatoxin B1. *Toxins* **2017**, *9*, 371. [CrossRef] [PubMed]
84. Popescu, R.G.; Rădulescu, A.L.; Georgescu, S.E.; Dinischiotu, A. Aflatoxins in feed: Types, metabolism, health consequences in swine and mitigation strategies. *Toxins* **2022**, *14*, 853. [CrossRef] [PubMed]
85. Li, C.; Liu, X.; Wu, J.; Ji, X.; Xu, Q. Research progress in toxicological effects and mechanism of aflatoxin B(1) toxin. *PeerJ* **2022**, *10*, e13850. [CrossRef]
86. Hou, L.; Qiu, H.; Li, A.; Dong, J.; Zhu, L.; Liu, G.; Chen, F. Effects of aflatoxin B1 on growth performance, antioxidant status, immune response, and pro-inflammatory cytokine mrna expression in isa chicks. *Front. Vet. Sci.* **2022**, *9*, 993039. [CrossRef] [PubMed]

87. Lin, L.; Fu, P.; Chen, N.; Gao, N.; Cao, Q.; Yue, K.; Xu, T.; Zhang, C.; Zhang, C.; Liu, F.; et al. Total flavonoids of rhizoma drynariae protect hepatocytes against aflatoxin B1-induced oxidative stress and apoptosis in broiler chickens. *Ecotoxicol. Environ. Saf.* **2022**, *230*, 113148. [CrossRef] [PubMed]
88. Liu, T.; Ma, Q.; Zhao, L.; Jia, R.; Zhang, J.; Ji, C.; Wang, X. Protective effects of sporoderm-broken spores of ganderma iucidum on growth performance, antioxidant capacity and immune function of broiler chickens exposed to low level of aflatoxin B1. *Toxins* **2016**, *8*, 278. [CrossRef]
89. Jurišić, N.; Schwartz-Zimmermann, H.E.; Kunz-Vekiru, E.; Moll, W.D.; Schweiger, W.; Fowler, J.; Berthiller, F. Determination of aflatoxin biomarkers in excreta and ileal content of chickens. *Poult. Sci.* **2019**, *98*, 5551–5561. [CrossRef]
90. Kim, S.W.; Duarte, M.E. Understanding intestinal health in nursery pigs and the relevant nutritional strategies. *Anim. Biosci.* **2021**, *34*, 338–344. [CrossRef]
91. Szabó, C.; Kachungwa Lugata, J.; Ortega, A. Gut health and influencing factors in pigs. *Animals* **2023**, *13*, 1350. [CrossRef]
92. Liu, N.; Wang, J.Q.; Liu, Z.Y.; Chen, Y.K.; Wang, J.P. Effect of cysteamine hydrochloride supplementation on the growth performance, enterotoxic status, and glutathione turnover of broilers fed aflatoxin B1 contaminated diets. *Poult. Sci.* **2018**, *97*, 3594–3600. [CrossRef]
93. El-Katcha, M.I.; Soltan, M.A.; El-Shobokshy, S.A.; Shokry, A. Protective effect of chemical and biological mycotoxin binder on growth performance, serum biochemistry and carcass traits in broiler chicks fed on aflatoxin contaminated diet. *Alex. J. Vet. Sci.* **2017**, *55*, 180–197.
94. Zou, Y.; Liu, S.-B.; Zhang, Q.; Tan, H.-Z. Effects of aflatoxin B1 on growth performance, carcass traits, organ index, blood biochemistry and oxidative status in chinese yellow chickens. *J. Vet. Med. Sci.* **2023**, *85*, 1015–1022. [CrossRef] [PubMed]
95. Zabiulla, I.; Malathi, V.; Swamy, H.V.L.N.; Naik, J.; Pineda, L.; Han, Y. The efficacy of a smectite-based mycotoxin binder in reducing aflatoxin B1 toxicity on performance, health and histopathology of broiler chickens. *Toxins* **2021**, *13*, 856. [CrossRef]
96. Pasha, T.N.; Farooq, M.U.; Khattak, F.M.; Jabbar, M.A.; Khan, A.D. Effectiveness of sodium bentonite and two commercial products as aflatoxin absorbents in diets for broiler chickens. *Anim. Feed Sci. Technol.* **2007**, *132*, 103–110. [CrossRef]
97. Nemati, Z.; Karimi, A.; Besharati, M. Effects of aflatoxin B1 and yeast cell wall supplementation on the growth performance of broilers. *Int. Conf. Innovations Chem. Agric. Eng.* **2015**, *8*, 117–120.
98. Huff, W.E.; Kubena, L.F.; Harvey, R.B.; Hagler Jr, W.M.; Swanson, S.P.; Phillips, T.D.; Creger, C.R. Individual and combined effects of aflatoxin and deoxynivalenol (DON, vomitoxin) in broiler chickens. *Poult. Sci.* **1986**, *65*, 1291–1298. [CrossRef] [PubMed]
99. Verma, J.; Johri, T.S.; Swain, B.K.; Ameena, S. Effect of graded levels of aflatoxin, ochratoxin and their combinations on the performance and immune response of broilers. *Br. Poult. Sci.* **2004**, *45*, 512–518. [CrossRef]
100. Ma, H.; Chen, Q.; Yang, H.; Wan, X. Effects of lycopene on the growth performance, meat quality, and antioxidant capacity of broiler chickens challenged with aflatoxin B1. *J. Food Sci.* **2024**, *89*, 96–103. [CrossRef]
101. Cravens, R.L.; Goss, G.R.; Chi, F.; De Boer, E.D.; Davis, S.W.; Hendrix, S.M.; Richardson, J.A.; Johnston, S.L. The effects of necrotic enteritis, aflatoxin B1, and virginiamycin on growth performance, necrotic enteritis lesion scores, and mortality in young broilers. *Poult. Sci.* **2013**, *92*, 1997–2004. [CrossRef] [PubMed]
102. Al-Shawabkeh, K.; Herzallah, S.; Al-Fataftah, A.; Zakaria, H. Effect of aflatoxin B1 contaminated feed on broiler chickens performance and meat content of conjugated linoleic acid. *J. Agric. Sci.* **2009**, *5*, 314–323.
103. Bintvihok, A.; Kositcharoenkul, S. Effect of dietary calcium propionate on performance, hepatic enzyme activities and aflatoxin residues in broilers fed a diet containing low levels of aflatoxin B1. *Toxicon* **2006**, *47*, 41–46. [CrossRef]
104. Saei, M.M.; Sadeghi, A.A.; Ahmadvand, H. The effect of myrtus communis oil extract on growth performance, serum biochemistry and humoral immune responses in broiler chicks fed diet containing aflatoxin B1. *Arch. Anim. Breed.* **2013**, *56*, 842–850. [CrossRef]
105. El-Sayed, H.G.M.; Elwan, H.A.M.; Aboelhassan, A.F.A.; Abdel-Wareth, A.A.A. Effects of probiotic feed additives on the growth performance and carcass criteria of broiler chickens exposed to an aflatoxin B1 challenge. *SVU Int. J. Agric. Sci.* **2022**, *4*, 29–36. [CrossRef]
106. Rashidi, N.; Khatibjoo, A.; Taherpour, K.; Akbari-Gharaei, M.; Shirzadi, H. Effects of licorice extract, probiotic, toxin binder and poultry litter biochar on performance, immune function, blood indices and liver histopathology of broilers exposed to aflatoxin-B1. *Poult. Sci.* **2020**, *99*, 5896–5906. [CrossRef] [PubMed]
107. Raj, J.; Vasiljević, M.; Tassis, P.; Farkaš, H.; Bošnjak-Neumüller, J.; Männer, K. Effects of a modified clinoptilolite zeolite on growth performance, health status and detoxification of aflatoxin B1 and ochratoxin a in male broiler chickens. *Br. Poult. Sci.* **2021**, *62*, 601–610. [CrossRef] [PubMed]
108. Fan, Y.; Zhao, L.; Ma, Q.; Li, X.; Shi, H.; Zhou, T.; Zhang, J.; Ji, C. Effects of *Bacillus subtilis* ANSB060 on growth performance, meat quality and aflatoxin residues in broilers fed moldy peanut meal naturally contaminated with aflatoxins. *Food Chem. Toxicol.* **2013**, *59*, 748–753. [CrossRef] [PubMed]
109. Saminathan, M.; Selamat, J.; Abbasi Pirouz, A.; Abdullah, N.; Zulkifli, I. Effects of nano-composite adsorbents on the growth performance, serum biochemistry, and organ weights of broilers fed with aflatoxin-contaminated feed. *Toxins* **2018**, *10*, 345. [CrossRef] [PubMed]

110. Shang, Q.H.; Yang, Z.B.; Yang, W.R.; Li, Z.; Zhang, G.G.; Jiang, S.Z. Toxicity of mycotoxins from contaminated corn with or withoutyeast cell wall adsorbent on broiler chickens. *Asian-Australas. J. Anim. Sci.* **2015**, *29*, 674–680. [CrossRef] [PubMed]
111. Coffey, M.T.; Hagler Jr, W.M.; Cullen, J.M. Influence of dietary protein, fat or amino acids on the response of weanling swine to aflatoxin B1. *J. Anim. Sci.* **1989**, *67*, 465–472. [CrossRef] [PubMed]
112. Dazuk, V.; Tarasconi, L.; Molosse, V.L.; Cécere, B.G.O.; Deolindo, G.L.; Strapazzon, J.V.; Bottari, N.B.; Bissacotti, B.F.; Schetinger, M.R.C.; Sareta, L. Can the inclusion of a vegetable biocholine additive in pig feed contaminated with aflatoxin reduce toxicological impacts on animal health and performance? *Animals* **2023**, *13*, 3010. [CrossRef]
113. Schell, T.C.; Lindemann, M.D.; Kornegay, E.T.; Blodgett, D.J. Effects of feeding aflatoxin-contaminated diets with and without clay to weanling and growing pigs on performance, liver function, and mineral metabolism. *J. Anim. Sci.* **1993**, *71*, 1209–1218. [CrossRef]
114. Lindemann, M.D.; Blodgett, D.J.; Kornegay, E.T.; Schurig, G.G. Potential ameliorators of aflatoxicosis in weanling/growing swine. *J. Anim. Sci.* **1993**, *71*, 171–178. [CrossRef]
115. Fu, J.-C.; Chen, Q.; Du, J.; Shi, B.-M.; Shan, A.-S. Effectiveness of maifanite in reducing the detrimental effects of aflatoxin B1 on hematology, aflatoxin B1 residues, and antioxidant enzymes activities of weanling piglets. *Livest. Sci.* **2013**, *157*, 218–224. [CrossRef]
116. Harper, A.F.; Estienne, M.J.; Meldrum, J.B.; Harrell, R.J.; Diaz, D.E. Assessment of a hydrated sodium calcium aluminosilicate agent and antioxidant blend for mitigation of aflatoxin-induced physiological alterations in pigs. *J. Swine Health Prod.* **2010**, *18*, 282–289. [CrossRef] [PubMed]
117. Wang, D.; Lindemann, M.D.; Estienne, M.J. Effect of folic acid supplementation and dietary protein level on growth performance, serum chemistry and immune response in weanling piglets fed differing concentrations of aflatoxin. *Toxins* **2020**, *12*, 651. [CrossRef] [PubMed]
118. Thieu, N.Q.; Ogle, B.; Pettersson, H. Efficacy of bentonite clay in ameliorating aflatoxicosis in piglets fed aflatoxin contaminated diets. *Trop. Anim Health Prod.* **2008**, *40*, 649–656. [CrossRef]
119. Marin, D.E.; Taranu, I.; Bunaciu, R.P.; Pascale, F.; Tudor, D.S.; Avram, N.; Sarca, M.; Cureu, I.; Criste, R.D.; Suta, V. Changes in performance, blood parameters, humoral and cellular immune responses in weanling piglets exposed to low doses of aflatoxin. *J. Anim. Sci.* **2002**, *80*, 1250–1257. [CrossRef] [PubMed]
120. Meissonnier, G.M.; Pinton, P.; Laffitte, J.; Cossalter, A.-M.; Gong, Y.Y.; Wild, C.P.; Bertin, G.; Galtier, P.; Oswald, I.P. Immunotoxicity of aflatoxin B1: Impairment of the cell-mediated response to vaccine antigen and modulation of cytokine expression. *Toxicol. Appl. Pharmacol.* **2008**, *231*, 142–149. [CrossRef] [PubMed]
121. Harvey, R.B.; Kubena, L.F.; Huff, W.E.; Corrier, D.E.; Rottinghaus, G.E.; Phillips, T.D. Effects of treatment of growing swine with aflatoxin and T-2 toxin. *Am. J. Vet. Res.* **1990**, *51*, 1688–1693. [CrossRef]
122. Shi, Y.H.; Xu, Z.R.; Feng, J.L.; Xia, M.S.; Hu, C.H. Effects of modified montmorillonite nanocomposite on growing/finishing pigs during aflatoxicosis. *Asian-Australas. J. Anim. Sci.* **2005**, *18*, 1305–1309. [CrossRef]
123. Sun, Y.; Park, I.; Guo, J.; Weaver, A.C.; Kim, S.W. Impacts of low level aflatoxin in feed and the use of modified yeast cell wall extract on growth and health of nursery pigs. *Anim. Nutr.* **2015**, *1*, 177–183. [CrossRef]
124. Weaver, A.C.; See, M.T.; Hansen, J.A.; Kim, Y.B.; De Souza, A.L.; Middleton, T.F.; Kim, S.W. The use of feed additives to reduce the effects of aflatoxin and deoxynivalenol on pig growth, organ health and immune status during chronic exposure. *Toxins* **2013**, *5*, 1261–1281. [CrossRef] [PubMed]
125. Chaytor, A.C.; See, M.T.; Hansen, J.A.; De Souza, A.L.P.; Middleton, T.F.; Kim, S. Effects of chronic exposure of diets with reduced concentrations of aflatoxin and deoxynivalenol on growth and immune status of pigs. *J. Anim. Sci.* **2011**, *89*, 124–135. [CrossRef]
126. Boudergue, C.; Burel, C.; Dragacci, S.; Favrot, M.C.; Fremy, J.M.; Massimi, C.; Prigent, P.; Debongnie, P.; Pussemier, L.; Boudra, H. Review of mycotoxin-detoxifying agents used as feed additives: Mode of action, efficacy and feed/food safety. *EFSA Support. Publ.* **2009**, *6*, 22E. [CrossRef]
127. Weaver, A.C.; See, M.T.; Kim, S.W. Protective effect of two yeast based feed additives on pigs chronically exposed to deoxynivalenol and zearalenone. *Toxins* **2014**, *6*, 3336–3353. [CrossRef]
128. Holanda, D.M.; Kim, S.W. Investigation of the efficacy of mycotoxin-detoxifying additive on health and growth of newly-weaned pigs under deoxynivalenol challenges. *Asian-Australas. J. Anim. Sci.* **2021**, *34*, 405–416. [CrossRef]
129. Rasheed, U.; Ain, Q.U.; Yaseen, M.; Yao, X.; Liu, B. Synthesis and characterization of tannic acid pillared bentonite composite for the efficient adsorption of aflatoxins. *Colloids Surf. B Biointerfaces.* **2021**, *202*, 111679. [CrossRef] [PubMed]
130. Rao, Z.X.; Tokach, M.D.; Woodworth, J.C.; DeRouchey, J.M.; Goodband, R.D.; Gebhardt, J.T. Evaluation of selenium source on nursery pig growth performance, serum and tissue selenium concentrations, and serum antioxidant status. *Transl. Anim. Sci.* **2023**, *7*, txad049. [CrossRef] [PubMed]
131. Mao, J.; Zhou, Y.; Lv, G.; Zhou, R. Simultaneous detoxification of aflatoxin B1, zearalenone and deoxynivalenol by modified montmorillonites. *Molecules* **2022**, *27*, 315. [CrossRef]

132. Kong, C.; Shin, S.Y.; Kim, B.G. Evaluation of mycotoxin sequestering agents for aflatoxin and deoxynivalenol: An in vitro approach. *SpringerPlus* **2014**, *3*, 1–4. [CrossRef]
133. Wijesooriya, M.M.; Wijesekara, H.; Bolan, N.; Rajapaksha, A.U.; Vithanage, M. Clays and clay minerals: Long-lasting applications in environmental remediation. In *Clay composites: Environmental applications*; Springer: Singapore, 2023; pp. 3–28.
134. Zhang, N.; Han, X.; Zhao, Y.; Li, Y.; Meng, J.; Zhang, H.; Liang, J. Removal of aflatoxin B1 and zearalenone by clay mineral materials: In the animal industry and environment. *Appl. Clay Sci.* **2022**, *228*, 106614. [CrossRef]
135. Ahn, J.Y.; Kim, J.; Cheong, D.H.; Hong, H.; Jeong, J.Y.; Kim, B.G. An in vitro study on the efficacy of mycotoxin sequestering agents for aflatoxin B1, deoxynivalenol, and zearalenone. *Animals* **2022**, *12*, 333. [CrossRef] [PubMed]
136. Huwig, A.; Freimund, S.; Käppeli, O.; Dutler, H. Mycotoxin detoxication of animal feed by different adsorbents. *Toxicol. Lett.* **2001**, *122*, 179–188. [CrossRef] [PubMed]
137. Heathcote, J.G.; Hibbert, J.R. Biological activity and electronic structure of the aflatoxins. *Br. J. Cancer* **1974**, *29*, 470–476. [CrossRef] [PubMed]
138. Dallies, N.; François, J.; Paquet, V. A new method for quantitative determination of polysaccharides in the yeast cell wall. Application to the cell wall defective mutants of *Saccharomyces cerevisiae*. *Yeast* **1998**, *14*, 1297–1306. [CrossRef]
139. Choi, H.; Kim, S.W. Characterization of β-glucans from cereal and microbial sources and their roles in feeds for intestinal health and growth of nursery pigs. *Animals* **2023**, *13*, 2236. [CrossRef]
140. Chlebicz, A.; Śliżewska, K. In vitro detoxification of aflatoxin B1, deoxynivalenol, fumonisins, T-2 toxin and zearalenone by probiotic bacteria from genus lactobacillus and saccharomyces cerevisiae yeast. *Probiotics Antimicro. Proteins* **2020**, *12*, 289–301. [CrossRef] [PubMed]
141. Drummond, R.A.; Brown, G.D. The role of dectin-1 in the host defence against fungal infections. *Curr. Opin. Microbiol.* **2011**, *14*, 392–399. [CrossRef]
142. Amer, S.A.; Attia, G.A.; Aljahmany, A.A.; Mohamed, A.K.; Ali, A.A.; Gouda, A.; Alagmy, G.N.; Megahed, H.M.; Saber, T.; Farahat, M. Effect of 1,3-beta glucans dietary addition on the growth, intestinal histology, blood biochemical parameters, immune response, and immune expression of CD3 and CD20 in broiler chickens. *Animals* **2022**, *12*, 3197. [CrossRef] [PubMed]
143. Peltonen, K.; El-Nezami, H.; Haskard, C.; Ahokas, J.; Salminen, S. Aflatoxin B1 binding by dairy strains of lactic acid bacteria and bifidobacteria. *J. Dairy Sci.* **2001**, *84*, 2152–2156. [CrossRef]
144. El Khoury, A.; Atoui, A.; Yaghi, J. Analysis of aflatoxin M1 in milk and yogurt and AFM1 reduction by lactic acid bacteria used in lebanese industry. *Food Control* **2011**, *22*, 1695–1699. [CrossRef]
145. Armanini, E.H.; Boiago, M.M.; de Oliveira, P.V.; Roscamp, E.; Strapazzon, J.V.; de Lima, A.G.; Copetti, P.M.; Morsch, V.M.; de Oliveira, F.C.; Wagner, R. Inclusion of a phytogenic bend in broiler diet as a performance enhancer and anti-aflatoxin agent: Impacts on health, performance, and meat quality. *Res. Vet. Sci.* **2021**, *137*, 186–193. [CrossRef] [PubMed]
146. Lindemann, M.D.; Blodgett, D.J.; Harper, A.F.; Kornegay, E.T.; Doerr, J.A. Appraisal of the value of selected clays and minerals in diets with and without aflatoxin-contaminated maize fed to young pigs. *J. Anim. Feed Sci.* **1997**, *6*, 507–519. [CrossRef]
147. Shu, X.; Wang, Y.; Zhou, Q.; Li, M.; Hu, H.; Ma, Y.; Chen, X.; Ni, J.; Zhao, W.; Huang, S. Biological degradation of aflatoxin B1 by cell-free extracts of bacillus velezensis DY3108 with broad pH stability and excellent thermostability. *Toxins* **2018**, *10*, 330. [CrossRef]
148. Xiong, D.; Wen, J.; Lu, G.; Li, T.; Long, M. Isolation, purification, and characterization of a laccase-degrading aflatoxin B1 from *Bacillus amyloliquefaciens* B10. *Toxins* **2022**, *14*, 250. [CrossRef] [PubMed]
149. Cheng, S.; Wu, T.; Zhang, H.; Sun, Z.; Mwabulili, F.; Xie, Y.; Sun, S.; Ma, W.; Li, Q.; Yang, Y. Mining lactonase gene from aflatoxin B1-degrading strain bacillus megaterium and degrading properties of the recombinant enzyme. *J. Agric. Food Chem.* **2023**, *71*, 20762–20771. [CrossRef] [PubMed]
150. Subagia, R.; Schweiger, W.; Kunz-Vekiru, E.; Wolfsberger, D.; Schatzmayr, G.; Ribitsch, D.; Guebitz, G.M. Detoxification of aflatoxin B1 by a *Bacillus subtilis* spore coat protein through formation of the main metabolites AFQ1 and epi-AFQ1. *Front. Microbiol.* **2024**, *15*, 1406707. [CrossRef] [PubMed]
151. Loi, M.; Renaud, J.B.; Rosini, E.; Pollegioni, L.; Vignali, E.; Haidukowski, M.; Sumarah, M.W.; Logrieco, A.F.; Mulè, G. Enzymatic transformation of aflatoxin B1 by Rh_Dypb peroxidase and characterization of the reaction products. *Chemosphere* **2020**, *250*, 126296. [CrossRef] [PubMed]
152. Pereyra, M.L.G.; Martínez, M.P.; Cavaglieri, L.R. Presence of aiia homologue genes encoding for n-acyl homoserine lactone-degrading enzyme in aflatoxin B1-decontaminating *Bacillus* strains with potential use as feed additives. *Food Chem. Toxicol.* **2019**, *124*, 316–323. [CrossRef] [PubMed]
153. El-Deeb, B.; Altalhi, A.; Khiralla, G.; Hassan, S.; Gherbawy, Y. Isolation and characterization of endophytic bacilli bacterium from maize grains able to detoxify aflatoxin B1. *Food Biotechnol.* **2013**, *27*, 199–212. [CrossRef]
154. Farzaneh, M.; Shi, Z.-Q.; Ghassempour, A.; Sedaghat, N.; Ahmadzadeh, M.; Mirabolfathy, M.; Javan-Nikkhah, M. Aflatoxin B1 degradation by *Bacillus subtilis* UTBSP1 isolated from pistachio nuts of iran. *Food control* **2012**, *23*, 100–106. [CrossRef]

155. Li, G.; Tong, Y.; Xiao, Y.; Huang, S.; Zhao, T.; Xia, X. Probiotic *Bacillus subtilis* contributes to the modulation of gut microbiota and blood metabolic profile of hosts. *Comp. Biochem. Physiol. C Toxicol. Pharmacol.* **2023**, *272*, 109712. [CrossRef] [PubMed]
156. Bai, K.; Feng, C.; Jiang, L.; Zhang, L.; Zhang, J.; Zhang, L.; Wang, T. Dietary effects of *Bacillus subtilis* fmbj on growth performance, small intestinal morphology, and its antioxidant capacity of broilers. *Poult. Sci.* **2018**, *97*, 2312–2321. [CrossRef]
157. Ciegler, A.; Lillehoj, E.B.; Peterson, R.E.; Hall, H.H. Microbial detoxification of aflatoxin. *Appl. Microbiol.* **1966**, *14*, 934–939. [CrossRef]
158. Teniola, O.D.; Addo, P.A.; Brost, I.M.; Färber, P.; Jany, K.-D.; Alberts, J.F.; Van Zyl, W.H.; Steyn, P.S.; Holzapfel, W.H. Degradation of aflatoxin B1 by cell-free extracts of *Rhodococcus erythropolis* and *Mycobacterium fluoranthenivorans* sp. *Nov. DSM44556T. Int. J. Food Microbiol.* **2005**, *105*, 111–117. [CrossRef] [PubMed]
159. Tejada-Castañeda, Z.I.; Avila-Gonzalez, E.; Casaubon-Huguenin, M.T.; Cervantes-Olivares, R.A.; Vásquez-Peláez, C.; Hernández-Baumgarten, E.M.; Moreno-Martínez, E. Biodetoxification of aflatoxin-contaminated chick feed. *Poult. Sci.* **2008**, *87*, 1569–1576. [CrossRef]
160. Papatsiros, V.G.; Eliopoulos, C.; Voulgarakis, N.; Arapoglou, D.; Riahi, I.; Sadurní, M.; Papakonstantinou, G.I. Effects of a multi-component mycotoxin-detoxifying agent on oxidative stress, health and performance of sows. *Toxins* **2023**, *15*, 580. [CrossRef] [PubMed]
161. Sabater-Vilar, M.; Malekinejad, H.; Selman, M.; Van Der Doelen, M.; Fink-Gremmels, J. In vitro assessment of adsorbents aiming to prevent deoxynivalenol and zearalenone mycotoxicoses. *Mycopathologia* **2007**, *163*, 81–90. [CrossRef] [PubMed]
162. Ao, X.; Kim, I.H. Effects of grape seed extract on performance, immunity, antioxidant capacity, and meat quality in pekin ducks. *Poult. Sci.* **2020**, *99*, 2078–2086. [CrossRef]
163. Nallathambi, R.; Poulev, A.; Zuk, J.B.; Raskin, I. Proanthocyanidin-rich grape seed extract reduces inflammation and oxidative stress and restores tight junction barrier function in Caco-2 colon cells. *Nutrients* **2020**, *12*, 1623. [CrossRef]
164. González-Quilen, C.; Gil-Cardoso, K.; Ginés, I.; Beltrán-Debón, R.; Pinent, M.; Ardévol, A.; Terra, X.; Blay, M.T. Grape-seed proanthocyanidins are able to reverse intestinal dysfunction and metabolic endotoxemia induced by a cafeteria diet in wistar rats. *Nutrients* **2019**, *11*, 979. [CrossRef] [PubMed]
165. Meng, Q.; Zhang, Y.; Li, J.; Shi, B.; Ma, Q.; Shan, A. Lycopene affects intestinal barrier function and the gut microbiota in weaned piglets via antioxidant signaling regulation. *J. Nutr.* **2022**, *152*, 2396–2408. [CrossRef] [PubMed]
166. Sun, L.-H.; Zhang, N.-Y.; Zhu, M.-K.; Zhao, L.; Zhou, J.-C.; Qi, D.-S. Prevention of aflatoxin B1 hepatoxicity by dietary selenium is associated with inhibition of cytochrome P450 isozymes and up-regulation of 6 selenoprotein genes in chick liver. *J. Nutr.* **2016**, *146*, 655–661. [CrossRef]
167. Liao, S.; Shi, D.; Clemons-Chevis, C.L.; Guo, S.; Su, R.; Qiang, P.; Tang, Z. Protective role of selenium on aflatoxin B1-induced hepatic dysfunction and apoptosis of liver in ducklings. *Biol. Trace Elem. Res.* **2014**, *162*, 296–301. [CrossRef] [PubMed]
168. Purushothaman, A.; Teena Rose, K.S.; Jacob, J.M.; Varatharaj, R.; Shashikala, K.; Janardanan, D. Curcumin analogues with improved antioxidant properties: A theoretical exploration. *Food Chem.* **2022**, *373*, 131499. [CrossRef] [PubMed]
169. Wang, Y.; Liu, F.; Liu, M.; Zhou, X.; Wang, M.; Cao, K.; Jin, S.; Shan, A.; Feng, X. Curcumin mitigates aflatoxin B1-induced liver injury via regulating the NLRP3 inflammasome and Nrf2 signaling pathway. *Food Chem. Toxicol.* **2022**, *161*, 112823. [CrossRef]
170. Gao, X.; Kuo, J.; Jiang, H.; Deeb, D.; Liu, Y.; Divine, G.; Chapman, R.A.; Dulchavsky, S.A.; Gautam, S.C. Immunomodulatory activity of curcumin: Suppression of lymphocyte proliferation, development of cell-mediated cytotoxicity, and cytokine production in vitro. *Biochem. Pharmacol.* **2004**, *68*, 51–61. [CrossRef]
171. Solís-Cruz, B.; Hernandez-Patlan, D.; Beyssac, E.; Latorre, J.D.; Hernandez-Velasco, X.; Merino-Guzman, R.; Tellez, G.; López-Arellano, R. Evaluation of chitosan and cellulosic polymers as binding adsorbent materials to prevent aflatoxin B1, fumonisin B1, ochratoxin, trichothecene, deoxynivalenol, and zearalenone mycotoxicoses through an in vitro gastrointestinal model for poultry. *Polymers* **2017**, *9*, 529. [CrossRef]
172. Huang, W.; Cao, Z.; Cui, Y.; Huo, S.; Shao, B.; Song, M.; Cheng, P.; Li, Y. Lycopene ameliorates aflatoxin B1-induced testicular lesion by attenuating oxidative stress and mitochondrial damage with nrf2 activation in mice. *Ecotoxicol. Environ. Saf.* **2023**, *256*, 114846. [CrossRef]
173. Sarker, M.T.; Wan, X.; Yang, H.; Wang, Z. Dietary lycopene supplementation could alleviate aflatoxin B1 induced intestinal damage through improving immune function and anti-oxidant capacity in broilers. *Animals* **2021**, *11*, 3165. [CrossRef]

Disclaimer/Publisher's Note: The statements, opinions and data contained in all publications are solely those of the individual author(s) and contributor(s) and not of MDPI and/or the editor(s). MDPI and/or the editor(s) disclaim responsibility for any injury to people or property resulting from any ideas, methods, instructions or products referred to in the content.

toxins

MDPI

Article

Innovative Mycotoxin Detoxifying Agents Decrease the Absorption Rate of Aflatoxin B1 and Counteract the Oxidative Stress in Broiler Chickens Exposed to Low Dietary Levels of the Mycotoxin

Matteo Cuccato [1], Neenu Amminikutty [1], Veronica Spalenza [1,†], Vanessa Conte [1], Stefano Bagatella [1], Donato Greco [2], Vito D'Ascanio [2], Francesco Gai [3], Achille Schiavone [1], Giuseppina Avantaggiato [2], Carlo Nebbia [1,*] and Flavia Girolami [1]

[1] Department of Veterinary Sciences, University of Torino, 10095 Grugliasco, Italy; matteo.cuccato@unito.it (M.C.); neenu.amminikutty@unito.it (N.A.); spalenza.veronica@gmail.com (V.S.); vanessa.conte@unito.it (V.C.); stefano.bagatella@unito.it (S.B.); achille.schiavone@unito.it (A.S.); flavia.girolami@unito.it (F.G.)

[2] Institute of Sciences of Food Production, Italian National Research Council, 70126 Bari, Italy; donato.greco@ispa.cnr.it (D.G.); vito.dascanio@ispa.cnr.it (V.D.); giuseppina.avantaggiato@ispa.cnr.it (G.A.)

[3] Institute of Sciences of Food Production, National Research Council, 10095 Grugliasco, Italy; francesco.gai@ispa.cnr.it

* Correspondence: carlo.nebbia@unito.it

† Current address: ANABORAPI—Associazione Nazionale Allevatori Bovini Razza Piemontese, 12061 Carrù, Italy.

Abstract: Aflatoxin B1 (AFB1) can impair the growth of chickens and reduce the quality of eggs and meat, resulting in significant economic losses. The inclusion of mycotoxin detoxifying agents (MyDA) with binding properties in the diet is an efficient tool to reduce their absorption rate in the gastrointestinal tract. Our aim was to investigate the ability of two innovative MyDA (SeOX, a feed additive featuring a tri-octahedral smectite mixed with lignocellulose, and CHS, a di-octahedral smectite functionalized with an organic non-toxic modifier) in both reducing the bio-accessibility and mitigating the adverse effects of AFB1 in broilers exposed for 10 days to concentrations approaching the European Union maximum limits in feed (0.02 mg/kg). The amount of AFB1 in the excreta of birds, collected over four consecutive days (starting on day 7), was significantly lower ($p < 0.001$) in the group exposed to AFB1 alone compared to the groups treated with either SeOX or CHS. The calculated bio-accessibility was decreased by nearly 30% with both MyDA. This positive effect was reflected by a significant reduction ($p < 0.001$) in the oxidative stress (measured as serum antioxidant capacity and hepatic lipid peroxidation) induced by AFB1. Although antioxidant enzyme activities and glutathione levels were unaffected by any treatment, AFB1 significantly induced ($p < 0.001$) the upregulation of CYP2A6 and the downregulation of Nrf2; the latter was reverted by each MyDA. Overall, these results demonstrate that the selected MyDA are effective in limiting the AFB1 absorption rate, thereby mitigating or even reverting the oxidative stress induced by AFB1 in broilers.

Keywords: aflatoxin B1; mycotoxin detoxifying agents; oxidative stress; liver; chicken; drug-metabolizing enzymes

Key Contribution: Two innovative additives with mycotoxin binding properties reduce the bio-accessibility of AFB1 in broilers exposed to concentrations approaching the EU maximum limits, thus mitigating the mycotoxin adverse effects.

1. Introduction

Aflatoxins (AF) are secondary metabolites mainly produced by mycotoxigenic fungi of the genus Aspergillus [1]. AFB1 is one of the most extensively studied mycotoxins in humans and in a diverse range of animal species. It is widely acknowledged for its hepatotoxic and immunotoxic properties, and it has been classified as a Group I genotoxic carcinogen by the International Agency for Research on Cancer (IARC) [2]. In sub-tropical and tropical countries, AFB1 contamination in food and feed commodities has long been recognized as a significant public health issue [3]. However, the combination of rising temperatures and increased humidity caused by climate change in tempered zones has escalated the contamination into a global emergency [4]. In the poultry industry, feed contamination by AFB1 poses a substantial challenge [5]. Although chickens are relatively less sensitive to aflatoxicosis compared to the other avian species raised for food production (e.g., ducks and turkeys), the prolonged exposure of broilers to AFB1 can impair their growth performances, with reduced weight gain, feed intake, and feed conversion ratio [6,7]. Experimental studies have demonstrated that AFB1 can induce oxidative stress (OS) in various tissues and organs, thereby compromising the antioxidant defense system [8–10]. The main driver in the AFB1-mediated OS is aflatoxin-8,9-epoxide (AFBO), a bioactivated metabolite primarily generated by the phase I enzyme CYP2A6 ortholog in chickens, with CYP1A1 playing a secondary role. AFBO can be detoxified by reduced glutathione (GSH) via glutathione S-transferase (GST) and subsequently excreted as a mercapturic acid derivative. The phase I metabolism of AFB1 produces other less toxic oxidated water-soluble metabolites (i.e., AFM1, AFB2a, AFQ1, and AFP1) and a reduced form, aflatoxicol (AFL), which can be easily re-converted to the parent compound in liver [11].

Given the significant health risks posed by AF, regulatory bodies worldwide have established maximum permitted levels in food and feed. In the European Union (EU), the limits in both complementary and complete feed for farm animals range from 0.005 mg/kg for dairy ruminants to 0.02 mg/kg for the other food-producing species (EC Regulation 574/2011). However, similar regulatory restrictions have been implemented only in the USA and in a few African countries, and a homogenous global regulatory framework is still lacking [3]. Thus, minimizing the formation and accumulation of mycotoxins in crops remains crucial. AFB1 contamination can occur at any stage of the feed and food chain, from the field production to the final use of various plant products. Several control methods have proven effective in mitigating AFB1 accumulation in cereals from harvest to storage, ensuring compliance with EU regulatory limits [12]. Additionally, the inclusion of low-cost and easy-to-use technological feed additives, known as mycotoxin adsorbents or binding agents (e.g., clays, silicates, yeast cell walls, and probiotic bacteria), represents the prevalent approach to reduce the gut absorption of mycotoxins, thus counteracting their toxic effects [13].

According to a Scientific Opinion of the European Food Safety Authority (EFSA), mycotoxin-detoxifying agents (MyDA) are defined as compounds either capable of adsorbing mycotoxins, binding them to their surfaces, or transforming them into less toxic derivatives through biotransformation [13]. From a regulatory perspective, mycotoxin binders are classified as technological feed additives, and their use is authorized for food-producing animals (EC Regulations 1831/2003 and 786/2015). Currently, bentonite is the only authorized mycotoxin binder for reducing AFB1 absorption in pigs, poultry, and cattle (EC Regulation 1060/2013). According to the recently published literature, several commercial feed additives containing bentonite as the main compound have been shown to effectively sequester AFB1 and tackle its toxicity in chickens [14–18]. However, such inorganic binder rarely shows activity against other mycotoxins, such as T-2 toxins, ochratoxin-A, deoxynivalenol, and fumonisins [12]. In recent years, several authorized and experimental additives have been investigated for their ability to function as MyDA and mitigate the harmful effects of mycotoxins in poultry [19–23]. The most studied ones are typically based on mineral clays or other inorganic compounds that can trap mycotoxins from contaminated feed, thereby reducing their absorption in the gastrointestinal system and facilitating their fecal elimination [12]. Conversely, organic additives such as yeast cell wall, probiotic Lactobacilli strains, or *Saccharomyces cerevisiae* act by modifying the chemical structure of the mycotoxins and alleviating their adverse effects [24,25].

In previous published studies, broilers dietarily exposed to AFB1 at levels around the EU regulatory limits (0.02 mg/kg) showed significant hepatic, renal, and systemic OS [9,10]. Of note, dietary supplementation with turmeric powder reverted such effects and counteracted the negative modulation of selected target genes induced by AFB1. The concomitant reduction in hepatic AFB1 residues and the positive gene regulation of intestinal and hepatic transporters suggest the potential ability of turmeric powder to reduce intestinal AFB1 absorption or to increase its excretion. In general, the use of feed exceeding the EU limit for AFB1, even with the supplementation of an authorized MyDA, is not permitted in the EU Member States. To the best of our knowledge, most of the research on chickens has investigated the efficacy of commercial or innovative MyDA in counteracting the negative effects of feed contaminated by AFB1 at concentrations ranging from 0.1 to 10 mg/kg [20,22,26,27], thus always exceeding the EU regulatory limit for poultry feed. It remains to be established whether the use of MyDA in feed contaminated by AFB1 concentrations compliant with the EU limits may be effective in ameliorating the OS caused by such low levels of the mycotoxin in chickens [10].

Therefore, the aim of this study was to investigate the potential ability of two innovative MyDA, a smectite-based material (SeOX) and a bio-organoclay (CHS), to limit both the bio-accessibility and the adverse effects of AFB1 in broilers exposed to concentrations approaching the EU maximum limits in feed (0.02 mg/kg). SeOX was obtained by combining a tri-octahedral smectite with a lignocellulose-based material showing antioxidant properties, as described by Greco et al. [28]. The CHS was developed by functionalization of a di-octahedral smectite with an organic, non-toxic modifier. This product can be considered safe, as it was obtained using agents listed in the EU Register of Feed Additives (EC Regulation, No.1831/2003). Since the primary target organ of AFB1 toxicity is the liver [29], the hepatic effects of the exposure to the mycotoxin in the presence or absence of each MyDa were explored.

2. Results

2.1. Effects of AFB1 and MyDA on Zootechnical Parameters and Liver Histology

Throughout the trial, the birds of each experimental group maintained good health, and no mortality occurred. No statistically significant differences in the zootechnical parameters were registered as the result of the exposure to AFB1 alone or in combination with either SeOX or CHS; this was also true for chicks treated with each MyDA alone (Table 1). Additionally, no effects on the dry matter (DM) digestibility were observed in any of the treated groups, with an average value of 64.3%.

No significant histological changes were observed in the hepatic tissue from animals of every experimental group. Mild to moderate multifocal lymphoplasmacytic hepatitis, accompanied by a multifocal lymphoid hyperplasia, and/or mild vacuolar degeneration of the hepatocytes were randomly distributed among birds (Figure 1). No significant differences were observed between broilers from the control and treated groups.

Table 1. Growth performance and productive parameters of broiler chickens fed the experimental diets (n = 4; mean ± SEM).

Parameters	Treatments						p-Value
	BD	SeOX	CHS	AFB1	AFB1 + SeOX	AFB1 + CHS	
Initial body weight (d23) (g)	1011 ± 22	1004 ± 15	999 ± 25	987 ± 29	996 ± 22	997 ± 34	0.802
Final body weight (d32) (g)	1944 ± 41	1965 ± 33	1954 ± 42	1944 ± 66	1976 ± 28	1949 ± 43	0.841
ADG 23–32 days of age (g)	118 ± 4.1	121 ± 2.9	120 ± 3.8	120 ± 3.6	123 ± 1.6	120 ± 5.2	0.574
ADFI 23–32 days of age (g)	144 ± 5.8	144 ± 2.5	145 ± 7.5	144 ± 7.1	149 ± 8.1	144 ± 11	0.776
FCR 23–32 days of age	1.53 ± 0.02	1.50 ± 0.03	1.51 ± 0.05	1.52 ± 0.02	1.52 ± 0.07	1.51 ± 0.08	0.936

ADG, average daily gain; ADFI, average daily feed intake; FCR, feed conversion ratio; BD, basal diet; AFB1, aflatoxin B1 group; SeOX, SeOX MyDA group; CHS, CHS MyDA group; AFB1 + SeOX, aflatoxin B1 plus SeOX group; AFB1 + CHS, aflatoxin B1 plus CHS group.

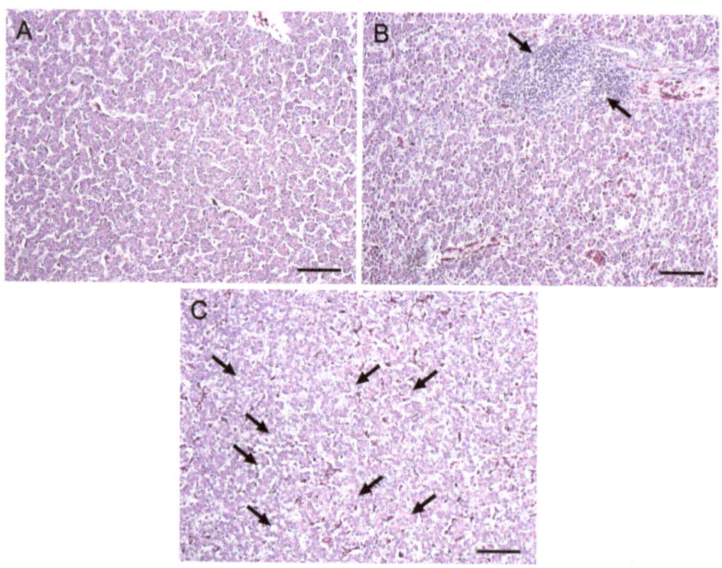

Figure 1. Representative histopathological examination of photomicrographs of broiler livers. (**A**) Absence of lesions from aflatoxin B1 (AFB1) group; (**B**) focal, moderate, and lymphoplasmacytic inflammation (arrows) from basal diet (BD) group; (**C**) mild vacuolar degeneration of hepatocytes (arrows) from AFB1 + SeOX group. Hematoxylin-eosin, 200X; scale bars: 50 μm.

2.2. *Effects of MyDA on the AF Content in Broiler Excreta and on the AF Absorption Rate*

The content of AF (i.e., AFB1, AFB2, AFG1, and AFG2) and their metabolites (i.e., AFM1, AFM2, AFQ1, and AFL) was measured in the broiler excreta collected in the last 4 days of the treatment period by means of an optimized high-performance liquid chromatography with fluorescence detection (HPLC-FLD) method. The average amounts of AFB1 and AFB2 in the different experimental groups are shown in Figure 2. The excreta of both the SeOX and CHS-supplemented groups exhibited a 7- to 9-fold higher AFB1 concentration ($p < 0.001$) compared to the broilers that received the mycotoxin alone (2.3 ± 0.3 ng/g). Moreover, SeOX treatment proved to be slightly more efficient in reducing the absorption of AFB1 than CHS ($p < 0.01$). Despite the minimal amount of AFB2 present in the contaminated feed (1.4 ± 0.2 μg/kg), the toxin was detected in all samples. The excreta collected from the SeOX- or CHS-supplemented group exhibited 5- to 6-fold higher AFB2 values compared with chickens fed the contaminated diet alone (0.3 ± 0.04 ng AFB2/g), with no statistical differences between each MyDA. With regard to the remaining molecules, AFG1, AFG2, AFM2, and AFQ1 were not detected in any of the samples, whereas AFM1 and AFL were present at relatively low levels (below the limit of quantification, LOQ).

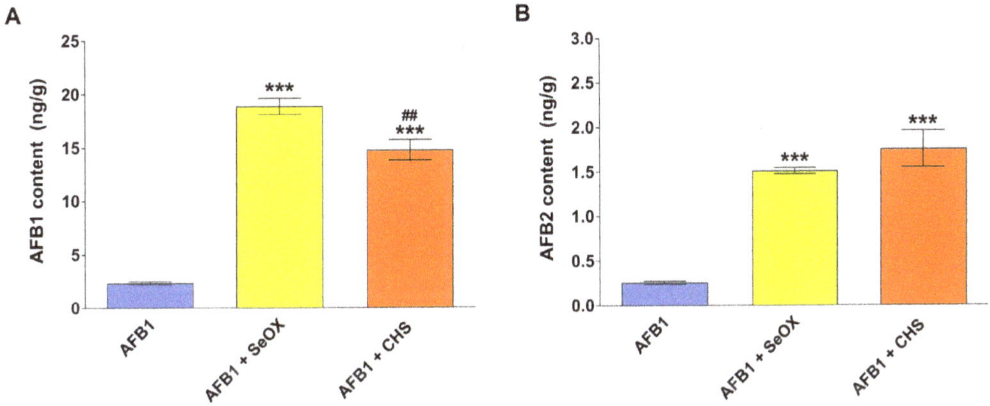

Figure 2. Concentrations of AFB1 (**A**) and AFB2 (**B**) in the excreta of broilers during the last 4 days of treatment. AFB1, aflatoxin B1 group; AFB1 + SeOX, aflatoxin B1 plus SeOX group; AFB1 + CHS, aflatoxin B1 plus CHS group. Each value represents the mean (±SEM) of four replicates analyzed in duplicate (n = 8) (*** $p < 0.001$ vs. AFB1; ## $p < 0.01$ vs. AFB1 + SeOX).

The mycotoxin absorption rate is depicted in Figure 3. In the AFB1 group, there was an extensive absorption rate for both AFB1 and AFB2 (close to 95%), which was significantly reduced (by about 30%, $p< 0.001$) in both AFB1 + SeOX and AFB1 + CHS groups. No statistically significant differences in the ability to reduce the AFB1 absorption rate were detected between SeOX and CHS.

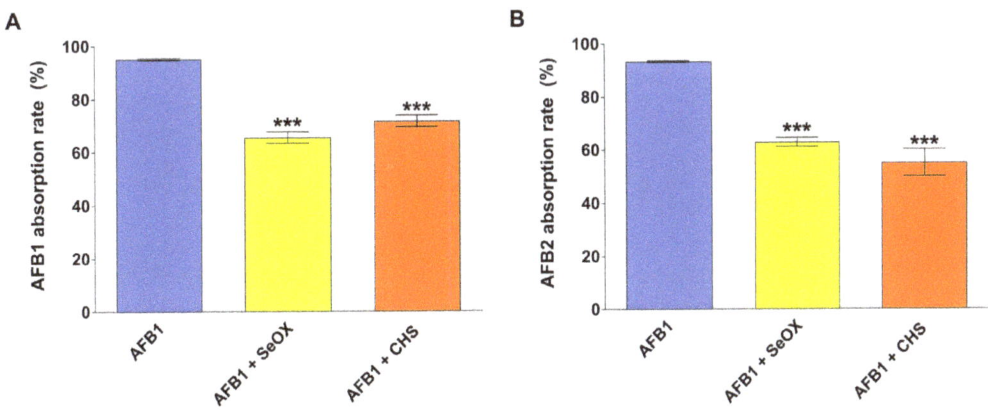

Figure 3. Absorption rate of AFB1 (**A**) and AFB2 (**B**) calculated on the dry matter (DM) and AF concentration of both feed intake and excreta (see details in the M&M section). AFB1, aflatoxin B1 group; AFB1 + SeOX, aflatoxin B1 plus SeOX group; AFB1 + CHS, aflatoxin B1 plus CHS group. Data are expressed as mean ± SEM, n = 8 (*** $p < 0.001$ vs. AFB1).

2.3. Effects of AFB1 and MyDA on Serum Antioxidant Capacity (SAC) and Hepatic Lipid Peroxidation

The measurement of SAC by the OXY test was performed to assess the systemic antioxidant response of broilers to the different treatments. As shown in Figure 4A, a 10-day exposure to AFB1-contaminated feed reduced the SAC by approximately 30% in comparison with the control group ($p < 0.01$). Neither the binder SeOX nor CHS alone affected the SAC of broilers. Conversely, when AFB1 was administered together with either of the two MyDA, the co-treatment induced a statistically significant increase in SAC ($p < 0.001$), reverting the pro-oxidant effect of AFB1 and restoring the antioxidant defense of the birds to control or even higher values (i.e., in the presence of CHS).

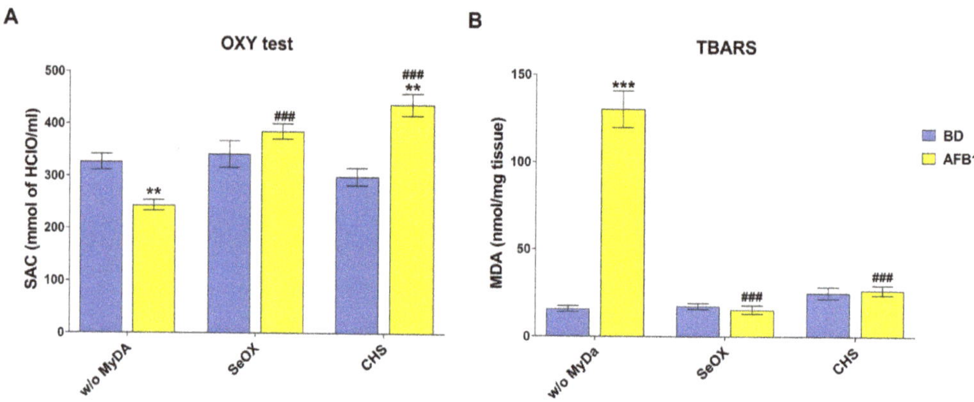

Figure 4. Effects of AFB1 and MyDA on SAC (**A**) and hepatic lipid peroxidation (**B**) in broilers at the end of the treatment (10 days). BD, basal diet; AFB1, BD + aflatoxin B1; w/o MyDA, diet not supplemented with MyDa; SeOX, SeOX-supplemented diet; CHS, CHS-supplemented diet. Data are expressed as mean ± SEM, n = 8 (** $p < 0.01$ vs. BD w/o MyDA; *** $p < 0.001$ vs. BD w/o MyDA; ### $p < 0.001$ vs. BD + AFB1 w/o MyDA).

MDA levels were assessed in liver samples by the Thiobarbituric Reactive Substances (TBARS) test to investigate the lipid peroxidation. An increase in MDA content by approximately 6-fold was observed in AFB1-treated broilers in comparison with the control group (Figure 4B, $p < 0.001$). Neither of the two MyDA alone affected the hepatic MDA levels. In contrast, the administration of AFB1 with either SeOX or CHS resulted in a statistically significant reduction in MDA content to control values, reverting the negative effects of the mycotoxin ($p < 0.001$).

2.4. Effects of AFB1 and MyDA on the Expression of Selected Hepatic Genes

The analysis of mRNA expression in liver was conducted on a panel of genes encoding for (i) enzymes involved in AFB1 metabolism (i.e., CYP1A1, CYP1A2, CYP2A6, CYP2H1, CYP3A4, EPHX1, EPHX2, all isoforms of GSTA, and GSTM2), (ii) the antioxidant defense pathway (i.e., Nrf2, CAT, GPX1, SOD1, and SOD 2), and (iii) drug transporters (DT) responsible for AFB1 efflux (i.e., ABCB1, ABCC2, and ABCG2). While all genes were expressed across all samples, only CYP2A6 and Nrf2 were significantly modulated in response to the treatments (Figure 5). Specifically, Nrf2 was downregulated by approximately 5-fold in the AFB1 group, but the concomitant administration of each MyDA successfully restored Nrf2 expression to control levels (Figure 5A). Conversely, CYP2A6 was upregulated by approximately 2.5-fold by AFB1 in comparison with the control group; however, the co-treatment with either SeOX or CHS was not able to mitigate such effect (Figure 5B).

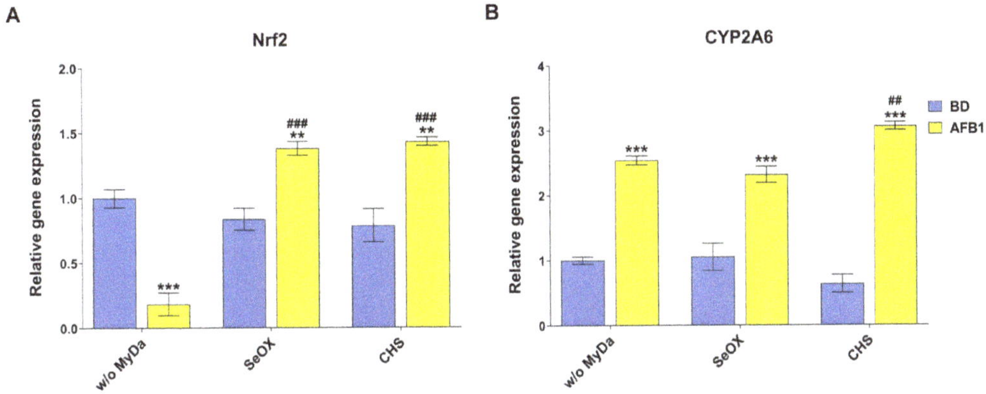

Figure 5. Effects of AFB1 and MyDA on the mRNA expression of Nrf2 (**A**) and CYP2A6 (**B**) in the liver of broilers at the end of the treatment period (10 days), measured using qRT-PCR. BD, basal diet; AFB1, BD + aflatoxin B1; w/o MyDA, diet not supplemented with MyDa; SeOX, SeOX-supplemented diet; CHS, CHS-supplemented diet. Data are expressed as mean ± SEM, n = 8 (** $p < 0.01$ vs. BD w/o MyDA; *** $p < 0.001$ vs. BD w/o MyDA; ## $p < 0.01$ vs. BD + AFB1 w/o MyDA; ### $p < 0.001$ vs. BD + AFB1 w/o MyDA).

2.5. Effects of AFB1 and MyDA on Hepatic GSH Content and Enzymatic Activities

Determination of GSH content and other enzymatic activities (total GST, μ-class GST, α-class GST, and DT-diaphorase) was conducted in liver to investigate the role of AFB1 and the selected MyDA in the modulation of the main detoxification pathways involved in AFB1 metabolism. None of the investigated enzymatic activities and the GSH content, shown in Table 2, were modulated either by AFB1 or by the other treatments.

Table 2. GSH content and enzymatic activities of total GST, GST α-class, GST μ-class, and DT-diaphorase in liver samples from experimental broilers.

Parameter	Treatments					
	BD	**SeOX**	**CHS**	**AFB1**	**AFB1 + SeOX**	**AFB1 + CHS**
GSH (μmol/g liver)	2.23 ± 0.24	2.29 ± 0.16	2.38 ± 0.27	2.15 ± 0.17	2.51 ± 0.13	1.77 ± 0.11
Total GST (nmol/min/mg protein)	504 ± 26.84	595.8 ± 25.7	586 ± 38.82	563 ± 26.28	551 ± 33.48	545.2 ± 25.26
α-class GST (nmol/min/mg protein)	45.1 ± 2.86	40.75 ± 1.04	42.57 ± 1.37	40.3 ± 2.86	43.29 ± 1.8	39.52 ± 1.77
μ-class GST (nmol/min/mg protein)	1.04 ± 0.06	1.43 ± 0.11	1.24 ± 0.15	1.50 ± 0.22	1.19 ± 0.18	1.14 ± 0.17
DT-diaphorase (nmol/min/mg protein)	367.8 ± 14.69	390.4 ± 6.42	404 ± 13.56	371.4 ± 18.71	396.5 ± 23.01	376.2 ± 16.88

BD, basal diet; AFB1, aflatoxin B1 group; SeOX, SeOX MyDA group; CHS, CHS MyDA group; AFB1 + SeOX, aflatoxin B1 plus SeOX group; AFB1 + CHS, aflatoxin B1 plus CHS group. Data are expressed as mean ± SEM, n = 8.

3. Discussion

In the poultry industry, MyDA are widely used to mitigate the negative effects of mycotoxin feed contamination. These agents often trap mycotoxins in the gastrointestinal tract, thereby reducing their bio-accessibility and preventing or limiting their toxic effects. AFB1 has been demonstrated to induce OS and other adverse effects in the liver and kidney of broilers exposed to levels around the EU regulatory limit [9,10]. Most published studies have investigated the efficacy of MyDA using much higher dietary concentrations of AFB1, typically ranging from 0.1 to 10 mg/kg [26]. According to the European legislative framework, the abovementioned scenario is unrealistic since the use of feed with such AFB1 contamination is not allowed in routine poultry farming. Therefore, this study aimed to investigate the efficacy of two innovative MyDA in limiting both the bio-accessibility and the adverse effects of AFB1 in broilers exposed to concentrations approaching the EU maximum limits for chicken feed.

Smectite-based MyDA have long been extensively employed to promote AFB1 sequestration in animal feed [30]. The efficacy of SeOX as a multi-mycotoxin adsorbing agent (i.e., AFB1, fumonisin B1, ochratoxin A, zearalenone, and T-2 toxin) has been previously demonstrated through a series of in vitro studies including adsorption isotherms. The additive proved effective to almost completely adsorb AFB1 at both gastric and intestinal pH values [28]. Likewise, the CHS MyDA at low concentrations (0.25–0.5% w/v) successfully sequestered in vitro more than 95% of AFB1 and other mycotoxins in a large range of pH values (3–9) (unpublished data).

As previously reported [9,31,32], the schedule of AFB1 administration used in this trial (0.02 mg/kg diet for 10 days) neither caused mortality nor affected zootechnical performances and diet digestibility. In addition, it did not induce any microscopical changes in hepatic tissue [10] despite the occurrence of OS as evidenced by the increase in liver TBARS. To the best of our knowledge, the treatment protocol that can induce initial histological changes in chicken liver is 0.05 mg/kg diet of AFB1 for 28 days [7]. Nevertheless, it is conceivable that the OS generated by the low doses used in this study may result in overt histological changes upon a longer exposure [33]. It is also noteworthy that neither SeOX nor CHS alone (0.5% dietary supplementation) affected the growth performances or health of broilers, in line with a substantial body of data indicating that smectite-based MyDA are free of negative effects in poultry when applied at dietary concentrations like those used in this study [30].

Most of the available literature on the efficacy of binders focuses on their ability to mitigate the adverse effects (endpoints) caused by the mycotoxins. For AFB1, the most common endpoint is the impairment of growth performances [30], which in our study was not affected by the treatment schedule. Another method to demonstrate the in vivo efficacy of a MyDA acting as a binder is the measurement of the mycotoxin concentration in the excreta in the absence or presence of the additive [34]. In a trial where chickens were fed a diet containing AFB1 at concentrations similar to those used in our study (0.018 mg/kg), the AFB1 and AFB2 content in the excreta after 14 days was superimposable to our study, amounting to 2.3 ± 1.6 ng/g and 0.2 ± 0.1 ng/g, respectively [35]. The significant increase in the content of both AFB1 (7- to 9-fold) and AFB2 (5- to 6-fold) that we observed in the excreta from SeOX- or CHS-supplemented chickens aligns with the sequestration of the AF in the gastrointestinal tract promoted by both additives. When administered at low levels to chickens, AFB1 is extensively bio-transformed into several more polar metabolites (e.g., AFM1, AFBQ1, AFBP1, and AFL), which, after undergoing phase II metabolism, may be excreted via the biliary and urinary routes [36]. After absorption, AF cannot be excreted as such in avian and mammalian species unless they undergo both oxidative and conjugative metabolism [37]. Therefore, it may be assumed that the amount of AFB1/AFB2 recovered in the excreta likely represents the unabsorbed toxins. Based on these data, the supplementation of SeOX or CHS at inclusion levels of 0.5% led to a decrease in AFB1/AFB2 bio-accessibility up to 30%, promoting their elimination via the excreta. Our findings are difficult to compare with the existing literature due to limited available data. For instance, a mixture of bentonite and sepiolite fed to chicks at an inclusion level of 0.1% resulted in a two-fold increase in the content of AFB1 in the excreta with respect to chicks exposed to the mycotoxin alone (0.01 mg/kg diet) [34].

Only sparse information is available on the effects of MyDA on AFB1-mediated OS. As regards SAC, the increase to control values elicited by either SeOX or CHS in the AFB1 co-treated groups is similar to the findings of Khanian and colleagues [38], who described the same effect using a *Lactobacillus plantarum 299v* probiotic strain in broilers fed a 0.2 or 2 mg AFB1/kg contaminated diet for 42 days. The use of natural antioxidants to restore SAC was also successfully exploited by turmeric powder in broilers exposed to 0.02 mg AFB1/kg feed for 10 days [9] or by vitamin E [26]; in the latter case, the restoration of blood oxidative resistance in laying hens treated for 3 weeks with increasing doses of AFB1 (from 0.1 to 10 mg AFB1/kg diet) occurred only at AFB1 concentrations lower than 3 mg/kg.

Further protective effects of SeOX and CHS against AFB1-mediated OS were also observed for the TBARS assay, with the hepatic MDA levels being comparable to control values. Liver MDA values similar to controls were also documented in broilers receiving a diet containing 0.25 mg/kg of AFB1 for 42 days, supplemented with a mixture of aluminosilicates and glucomannans [39]. The increase in MDA content detected in the blood and kidney but not in the liver of broilers treated with diets containing 0.1 mg/kg AFB1 for 21 days was fully counteracted by the addition of zeolite [20].

For a more comprehensive understanding of the interactions between AFB1 and the selected MyDA, a gene expression analysis was conducted on genes encoding for enzymes involved in AFB1 bioactivation/detoxification but also related to OS modulation and drug transport. In the current study, none of the investigated genes (Table S1) were modulated by any of the different treatments except for CYP2A6 and Nrf2. As regards CYPs, most of the published literature on avian species focused on the effects of AFB1 and MyDA on liver CYP1A1, CYP1A2, CYP2H1, and CYP3A4. Studies in chickens and ducks generally suggest an AFB1-mediated downregulation of these CYPs, which is partially restored upon MyDA supplementation [15,39,40]. Although these CYPs contribute to some extent to AFB1 oxidative bio-transformations, CYP2A6 is consistently identified as the key enzyme involved in AFBO generation in chickens, which leads to pro-oxidant and cytotoxic effects [11]. Despite the low AFB1 exposure level and the short treatment duration, we observed an almost 2.5-fold upregulation of CYP2A6, in line with other studies performed with the administration of higher AFB1 dosages. For instance, the inclusion of either 0.5 mg/kg or 5 mg/kg of AFB1 in broiler feed for 21 or 28 days increased CYP2A6 expression by approximately 1.5- and 2.5-fold, respectively [41,42]. The AFB1-mediated increased expression of liver CYP2A6 is expected to enhance the generation of AFBO, triggering an overall worsening of the adverse effects in exposed birds. Contrary to what was observed for other parameters, the supplementation of either SeOX or CHS failed to counteract the increase in hepatic CYP2A6 expression caused by AFB1. These results are difficult to interpret and suggest that liver CYP2A6 induction is one of the most sensitive biomarkers of AFB1 exposure in chickens. It is worth noting that turmeric powder, curcumin, and curcuminoids were effective in counteracting the AFB1-related increase in hepatic CYP2A6 in several in vivo experiments performed in chickens [42–44].

Nrf2, a key transcription factor regulating the cell response to OS, underwent downregulation by AFB1, which is considered a feature of the pro-oxidant mechanisms displayed by the mycotoxin in different species, tissues, and cell systems [45]. However, in our study, the inclusion of each mycotoxin binder in the AFB1-contaminated diet resulted in an increase in Nrf2 expression, with levels even slightly higher than those observed in control animals. For comparison, the hepatic Nrf2 downregulation detected in broilers treated with 0.1 mg

AFB1/kg diet for 14 days was not reversed by the addition of zeolite [20]. It is noteworthy that when SeOX or CHS were added alone to the control diet, they failed to modulate Nrf2 expression, pointing to a lack of systemic effects of the tested MyDA. In our study, the AFB1-mediated Nrf2 downregulation was not associated with a modulation of OS genes (i.e., SOD1, SOD2, CAT, or GPX1), as previously documented in chickens treated with higher doses of AFB1 (0.33 mg/kg for 21 days [46], 0.75 mg/kg for 30 days [47], or 1 mg/kg for 21 days [48]) According to the published literature, the consequences of Nrf2 downregulation in chickens are not yet fully understood. Given the complexity of the Nrf2 pathway, alternative genes may be regulated instead of the major ones commonly investigated, such as SOD, CAT, and GPX [49].

GST enzymes, especially those belonging to the α- and μ-class families, are responsible for AFB1 detoxification through the conjugation of AFBO with GSH [11]. An additional enzyme involved in the regulation of AFB1-mediated OS is DT-diaphorase, which has been shown to be downregulated by AFB1 in the liver of various animal species, including chickens, ducks, and rabbits [50–52]. Limited data exist on the hepatic modulation of GSTs and DT-diaphorase activities in response to low AFB1 exposure [31], and there is scant information on how MyDA affect these parameters. In our study, liver GSH content, GST activities related to α- and μ-class families, and DT-diaphorase activities were not modulated in broilers exposed to 0.02 mg AFB1/kg diet for 10 days. None of the above parameters were affected by either SeOX and CHS alone or in combination with AFB1. Our findings are consistent with previous investigations, which also found that AFB1 (0.1 mg/kg diet for 21 days) did not change hepatic GSH content and GSH-peroxidase activity in broilers even in the presence of zeolite [20].

In our study, the observed increase in liver TBARS did not coincide with a reduction in GSH content. Although these results are difficult to interpret, controversial effects of AFB1 on liver GSH and TBARS have been reported in the limited studies conducted on chickens. The increase in liver TBARS content was not matched by a decrease in GSH content, which remained unaltered, in broilers exposed to 17 μg AFB1/kg diet for 21 days [20]. In another study, the treatment of broilers with higher AFB1 dietary concentrations (100 μg/kg diet for 14 days) resulted in a limited decrease in hepatic GSH content (−30%) along with an increase in TBARS [44]. In contrast, neither liver TBARS nor liver GSH were altered by the repeated exposure of broilers even to higher dosages (150 μg/kg diet for 14 days) [53].

A number of potential limitations of this study should be mentioned. In some instances (CYPs, Nrf2, and some antioxidant enzymes), only the effects on gene expression were investigated. However, it should be noted that the main aim of this study was to assess the ability of two different MyDA to reduce the bio-accessibility and the enteric uptake of AF (as it was the case), thereby positively affecting the expression of target genes known as biomarkers of AFB1 toxicity. Secondly, the trial was conducted using a single dietary concentration of both AF and MyDA. It might be interesting, for instance, to test the efficacy of lower SeOX or CHS concentrations in the presence of the same AF dietary levels used in this study; alternatively, the binding ability of the additives might be assessed with AF concentrations in feed higher than those permitted in the EU, as is the case for most maize samples from many African countries [3].

4. Conclusions

The dietary exposure of chickens to the EU maximum allowed AF concentration in poultry feed (0.02 mg/kg) for 10 days did not affect either the zootechnical performances or the histological morphology of liver. However, the treatment elicited systemic and hepatic OS as well as pro-oxidant changes in liver (i.e., increase in CYP2A6 and decrease in Nrf2 gene expression). The dietary inclusion (0.5%) of either SeOX or CHS, two novel smectite-based MyDA, was able to mitigate or even reverse most of the AFB1-mediated adverse effects mentioned above, conceivably by decreasing the bio-accessibility of the mycotoxin up to 30%. Interestingly, when administered alone, no negative effects were detected on zootechnical or any of the other tested endpoints. Further research is warranted to investigate the efficacy of SeOX or CHS supplementation in the presence of higher dietary AF concentrations and to assess possible interferences with the supply of nutrients, micronutrients, or feed additives [54].

5. Materials and Methods

5.1. In Vivo Trial and Samples Collection

The in vivo trial was authorized by the Institutional Animal Care and Ethic Committee of the University of Turin, Italy (Approval number = 319508/2017-PR), and performed according to Directive 2007/43/EC in the farm facilities of the Department of Veterinary Sciences in Turin, as previously described [9,10]. In brief, a total of 48 male, 18-day-old broiler chickens (ROSS 308, average weight 752 ± 26 g) were randomly allocated to six experimental groups (n = 8 animals/group; 2 birds per cage): basal diet (BD) group; AFB1 group (BD + 0.02 mg/kg feed AFB1); SeOX group (BD + 0.5% SeOX); CHS group (BD + 0.5% CHS); AFB1 + SeOX group (BD + 0.02 mg/kg feed AFB1 + 0.5% SeOX); and AFB1 + CHS group (BD + 0.02 mg/kg feed AFB1 + 0.5% CHS). Details on the preparation of the AFB1-contaminated feed can be found in Damiano et al. (2022) [9]. Briefly, a BD containing 5.0 ± 1.3 μg/kg of AFB1 and 0.9 ± 0.1 μg/kg of AFB2 (mean ± SD, n = 3) was artificially contaminated to reach 18.8 ± 3.7 μg/kg of AFB1 and 1.4 ± 0.2 μg/kg of AFB2 (mean ± SD, n = 5). The different diets were administered for 10 consecutive days, from 23 to 32 days of age. Water and feed were provided ad libitum, and overall growth performances were recorded. During the last 4 days of the trial, 24 h excreta samples were collected from each cage, according to Schiavone et al. [55]. Then, the excreta were cleaned from feathers and any other foreign matter and weighed. A single sample was prepared for each replicate (n = 2 animals) by pooling representative aliquots (20%) of the daily excreted weight over the last 4 days. Subsequently, a representative sample of the entire pool (cage) was frozen and freeze-dried before determining the toxin content. At the end of the treatment (day 10), blood from each bird was sampled from the brachial vein. Then, the animals were euthanatized by sodium pentobarbital overdosage, and liver samples were collected from each broiler, cut into aliquots and properly stored for the subsequent analyses. For the histological investigation, specimens were fixed in a 10% buffered formalin solution; for the determination of GSH content and lipid peroxidation and for the enzymatic activities assays, they were immediately frozen in liquid nitrogen and then stored at −80 °C. For the gene expression study, the samples were kept in RNAlater® stabilization solution (Sigma-Aldrich, Milan, Italy) for 24 h and then stored at −80 °C until analysis.

5.2. DM Digestibility and Mycotoxin Absorption Rate Calculation

The DM digestibility was calculated according to the standard protocol reported by Schiavone et al. [55] as given below:

$$\text{DM digestibility (\%)} = (\text{DM feed intake} - \text{DM excreta})/\text{DM feed intake} \times 100$$

Then, the mycotoxin absorption rate was calculated as follows:

$$\text{Mycotoxin absorption rate (\%)} = [(\text{DM feed intake} \times \text{Mycotoxin}_{\text{feed}}) - (\text{DM excreta} \times \text{Mycotoxin}_{\text{excreta}})]/(\text{DM feed intake} \times \text{Mycotoxin}_{\text{feed}}) \times 100$$

where $Mycotoxin_{feed}$ and $Mycotoxin_{exreta}$ represent the mycotoxin (AFB1 or AFB2) concentration (ng/g DM) in feed and excreta, respectively.

5.3. Histopathological Examination

The formalin-fixed liver samples were embedded in paraffin according to the standard protocol, cut into sections of 5 μm thickness, and stained with traditional hematoxylin–eosin dye. Two sections from each chick were blinded-analyzed by two different pathologists under a light microscope.

5.4. Chemical Analysis of AF and Their Metabolites in Broiler Excreta

Unless mentioned otherwise, all reagents were of analytical grade. Acetonitrile and methanol were purchased from Carlo Erba Reagents srl (Milan, Italy). The water was of Milli-Q® quality (Millipore, Bedford, MA, USA). The β-glucuronidase/aryl sulfatase enzyme solution (30/60 U/mL) was acquired from Merck KGaA (Darmstadt, Germany). AF standards in powder form were provided by Fermentek Ltd. (Jerusalem, Israel), with the exception of AFQ1 that was supplied by Cfm Oskar Tropitzsch GmbH (Marketredwitz, Germany). Stock solutions of each standard (1 μg/mL in acetonitrile) were kept in the dark at 4 °C. AflaTest™ WB immunoaffinity columns (IMA) were supplied by VICAM (Waters Corporation, Milford, MA, USA).

The method to analyze AF in broiler excreta was optimized and validated in-house. For the optimization study, "blank" chicken excreta, collected from animals fed AF-free diets, were artificially spiked with a known amount of AF and analyzed using various extraction procedures. These procedures differed in specific operational conditions, including (i) type and volume of solvents, (ii) extraction mode (blending and/or shaking), and (iii) extraction time. Furthermore, the clean-up process using immunoaffinity columns was enhanced by testing different types and volumes of washing solvents. The combination of operational parameters that provided the highest extraction yield was selected as the definitive method for determining AF in chicken excreta. This optimized method was validated by assessing the following parameters: selectivity, intermediate precision, recovery, linearity, limit of detection (LOD), and limit of quantification (LOQ). Under optimized experimental conditions, 5 g of lyophilized excreta were mixed with 1 g of NaCl and 50 mL of an extraction solvent consisting of acetonitrile and water in a 70:30 (*v/v*) ratio. The suspension was homogenized for 2 min using a blender and then shaken (170 rpm) at room temperature for 1 h. After AF extraction by shaking, the suspension was filtered through paper (Whatman® Grade 4, Maidstone, UK), diluted 10 times with ammonium acetate buffer (0.1 M, pH 4.8), and incubated overnight with 25 μL of the β-glucuronidase/arylsulfatase (30/60 U/mL) solution. The extract was filtered again using glass microfiber filters (Whatman® Grade GF/A) and cleaned up (20 mL) using AflaTest® WB columns. Following the extract elution (approximatively 1 drop per second), a washing step was performed by 30 mL of PBS to remove impurities. AF were eluted from the columns with 2 mL of methanol and collected into a 4 mL amber silanized vial. The cleaned-up sample was concentrated by evaporation under an air stream at 40° C, and the residual AF was reconstituted with 1 mL of a mixture containing methanol/water (50:50, *v*/*v*). A volume of 100 μL was injected into a HPLC-FLD system (Agilent 1100 series coupled with the UVE™ system for post-column photochemical derivatization) for the detection of AF according to the method described by Amminikutty et al. [10]. The AF content is expressed as ng/g excreta. The validation of the method was performed according to Malachová et al. [56]. The linearity of the method was evaluated by analyzing five standard calibrants (prepared in methanol/water, 50/50, *v/v*) containing the eight AF (i.e., AFB1, AFB2, AFG1, AFG2, AFM1, AFM2, AFQ1, and AFL) at concentrations ranging from 0.1 to 25 ng/mL, which correspond to the contamination levels of poultry excreta from 0.5 to 125 ng/g. Matrix-matched calibrants were prepared by spiking the methanolic eluate of uncontaminated excreta samples with different amount of AF stock solutions according to the desired concentration. Standard and matrix-matched calibrants (both analyzed in triplicate) were used to prepare the standard and the matrix-matched calibration curves, respectively. The two calibration curves were used to calculate the signal suppression/enhancement (SSE%) as described by Amminikutty et al. [10]. As shown in Table S2 the matrix-matched calibration curves showed good determination coefficients ($R^2 > 0.998$), although the goodness of fit (linearity) was assessed by the lack-of-fit test ($p > 0.05$). In addition, SEE% values between 92% and 102% confirmed the efficacy of the optimized protocol in separating interfering matrix compounds from AF. The LOD and the LOQ, expressed as ng AF/g excreta, were evaluated as a signal-to-noise ratio (S/N) equal to 3 and 10, respectively (Table S3). The low LOQ values indicated that the optimized method, in combination with the use of sensitive HPLC equipment, was highly sensitive in detecting AF in a complex matrix such as broiler excreta. Recovery values were determined by spiking uncontaminated fecal samples at 5 and 25 ng/g. The recovery and precision values (RSD%) were assessed by analyzing three replicates at each spiking level. Recovery values ranging from 101% to 112% were considered acceptable (Table S4). The calculated RSD% values were in the range of acceptability at ≤6% (EC Regulation 519/2014).

5.5. SAC Measurement

The SAC was assessed by the OXY-Adsorbent test from Diacron (Grosseto, Italy), according to the modified method previously described [57]. Briefly, 2 µL of serum (1:100 (*v*/*v*) dilution in MilliQ water) was incubated with 200 µL of HClO solution for 10 min at 37 °C in a 96-well plate. Then, 2 µL of N, N-diethyl-p-phenylenediamine was added, and the absorbance was read at 505 nm by a spectrophotometer. The intensity of the chromogenic complex is negatively correlated to the SAC. Data are expressed as mmol of HClO per mL.

5.6. TBARS Assay

The hepatic lipid peroxidation was determined by a modified version of the TBARS test [58]. Briefly, liver homogenates were obtained by disrupting the samples through the Tissue Lyzer LT (Qiagen, Hilden, Germany) for 5 min at 50 Hz in a buffer comprising 200 µL of NaCl (0.9% *w*/*v*), 200 µL of TCA (10% *w*/*v*), and 4 µL of butylated hydroxytoluene (BHT) (2% *w/v*). After centrifugation at 13,000× *g* for 15 min at 4 °C, supernatants, blanks, and/or MDA standard solutions were mixed with 400 µL of TBA solution (15% TCA (*w*/*v*) in glacial acetic acid, 0.38% TBA (*w*/*v*), and 0.25 N hydrochloric acid) and vortexed for 1 min. The development of MDA-TBA adducts was obtained by immersing the samples in a 95 °C water bath for 60 min and then cooling them in a cold-water bath for 10 min. After centrifugation at 13,000× g for 15 min, the absorbance of supernatants was read at 532 nm. All samples, blanks, and the MDA standard curve were analyzed in triplicate. Data are expressed as nmol of MDA per mg of tissue.

5.7. RNA Extraction and qRT-PCR

The gene expression analysis was performed as reported by Amminikutty et al. [10]. Briefly, total hepatic RNA was extracted using the Maxwell RSC simplyRNA Kit (Promega, Madison, WI, USA) following to the manufacturer's protocol, and RNA quantitation was obtained through the NanoDrop ND-2000 UV-Vis spectrophotometer (Thermo Fisher Scientific, Waltham, MA, USA). To check RNA purity, the 260 and 280 nm ratio was measured, and all analyzed samples had values > 1.9. An automated electrophoresis station (Experion Instrument, Bio-Rad, Hercules, CA, USA) was used to verify the RNA Integrity Number (RIN), which was >7 for all samples. Next, reverse transcription of 1 µg of total RNA from each sample in a final volume of 20 µL was performed through the iScriptTM cDNA Synthesis Kit (Bio-Rad, Hercules, CA, USA) following the manufacturer's instructions in the GeneAmp PCR System 9700 (Perkin Elmer, Waltham, MA, USA). The primers were picked on the *Gallus gallus* GenBank database and verified on the Ensembl mRNA sequences using Primer3 software (version 3.0, Applied Biosystems, Foster City, CA, USA). To avoid the amplification of contaminant genomic DNA, primers were selected at the exon/exon boundaries and tested for hairpin and dimer formation with the NetPrimer tool (available at http://www.premierbiosoft.com/netprimer/index.html, access date: 22 July 2024). Finally, the primers were validated through the BLAST tool (Basic local alignment search tool; https://blast.ncbi.nlm.nih.gov/Blast.cgi, access date: 17 June 2024). All primer information is reported in Table S1. PGK2 and RPS7 were used as housekeeping genes (HKG) according to Amminikutty et al. [10]. qRT-PCR was performed through the Bio-Rad CFX Opus 96 (Bio-Rad) at the following conditions: 30 s at 95 °C, 40 cycles of 5 s at 95 °C, and 30 s at 60 °C. Intra-assay variability was reduced by running each reaction in triplicate; water instead of cDNA was used as a blank control. An inter-run calibrator comprising a pool of cDNAs from control samples was included to check the inter-assay variation. Gene expression results were analyzed using the CFX Maestro software (Bio-Rad, version 5.2.008.0222), and data are expressed as relative mRNA quantity.

5.8. Total GSH Content Determination and Enzymatic Activity Assays

The determination of total GSH content and enzymatic activities was performed on hepatic cytosolic fractions obtained by differential ultracentrifugation [59]. Bovine serum albumin was used as reference standard to quantify protein content [60]. GSH was measured on TCA deproteinized samples using the Elmann's reagent as previously reported [61], and the results are expressed as µmol of GSH per g of liver. Total GST, GST µ-class, and GST α-class activities were tested by monitoring the continuous increase at $\lambda = 340$ nm of the GSH conjugates of 1 mM 1-chloro, 2,4-dinitrobenzene (CDNB), 1 mM 3,4-dichloronitrobenzene (DCNB) [62,63], and tert-butyl hydroperoxide 1.5 mM [64], respectively. The activity of DT-diaphorase was measured following the continuous reduction of 2,6-dichloroindophenol at 600 nm [65]. All assays were performed under Vmax conditions and were linear in terms of protein concentration and time. The results are expressed as nmol/min per mg of protein.

5.9. Statistical Analysis

Data are expressed as mean ± standard error of the mean (SEM). Statistical analysis was run on the GraphPad Prism 7.03 software (GraphPad Software, San Diego, CA, USA). The D'Agostino and Pearson normality omnibus test was applied to verify the normal distribution of the data (n = 8 for each experimental group). The one-way or the two-way analysis of variance (ANOVA), followed by the Tukey's post hoc test, was used to check the significant differences among groups. The one-way ANOVA was applied to the concentrations in the excreta and the calculated absorption rate of AFBs. The other parameters were analyzed through the two-way ANOVA, considering the exposure to the mycotoxin (AFB1) and the presence/absence of the MyDA as the factors. A two-sided p-value of less than 0.05 was considered statistically significant.

Supplementary Materials: The following supporting information can be downloaded at https://www.mdpi.com/article/10.3390/toxins17020082/s1. Table S1: Primers for quantitative real-time PCR (qRT-PCR); Table S2: Matrix-matched regression curves and related R2 and SSE (%) values; Table S3: Limit of detection (LOD, S/N = 3) and limit of quantification (LOQ, S/N = 10) calculated for AFB1, AFB2, AFG1, AFG2, AFM1, AFM2, AFQ1, and AFL in excreta samples; Table S4: Recoveries of AFB1,AFB2, AFG1, AFG2, AFM1, AFM2, AFQ1, and AFL in excreta samples spiked at two different concentration levels (5 and 25 ng/g).

Author Contributions: Conceptualization, G.A., C.N. and F.G. (Flavia Girolami); methodology, A.S., F.G. (Francesco Gai), D.G. and V.D.; validation, M.C., N.A., V.S., D.G. and V.D.; formal analysis, V.S. and F.G. (Flavia Girolami); investigation, M.C., N.A., V.S., S.B., A.S., V.C., A.S., D.G. and V.D.; resources, A.S., C.N. and F.G. (Flavia Girolami); writing—original draft preparation, M.C., A.S., D.G., V.D., G.A., C.N. and F.G. (Flavia Girolami); writing—review and editing, C.N., G.A. and F.G. (Flavia Girolami); visualization, F.G. (Flavia Girolami); supervision, G.A., C.N. and F.G. (Flavia Girolami); funding acquisition, A.S. and G.A. All authors have read and agreed to the published version of the manuscript.

Funding: This research was funded by H2020-EU.3.2. "Integrated and innovative key actions for mycotoxin management in the food and feed chain" (MycoKey) (ID n. 678781).

Institutional Review Board Statement: The animal study protocol was approved by the Institutional Animal Care and Ethic Committee of the University of Turin (Approval number = 319508/2017-PR; date of approval: 21 September 2017).

Informed Consent Statement: Not applicable.

Data Availability Statement: The original contributions presented in the study are included in the article/Supplementary Material, and further inquiries can be directed to the corresponding author/s.

Conflicts of Interest: The authors declare no conflicts of interest.

References

1. Caceres, I.; Khoury, A.A.; Khoury, R.E.; Lorber, S.; Oswald, I.P.; Khoury, A.E.; Atoui, A.; Puel, O.; Bailly, J.-D. Aflatoxin Biosynthesis and Genetic Regulation: A Review. *Toxins* **2020**, *12*, 150. [CrossRef] [PubMed]
2. Mughal, M.J.; Peng, X.; Kamboh, A.A.; Zhou, Y.; Fang, J. Aflatoxin B1 Induced Systemic Toxicity in Poultry and Rescue Effects of Selenium and Zinc. *Biol. Trace Elem. Res.* **2017**, *178*, 292–300. [CrossRef] [PubMed]
3. Ochieng, P.E.; Scippo, M.-L.; Kemboi, D.C.; Croubels, S.; Okoth, S.; Kang'ethe, E.K.; Doupovec, B.; Gathumbi, J.K.; Lindahl, J.F.; Antonissen, G. Mycotoxins in Poultry Feed and Feed Ingredients from Sub-Saharan Africa and Their Impact on the Production of Broiler and Layer Chickens: A Review. *Toxins* **2021**, *13*, 633. [CrossRef]
4. Shabeer, S.; Asad, S.; Jamal, A.; Ali, A. Aflatoxin Contamination, Its Impact and Management Strategies: An Updated Review. *Toxins* **2022**, *14*, 307. [CrossRef] [PubMed]
5. Guerre, P. Worldwide Mycotoxins Exposure in Pig and Poultry Feed Formulations. *Toxins* **2016**, *8*, 350. [CrossRef]
6. Rawal, S.; Kim, J.E.; Coulombe, R. Aflatoxin B1 in Poultry: Toxicology, Metabolism and Prevention. *Res. Vet. Sci.* **2010**, *89*, 325–331. [CrossRef] [PubMed]
7. Magnoli, A.P.; Monge, M.P.; Miazzo, R.D.; Cavaglieri, L.R.; Magnoli, C.E.; Merkis, C.I.; Cristofolini, A.L.; Dalcero, A.M.; Chiacchiera, S.M. Effect of low levels of aflatoxin B_1 on performance, biochemical parameters, and aflatoxin B_1 in broiler liver tissues in the presence of monensin and sodium bentonite. *Poult. Sci.* **2011**, *90*, 48–58. [CrossRef] [PubMed]
8. Lin, L.; Fu, P.; Chen, N.; Gao, N.; Cao, Q.; Yue, K.; Xu, T.; Zhang, C.; Zhang, C.; Liu, F.; et al. Total Flavonoids of *Rhizoma Drynariae* Protect Hepatocytes against Aflatoxin B1-Induced Oxidative Stress and Apoptosis in Broiler Chickens. *Ecotoxicol. Environ. Saf.* **2022**, *230*, 113148. [CrossRef] [PubMed]
9. Damiano, S.; Jarriyawattanachaikul, W.; Girolami, F.; Longobardi, C.; Nebbia, C.; Andretta, E.; Lauritano, C.; Dabbou, S.; Avantaggiato, G.; Schiavone, A.; et al. Curcumin Supplementation Protects Broiler Chickens Against the Renal Oxidative Stress Induced by the Dietary Exposure to Low Levels of Aflatoxin B1. *Front. Vet. Sci.* **2022**, *8*, 822227. [CrossRef] [PubMed]
10. Amminikutty, N.; Spalenza, V.; Jarriyawattanachaikul, W.; Badino, P.; Capucchio, M.T.; Colombino, E.; Schiavone, A.; Greco, D.; D'Ascanio, V.; Avantaggiato, G.; et al. Turmeric Powder Counteracts Oxidative Stress and Reduces AFB1 Content in the Liver of Broilers Exposed to the EU Maximum Levels of the Mycotoxin. *Toxins* **2023**, *15*, 687. [CrossRef] [PubMed]
11. Deng, J.; Zhao, L.; Zhang, N.-Y.; Karrow, N.A.; Krumm, C.S.; Qi, D.-S.; Sun, L.-H. Aflatoxin B1 Metabolism: Regulation by Phase I and II Metabolizing Enzymes and Chemoprotective Agents. *Mutat. Res. Rev. Mutat. Res.* **2018**, *778*, 79–89. [CrossRef] [PubMed]
12. Stoev, S.D. Natural Feed Additives and Bioactive Supplements versus Chemical Additives as a Safe and Practical Approach to Combat Foodborne Mycotoxicoses. *Front. Nutr.* **2024**, *11*, 1335779. [CrossRef] [PubMed]
13. Boudergue, C.; Burel, C.; Dragacci, S.; Favrot, M.-C.; Fremy, J.-M.; Massimi, C.; Prigent, P.; Debongnie, P.; Pussemier, L.; Boudra, H.; et al. Review of Mycotoxin-Detoxifying Agents Used as Feed Additives: Mode of Action, Efficacy and Feed/Food Safety. *EFSA Support. Publ.* **2009**, *6*, 22E. [CrossRef]

14. Ruan, M.-L.; Wang, J.; Xia, Z.-Y.; Li, X.-W.; Zhang, B.; Wang, G.-L.; Wu, Y.-Y.; Han, Y.; Deng, J.; Sun, L.-H. An Integrated Mycotoxin-Mitigating Agent Can Effectively Mitigate the Combined Toxicity of AFB1, DON and OTA on the Production Performance, Liver and Oviduct Health in Broiler Breeder Hens. *Food Chem. Toxicol.* **2023**, *182*, 114159. [CrossRef] [PubMed]
15. Lai, Y.; Sun, M.; He, Y.; Lei, J.; Han, Y.; Wu, Y.; Bai, D.; Guo, Y.; Zhang, B. Mycotoxins Binder Supplementation Alleviates Aflatoxin B1 Toxic Effects on the Immune Response and Intestinal Barrier Function in Broilers. *Poult. Sci.* **2022**, *101*, 101683. [CrossRef] [PubMed]
16. Mesgar, A.; Aghdam Shahryar, H.; Bailey, C.A.; Ebrahimnezhad, Y.; Mohan, A. Effect of Dietary L-Threonine and Toxin Binder on Performance, Blood Parameters, and Immune Response of Broilers Exposed to Aflatoxin B1. *Toxins* **2022**, *14*, 192. [CrossRef]
17. Zabiulla, I.; Malathi, V.; Swamy, H.V.L.N.; Naik, J.; Pineda, L.; Han, Y. The Efficacy of a Smectite-Based Mycotoxin Binder in Reducing Aflatoxin B1 Toxicity on Performance, Health and Histopathology of Broiler Chickens. *Toxins* **2021**, *13*, 856. [CrossRef]
18. Ochieng, P.E.; Croubels, S.; Kemboi, D.; Okoth, S.; De Baere, S.; Cavalier, E.; Kang'ethe, E.; Faas, J.; Doupovec, B.; Gathumbi, J.; et al. Effects of Aflatoxins and Fumonisins, Alone or in Combination, on Performance, Health, and Safety of Food Products of Broiler Chickens, and Mitigation Efficacy of Bentonite and Fumonisin Esterase. *J. Agric. Food Chem.* **2023**, *71*, 13462–13473. [CrossRef] [PubMed]
19. Nasiri Poroj, S.; Fazeli, M.R.; Larypoor, M.; Shariatmadari, F. Developing a New Biologic Toxin Binder for Reducing AFB1 Toxicity in Laying Hens. *Lett. Appl. Microbiol.* **2023**, *76*, ovad001. [CrossRef]
20. Kövesi, B.; Kulcsár, S.; Cserháti, M.; Erdélyi, M.; Ancsin, Z.; Zándoki, E.; Mézes, M.; Balogh, K. Modification of the Effects of Aflatoxin B1 on the Glutathione System and Its Regulatory Genes by Zeolite. *Acta Vet. Hung.* **2021**, *69*, 23–30. [CrossRef] [PubMed]
21. Rashidi, N.; Khatibjoo, A.; Taherpour, K.; Akbari-Gharaei, M.; Shirzadi, H. Effects of Licorice Extract, Probiotic, Toxin Binder and Poultry Litter Biochar on Performance, Immune Function, Blood Indices and Liver Histopathology of Broilers Exposed to Aflatoxin-B1. *Poult. Sci.* **2020**, *99*, 5896–5906. [CrossRef] [PubMed]
22. Tavangar, P.; Gharahveysi, S.; Rezaeipour, V.; Irani, M. Efficacy of Phytobiotic and Toxin Binder Feed Additives Individually or in Combination on the Growth Performance, Blood Biochemical Parameters, Intestinal Morphology, and Microbial Population in Broiler Chickens Exposed to Aflatoxin B1. *Trop. Anim. Health Prod.* **2021**, *53*, 335. [CrossRef] [PubMed]
23. Fouad, A.M.; Ruan, D.; El-Senousey, H.K.; Chen, W.; Jiang, S.; Zheng, C. Harmful Effects and Control Strategies of Aflatoxin B1 Produced by *Aspergillus flavus* and *Aspergillus parasiticus* Strains on Poultry: Review. *Toxins* **2019**, *11*, 176. [CrossRef]
24. Pfliegler, W.P.; Pusztahelyi, T.; Pócsi, I. Mycotoxins—Prevention and decontamination by yeasts. *J. Basic Microbiol.* **2015**, *55*, 805–818. [CrossRef]
25. Nešić, K.; Habschied, K.; Mastanjević, K. Possibilities for the Biological Control of Mycotoxins in Food and Feed. *Toxins* **2021**, *13*, 98. [CrossRef]
26. Khan, W.A.; Khan, M.Z.; Khan, A.; Hassan, Z.U.; Rafique, S.; Saleemi, M.K.; Ahad, A. Dietary Vitamin E in White Leghorn Layer Breeder Hens: A Strategy to Combat Aflatoxin B1-Induced Damage. *Avian Pathology* **2014**, *43*, 389–395. [CrossRef] [PubMed]
27. Allameh, A.; Khanian, M.; Karimi-Torshizi, M.-A.; Kalantari-Hesari, A. Hepatoprotective Effects of Lactobacillus Plantarum 299v Supplemented via Drinking Water against Aflatoxin-Induced Liver Damage. *Avian Pathol.* **2021**, *50*, 522–530. [CrossRef]
28. Greco, D.; D'Ascanio, V.; Abbasciano, M.; Santovito, E.; Garbetta, A.; Logrieco, A.F.; Avantaggiato, G. Simultaneous Removal of Mycotoxins by a New Feed Additive Containing a Tri-Octahedral Smectite Mixed with Lignocellulose. *Toxins* **2022**, *14*, 393. [CrossRef] [PubMed]
29. Wang, Y.; Wang, X.; Li, Q. Aflatoxin B1 in poultry liver: Toxic mechanism. *Toxicon* **2023**, *233*, 107262. [CrossRef]
30. Dixon, J.B.; Kannewischer, I.; Tenorio Arvide, M.G.; Barrientos Velazquez, A.L. Aflatoxin Sequestration in Animal Feeds by Quality-Labeled Smectite Clays: An Introductory Plan. *Appl. Clay Sci.* **2008**, *40*, 201–208. [CrossRef]
31. Zhang, Z.F.; Xi, Y.; Wang, S.T.; Zheng, L.Y.; Qi, Y.; Guo, S.S.; Ding, B.Y. Effects of Chinese Gallnut Tannic Acid on Growth Performance, Blood Parameters, Antioxidative Status, Intestinal Histomorphology, and Cecal Microbial Shedding in Broilers Challenged with Aflatoxin B1. *J. Anim. Sci.* **2022**, *100*, skac099. [CrossRef] [PubMed]
32. Raj, J.; Vasiljević, M.; Tassis, P.; Farkaš, H.; Bošnjak-Neumüller, J.; Männer, K. Effects of a Modified Clinoptilolite Zeolite on Growth Performance, Health Status and Detoxification of Aflatoxin B1 and Ochratoxin A in Male Broiler Chickens. *Br. Poult. Sci.* **2021**, *62*, 601–610. [CrossRef] [PubMed]
33. Allameh, A.; Niayesh-Mehr, R.; Aliarab, A.; Sebastiani, G.; Pantopoulos, K. Oxidative Stress in Liver Pathophysiology and Disease. *Antioxidants* **2023**, *12*, 1653. [CrossRef] [PubMed]
34. EFSA Panel on Additives and Products or Substances used in Animal Feed (FEEDAP) Scientific Opinion on the Safety and Efficacy of a Preparation of Bentonite-and Sepiolite (Toxfin® Dry) as Feed Additive for All Species. *EFSA J.* **2013**, *11*, 3179. [CrossRef]
35. Jurišić, N.; Schwartz-Zimmermann, H.E.; Kunz-Vekiru, E.; Moll, W.D.; Schweiger, W.; Fowler, J.; Berthiller, F. Determination of Aflatoxin Biomarkers in Excreta and Ileal Content of Chickens. *Poult. Sci.* **2019**, *98*, 5551–5561. [CrossRef] [PubMed]

36. Mabee, M.S.; Chipley, J.R. Tissue Distribution and Metabolism of Aflatoxin B1-14c in Layer Chickens. *J. Food Sci.* **1973**, *38*, 566–570. [CrossRef]
37. Dohnal, V.; Wu, Q.; Kuča, K. Metabolism of aflatoxins: Key enzymes and interindividual as well as interspecies differences. *Arch. Toxicol.* **2014**, *9*, 1635–1644. [CrossRef] [PubMed]
38. Khanian, M.; Karimi-Torshizi, M.-A.; Allameh, A. Alleviation of Aflatoxin-Related Oxidative Damage to Liver and Improvement of Growth Performance in Broiler Chickens Consumed *Lactobacillus Plantarum 299v* for Entire Growth Period. *Toxicon* **2019**, *158*, 57–62. [CrossRef]
39. Salem, R.; El-Habashi, N.; Fadl, S.E.; Sakr, O.A.; Elbialy, Z.I. Effect of Probiotic Supplement on Aflatoxicosis and Gene Expression in the Liver of Broiler Chicken. *Environ. Toxicol. Pharmacol.* **2018**, *60*, 118–127. [CrossRef] [PubMed]
40. Arak, H.; Karimi Torshizi, M.A.; Hedayati, M.; Rahimi, S. The Efficiency of Synthetic Polymers to Ameliorate the Adverse Effects of Aflatoxin on Plasma Biochemistry, Immune Responses, and Hepatic Genes Expression in Ducklings. *Toxicon* **2020**, *187*, 136–143. [CrossRef] [PubMed]
41. Sang, R.; Ge, B.; Li, H.; Zhou, H.; Yan, K.; Wang, W.; Cui, Q.; Zhang, X. Taraxasterol Alleviates Aflatoxin B1-Induced Liver Damage in Broiler Chickens via Regulation of Oxidative Stress, Apoptosis and Autophagy. *Ecotoxicol. Environ. Saf.* **2023**, *251*, 114546. [CrossRef]
42. Muhammad, I.; Wang, H.; Sun, X.; Wang, X.; Han, M.; Lu, Z.; Cheng, P.; Hussain, M.A.; Zhang, X. Dual Role of Dietary Curcumin Through Attenuating AFB1-Induced Oxidative Stress and Liver Injury via Modulating Liver Phase-I and Phase-II Enzymes Involved in AFB1 Bioactivation and Detoxification. *Front. Pharmacol.* **2018**, *9*, 554. [CrossRef]
43. Muhammad, I.; Sun, X.; Wang, H.; Li, W.; Wang, X.; Cheng, P.; Li, S.; Zhang, X.; Hamid, S. Curcumin Successfully Inhibited the Computationally Identified CYP2A6 Enzyme-Mediated Bioactivation of Aflatoxin B1 in Arbor Acres Broiler. *Front. Pharmacol.* **2017**, *8*, 143. [CrossRef] [PubMed]
44. Zhang, N.-Y.; Qi, M.; Zhao, L.; Zhu, M.-K.; Guo, J.; Liu, J.; Gu, C.-Q.; Rajput, S.A.; Krumm, C.S.; Qi, D.-S.; et al. Curcumin Prevents Aflatoxin B_1 Hepatoxicity by Inhibition of Cytochrome P450 Isozymes in Chick Liver. *Toxins* **2016**, *8*, 327. [CrossRef] [PubMed]
45. Kozieł, M.J.; Kowalska, K.; Piastowska-Ciesielska, A.W. Nrf2: A Main Responsive Element in Cells to Mycotoxin-Induced Toxicity. *Arch. Toxicol.* **2021**, *95*, 1521–1533. [CrossRef] [PubMed]
46. Ma, Q.; Li, Y.; Fan, Y.; Zhao, L.; Wei, H.; Ji, C.; Zhang, J. Molecular Mechanisms of Lipoic Acid Protection against Aflatoxin B1-Induced Liver Oxidative Damage and Inflammatory Responses in Broilers. *Toxins* **2015**, *7*, 5435–5447. [CrossRef]
47. Wang, Y.; Liu, F.; Liu, M.; Zhou, X.; Wang, M.; Cao, K.; Jin, S.; Shan, A.; Feng, X. Curcumin Mitigates Aflatoxin B1-Induced Liver Injury via Regulating the NLRP3 Inflammasome and Nrf2 Signaling Pathway. *Food Chem. Toxicol.* **2022**, *161*, 112823. [CrossRef] [PubMed]
48. Yarru, L.P.; Settivari, R.S.; Gowda, N.K.S.; Antoniou, E.; Ledoux, D.R.; Rottinghaus, G.E. Effects of Turmeric (*Curcuma longa*) on the Expression of Hepatic Genes Associated with Biotransformation, Antioxidant, and Immune Systems in Broiler Chicks Fed Aflatoxin. *Poult. Sci.* **2009**, *88*, 2620–2627. [CrossRef] [PubMed]
49. Ma, Q. Role of Nrf2 in Oxidative Stress and Toxicity. *Annu. Rev. Pharmacol. Toxicol.* **2013**, *53*, 401–426. [CrossRef]
50. Zhang, J.; Sun, X.; Chai, X.; Jiao, Y.; Sun, J.; Wang, S.; Yu, H.; Feng, X. Curcumin Mitigates Oxidative Damage in Broiler Liver and Ileum Caused by Aflatoxin B1-Contaminated Feed through Nrf2 Signaling Pathway. *Animals* **2024**, *14*, 409. [CrossRef] [PubMed]
51. Wang, P.; Wang, Y.; Feng, T.; Yan, Z.; Zhu, D.; Lin, H.; Iqbal, M.; Deng, D.; Kulyar, M.F.-E.-A.; Shen, Y. Hedyotis Diffusa Alleviate Aflatoxin B1-Induced Liver Injury in Ducks by Mediating Nrf2 Signaling Pathway. *Ecotoxicol. Environ. Saf.* **2023**, *249*, 114339. [CrossRef] [PubMed]
52. Liu, Y.; Wang, J.; Chang, Z.; Li, S.; Zhang, Z.; Liu, S.; Wang, S.; Wei, L.; Lv, Q.; Ding, K.; et al. SeMet Alleviates AFB1-Induced Oxidative Stress and Apoptosis in Rabbit Kidney by Regulating Nrf2//Keap1/NQO1 and PI3K/AKT Signaling Pathways. *Ecotoxicol. Environ. Saf.* **2024**, *269*, 115742. [CrossRef]
53. Balogh, K.; Kövesi, B.; Zándoki, E.; Kulcsár, S.; Ancsin, Z.; Erdélyi, M.; Dobolyi, C.; Bata-Vidács, I.; Inotai, K.; Szekeres, A.; et al. Effect of Sterigmatocystin or Aflatoxin Contaminated Feed on Lipid Peroxidation and Glutathione Redox System and Expression of Glutathione Redox System Regulatory Genes in Broiler Chicken. *Antioxidants* **2019**, *8*, 201. [CrossRef]
54. EFSA Panel on Additives and Products or Substances used in Animal Feed (FEEDAP). Statement on the Establishment of Guidelines for the Assessment of Additives from the Functional Group 'Substances for Reduction of the Contamination of Feed by Mycotoxins'. *EFSA J.* **2010**, *8*, 1693. [CrossRef]
55. Schiavone, A.; De Marco, M.; Martínez, S.; Dabbou, S.; Renna, M.; Madrid, J.; Hernandez, F.; Rotolo, L.; Costa, P.; Gai, F.; et al. Nutritional Value of a Partially Defatted and a Highly Defatted Black Soldier Fly Larvae (*Hermetia illucens* L.) Meal for Broiler Chickens: Apparent Nutrient Digestibility, Apparent Metabolizable Energy and Apparent Ileal Amino Acid Digestibility. *J. Anim. Sci. Biotechnol.* **2017**, *8*, 51. [CrossRef]
56. Malachová, A.; Sulyok, M.; Beltrán, E.; Berthiller, F.; Krska, R. Optimization and Validation of a Quantitative Liquid Chromatography–Tandem Mass Spectrometric Method Covering 295 Bacterial and Fungal Metabolites Including All Regulated Mycotoxins in Four Model Food Matrices. *J. Chromatogr. A* **2014**, *1362*, 145–156. [CrossRef]

57. Jansen, E.H.; Ruskovska, T. Comparative Analysis of Serum (Anti)Oxidative Status Parameters in Healthy Persons. *Int. J. Mol. Sci.* **2013**, *14*, 6106–6115. [CrossRef]
58. Espín, S.; Sánchez-Virosta, P.; García-Fernández, A.J.; Eeva, T. A microplate adaptation of the thiobarbituric acid reactive substances assay to determine lipid peroxidation fluorometrically in small sample volumes. *Rev. Toxicol.* **2017**, *34*, 94–98.
59. Nebbia, C.; Della Donna, L.; Carletti, M.; Balbo, A.; Barbarino, G.; Gardini, G. Use of Hepatic Protein Biomarkers for Tracing the Exposure of Veal Calves to Illegal Growth-Promoters: Investigations on Experimental Samples and Preliminary Application under Field Conditions. *J. Vet. Pharmacol. Ther.* **2008**, *31*, 272–275. [CrossRef]
60. Lowry, O.H.; Rosebrough, N.J.; Farr, A.L.; Randall, R.J. Protein measurement with the folin phenol reagent. *J. Biol. Chem.* **1951**, *193*, 265–275. [CrossRef] [PubMed]
61. Roberts, J.C.; Francetic, D.J. The importance of sample preparation and storage in glutathione analysis. *Anal. Biochem.* **1993**, *211*, 183–187. [CrossRef] [PubMed]
62. Habig, W.H.; Pabst, M.J.; Jakoby, W.B. Glutathione S-transferases. The first enzymatic step in mercapturic acid formation. *J. Biol. Chem.* **1974**, *249*, 7130–7139. [CrossRef] [PubMed]
63. Dai, H.; Edens, F.W.; Roe, R.M. Glutathione S-transferases in the Japanese quail: Tissue distribution and purification of the liver isozymes. *J. Biochem. Toxicol.* **1996**, *11*, 85–96. [CrossRef]
64. Reddy, C.C.; Scholz, R.W.; Massaro, E.J. Cadmium, Methylmercury, Mercury, and Lead Inhibition of Calf Liver Glutathione S-Transferase Exhibiting Selenium-Independent Glutathione Peroxidase Activity. *Toxicol. Appl. Pharmacol.* **1981**, *61*, 460–468. [CrossRef] [PubMed]
65. Lind, C.; Cadenas, E.; Hochstein, P.; Ernster, L. DT-diaphorase: Purification, properties, and function. *Meth. Enzymol.* **1990**, *186*, 287–301. [CrossRef]

Disclaimer/Publisher's Note: The statements, opinions and data contained in all publications are solely those of the individual author(s) and contributor(s) and not of MDPI and/or the editor(s). MDPI and/or the editor(s) disclaim responsibility for any injury to people or property resulting from any ideas, methods, instructions or products referred to in the content.

toxins

MDPI

Article

Acyclic Cucurbit[n]uril-Enabled Detection of Aflatoxin B1 via Host–Guest Chemistry and Bioluminescent Immunoassay

Shaowen Wu [1,†], Ke Feng [2,†], Jinlu Niu [1,†], Jintao Xu [1], Hualian Mo [1], Xiaoman She [3], Shang-Bo Yu [2,*], Zhan-Ting Li [2] and Shijuan Yan [1,*]

1 State Key Laboratory of Swine and Poultry Breeding Industry, Agro-Biological Gene Research Center, Guangdong Academy of Agricultural Sciences, Guangzhou 510640, China; wushaowen@agrogene.ac.cn (S.W.); niujinlu2021@163.com (J.N.); xujintao_1999@163.com (J.X.); mhl981585@163.com (H.M.)

2 State Key Laboratory of Organometallic Chemistry, Shanghai Institute of Organic Chemistry, Chinese Academy of Sciences, University of Chinese Academy of Sciences, 345 Lingling Lu, Shanghai 200032, China; fengke@sioc.ac.cn (K.F.); ztli@mail.sioc.ac.cn (Z.-T.L.)

3 Guangdong Provincial Key Laboratory of High Technology for Plant Protection, Institute of Plant Protection, Guangdong Academy of Agricultural Sciences, Guangzhou 510640, China; shexiaoman@gdppri.com

* Correspondence: yushbo@sioc.ac.cn (S.-B.Y.); yanshijuan@gdaas.cn (S.Y.)

† These authors contributed equally to this work.

Abstract: Aflatoxin B1 (AFB1), a highly toxic secondary metabolite produced by Aspergillus species, represents a significant health hazard due to its widespread contamination of agricultural products. The urgent need for sensitive and sustainable detection methods has driven the development of diverse analytical approaches, most of which heavily rely on organic solvents, posing environmental challenges for routine food safety analysis. Here, we introduce a supramolecular platform leveraging acyclic cucurbit[n]uril (acCB) as a host molecule for environmentally sustainable AFB1 detection. Screening various acCB derivatives identified acCB6 as a superior host capable of forming a stable 1:1 complex with AFB1 in an aqueous solution, exhibiting a high binding affinity. Proton nuclear magnetic resonance (1H NMR) spectroscopy confirmed that AFB1 was deeply encapsulated within the host cavity, with isothermal titration calorimetry (ITC) experiments and molecular dynamics simulations further substantiating the stability of the interaction, driven by enthalpic and entropic contributions. This supramolecular host was incorporated into a scaffold-assembly-based bioluminescent enzyme immunoassay (SA-BLEIA), providing a green detection platform that rivals the performance of traditional organic solvent-based assays. Our findings highlight the potential of supramolecular chemistry as a foundation for eco-friendly mycotoxin detection and offer valuable insights into designing environmentally sustainable analytical methods.

Keywords: aflatoxin B1; host–guest chemistry; acyclic cucurbit[n]urils; bioluminescent immunoassay; supramolecular recognition

Key Contribution: We developed a supramolecular platform utilizing acyclic cucurbit[6]uril (acCB6) as a host molecule for aflatoxin B1 (AFB1) detection. The acCB6-AFB1 complex demonstrates a stable 1:1 binding stoichiometry in aqueous solution with remarkably high binding affinity, showcasing the potential of supramolecular chemistry for environmentally friendly mycotoxin detection.

1. Introduction

Aflatoxins, particularly aflatoxin B1 (AFB1), are among the most toxic secondary metabolites produced by *Aspergillus* species, frequently contaminating staple agricultural commodities such as peanuts, corn, cereals, and their derivatives [1]. AFB1 has been classified as a Group 1 carcinogen by the International Agency for Research on Cancer due to its potent carcinogenic, mutagenic, and immunosuppressive effects, even at trace levels [1,2]. These serious health risks, combined with the widespread occurrence of AFB1, have prompted stringent regulatory limits on AFB1 concentrations in food and feed products, typically ranging from 2 to 20 μg/kg [3,4]. Consequently, developing sensitive, reliable, and scalable methods for AFB1 monitoring remains a critical priority.

Current AFB1 detection techniques predominantly rely on chromatographic methods, including high-performance liquid chromatography (HPLC) and liquid chromatography-tandem mass spectrometry (LC-MS), alongside immunological assays such as enzyme-linked immunosorbent assay (ELISA) [5–7]. While HPLC and LC-MS deliver excellent sensitivity and specificity, their complex instrumentation, high operational costs, and reliance on skilled technicians limit their scalability for routine applications. Immunoassays, on the other hand, are better suited for high-throughput analysis [8,9]. However, a shared limitation across these methodologies is their heavy dependence on organic solvents for sample preparation and standard solutions, which raises concerns about their environmental sustainability. Addressing these challenges requires the development of green alternatives that can balance sensitivity, reliability, and environmental impact.

Host–guest chemistry has emerged as a powerful tool for molecular recognition, enabling the sequestration and detection of toxins, drugs, and other contaminants [10–18]. One notable example is the inclusion–antagonism mechanism of sugammadex, which efficiently removes residual neuromuscular blocking agents in clinical applications. Supramolecular hosts such as cyclodextrins, calixarenes, cucurbiturils, and pillararenes exhibit excellent biocompatibility and form strong complexes with target molecules, making them suitable for in vivo and in vitro applications. Among these, acyclic cucurbit[n]urils (acCBs) are particularly promising due to their high water solubility and adaptable cavities, which can flexibly bind guests of varying sizes. Recent advancements have shown that modifying acCBs with exocyclic substituents can fine-tune their backbone curvature and steric properties, enhancing their binding versatility [16,19]. Such functional adaptability makes acCBs ideal candidates for various guest-binding applications, including toxin detection [20,21].

AFB1 possesses a difuranocoumarin-like structure characterized by a highly planar conjugated aromatic system (Figure 1a). The aromatic moieties of acCBs provide opportunities for π-π stacking interactions with AFB1, potentially forming tightly packed complexes. Building on the binding versatility of acCBs, this study investigates their application in AFB1 detection. Spectroscopic, thermodynamic, and computational analyses revealed that acCB6 forms a highly stable 1:1 complex with AFB1 in aqueous solution. Furthermore, integrating this host–guest system into a scaffold-assembly-based bioluminescent enzyme immunoassay (SA-BLEIA) enabled sensitive, environmentally sustainable AFB1 detection. This work not only establishes the utility of acCBs for mycotoxin detection but also underscores their potential for advancing green analytical methodologies through host–guest chemistry.

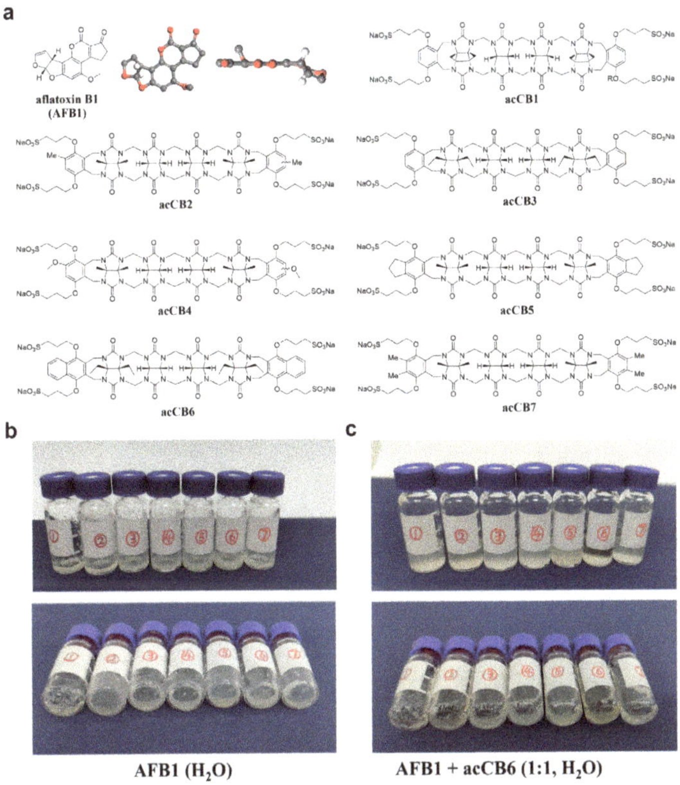

Figure 1. Chemical structures of acyclic cucurbit[n]urils and their ability to enhance the solubility of aflatoxin B1 (AFB1). (**a**) Chemical structure of AFB1 alongside the structures of acCB derivatives, illustrating their substituents and cavity dimensions. (**b**,**c**) Visual representation of AFB1 dissolution enhancement in aqueous solution. Sample vials containing AFB1 with various acCBs (vials 1–7 corresponding to acCB 1–7) at 0 molar equivalents (**b**) and 1.0 molar equivalents (**c**). Top and side views of the vials are provided.

2. Results

2.1. Dissolution Enhancement of AFB1 by acCBs

To evaluate the potential of acCBs as solubilizing agents for AFB1, we systematically screened a series of acCB derivatives with varying structure features (Figure 1a). While acCBs have previously demonstrated efficacy in enhancing the solubility of poorly soluble pharmaceuticals, their ability to dissolve AFB1 in water—given its planar spatial structure and extremely low solubility (10–30 µg/mL)—remains uncertain [22]. Initial solubility experiments were performed via visual inspection. AFB1 was dissolved in water in sample vials, followed by the addition of acCBs at 1.0 molar equivalents (Figure 1b,c). Among the derivatives tested, acCB6 exhibited exceptional dissolution efficacy, forming a completely transparent solution, indicative of full AFB1 either remaining as visible deposits on the vial walls or suspended in the solution. These observations suggest that the naphthalene substituents of acCB6 provide an optimal cavity size and aromatic surface for effective AFB1 encapsulation. This preliminary screening identified acCB6 as the most effective solubilizing agent for AFB1.

2.2. Spectroscopic Evidence for 1:1 Host–Guest Complex Formation

To elucidate the molecular interactions underpinning acCB6's superior solubilizing capability, we performed spectroscopic analyses. NMR titration experiments provided detailed insights into the binding mechanism and stoichiometry of the acCB6-AFB1 interaction. Stepwise addition of acCB6 (0.25–8.0 equivalents) to AFB1 led to systematic chemical shift changes in the diagnostic proton signals within the ^{1}H NMR spectra (Figure 2a). Significant upfield shifts were observed for signals A and B, indicative of AFB1 inclusion within the acCB6 cavity. Binding stoichiometry analysis revealed an inflection point at a 1:1 molar ratio (Figure 2b), supporting the formation of a well-defined 1:1 host–guest complex. Given AFB1's extremely low solubility in water, these NMR observations further support the fact that acCB6 successfully solubilized AFB1.

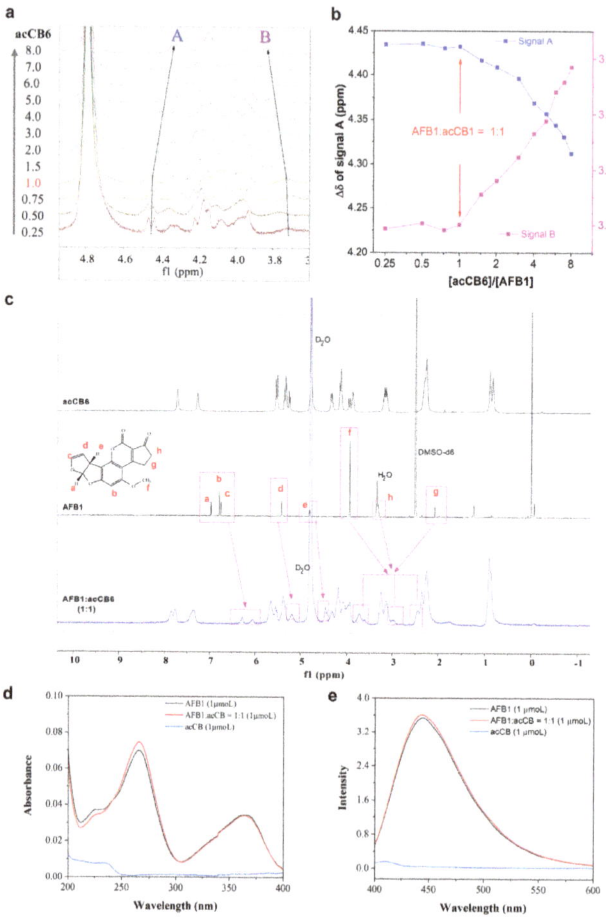

Figure 2. NMR, UV-vis, and fluorescence spectroscopic characterization of the acCB6-AFB1 complex. (**a**) ^{1}H NMR titration spectra showing the evolution of chemical shift changes in the 3.6–5.0 ppm region upon the stepwise addition of acCB6 (0.8–25.6 mM, 0.25–8.0 equivalents) to AFB1 (3.2 mM). (**b**) Signal shift analysis for A and B as a function of [acCB6]/[AFB1] ratio, highlighting a 1:1 binding stoichiometry. (**c**) Comparative ^{1}H NMR spectra of free acCB6 (3.2 mM, D_2O), free AFB1 (3.2 mM, DMSO-d6), and their 1:1 mixture (3.2 mM, D_2O) at 25 °C, with key proton assignments of AFB1 (a–g) indicated. (**d**) UV–visible absorption spectra and (**e**) fluorescence emission spectra of acCB6 (1 μM), AFB1 (1 μM), and their 1:1 complex (1 μM) in H_2O at 25 °C, demonstrating spectral changes upon complex formation.

Comparative ^{1}H NMR spectra of free acCB6, free AFB1, and their equimolar mixture further confirmed this interaction. Notable upfield shifts in AFB's H-a, H-b, H-c, and H-f proton signals (Figure 2c), suggest deep encapsulation of nearly the entire AFB1 molecule with acCB6 cavity. Complementary UV–visible and fluorescence spectroscopy analyses revealed slight increases in AFB1's absorbance and fluorescence upon complexation with acCB6 (Figure 2d,e), corroborating the host–guest complex formation. These spectroscopic findings establish acCB6 as a highly effective host molecule for AFB1 encapsulation in aqueous media.

2.3. Thermodynamic Analysis and Molecular Dynamics Simulations Reveal the Driving Forces of Complex Formation

To quantitatively investigate the driving forces of the acCB6-AFB1 interaction, we conducted isothermal titration calorimetry (ITC) experiments. Titration of acCB6 (0.1 mM) into AFB1 solution (0.01 mM) at 25 °C produced well-defined binding isotherms that fit a 1:1 binding model (Figure 3a). The binding constant (Ka) was determined to be $(3.02 \pm 4.7) \times 10^6$ M^{-1}, corresponding to a favorable Gibbs free energy change (ΔG) of −8.84 kcal/mol (Table 1). The interaction was driven by both enthalpic contributions ($\Delta H = -3.96 \pm 0.582$ kcal/mol) and entropic gains ($-T\Delta S = -4.88$ kcal/mol), highlighting the thermodynamic stability of the complex. The stoichiometry value ($N = 1.27 \pm 0.08$) further confirmed the 1:1 binding mode observed in NMR studies.

Table 1. Thermodynamic parameters for acCB6 and AFB1 binding at 25 °C measured by ITC.

Syringe	Cell	N	Ka (M^{-1})	ΔH (kcal/mol)	$-T\Delta S$ (kcal/mol)	ΔG (kcal/mol)
0.1 mM acCB6	0.01 mM AFB1	1.27 ± 0.08	$(3.02 \pm 4.7) \times 10^6$	-3.96 ± 0.582	−4.88	−8.84

To gain atomic-level insights into the binding mechanism of the acCB6-AFB1 complex, molecular dynamics (MD) simulations were performed. Three independent 1000 ns trajectories were analyzed to assess stability and binding modes (Figure 3b). Root-mean-square deviation (RMSD) analysis indicated rapid equilibration and stable complex formation, with fluctuations stabilizing at approximately 0.3 nm. The center-of-mass distance between acCB6 and AFB1 remained consistently close (~0.2 nm), confirming a stable host–guest association throughout the simulations. Energy component analysis revealed a total binding energy of −76,488.2 kJ/mol, consisting of potential energy (−92,910.53333 kJ/mol), Coulombic short-range interactions (−106,370.0 kJ/mol), and Lennard-Jones interactions (13,364.53333 kJ/mol). The large negative Coulomb energy highlights strong electrostatic interactions between the host and guest molecules, while the positive Lennard-Jones term reflects the balance between attractive van der Waals forces and repulsive steric interactions at close-contact distances (Table 2). These results align with the favorable energetics observed in ITC measurements and MD calculations, underscoring the structural stability of the complex.

Table 2. Energy components of the acCB6-AFB1 complex from molecular dynamics simulations.

Total Energy (kJ/mol)	Potential (kJ/mol)	Coulomb (SR) (kJ/mol)	LJ (SR) (kJ/mol)
−76,488.2	−92,910.53333	−10,6370.0	13,364.53333

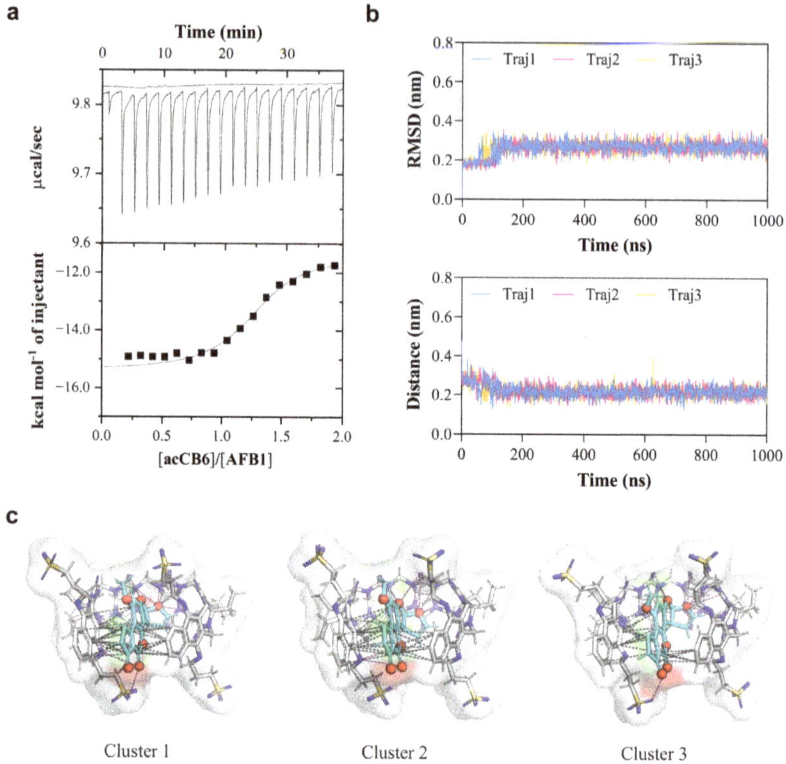

Figure 3. ITC and molecular dynamics analyses of acCB6-AFB1 interactions. (**a**) ITC titration curve showing raw calorimetric data and integrated heat changes for the addition of acCB6 (0.1 mM) to AFB1 (0.01 mM) at 25 °C. The solid line represents the best fit to a 1:1 binding model. (**b**) Molecular dynamics simulation results, including an RMSD plot of the acCB6-AFB1 complex (top) and the center-of-mass distance between acCB6 and AFB1 (bottom) over 1000 ns across three independent trajectories. (**c**) Representative structures from cluster analysis of MD trajectories showing three dominant binding modes (Clusters 1–3). AFB1 is depicted in cyan, acCB6 in gray, with polar interactions (magenta) and π-π stacking interactions (gray) shown as dashed lines. Polar O/N atoms are colored red in AFB1 and blue in acCB6. The host cavity surface is rendered in transparent white to illustrate the encapsulation of AFB1.

Cluster analysis of the MD trajectories was conducted to identify predominant binding modes. Representative structures from the top three clusters showed the AFB1 molecule consistently embedded deeply within the acCB6 cavity. In all clusters, the complex was stabilized by an intricate network of intermolecular interactions, visualized as polar contacts (magenta dashed lines, Figure 3c) and π-π stacking interactions (gray dashed lines). These interactions contribute directly to the observed Coulomb and Lennard-Jones energies and provide a molecular explanation for the favorable enthalpic contributions measured by ITC. The recurrent presence of these interactions across binding modes supports the specificity and strength of the acCB6-AFB1 recognition mechanism.

2.4. Development of acCB6-Enabled Aqueous Detection Assay for AFB1

With acCB6 established as a potent host for AFB1 solubilization, its potential as a green alternative to organic solvents in analytical detection was explored. Specifically, we investigated its application in a previously developed scaffold-assembly-based bioluminescent enzyme immunoassay (SA-BLEIA), comparing traditional methanol-dissolved AFB1 standards with the novel acCB6-dissolved AFB1 standards (Figure 4a) [9]. In this assay, the AFB1 present in the solution engages in a competitive binding interaction with hapten AFB1-BSA that is immobilized on the microplate surface, vying for attachment to nanobody–luciferase conjugates. As the concentration of AFB1 escalates, the bioluminescence signal, measured in Relative Light Units (RLUs), correspondingly diminishes. Initial experiments demonstrated that while the acCB6-based method produced detection curves comparable to the conventional approach, it exhibited reduced sensitivity. The IC_{50} value increased from 0.3730 ng/mL (methanol) to 0.7045 ng/mL (acCB6) (Figure 4b, Table 3). Given the MD simulations showing strong electrostatic interactions in the acCB6-AFB1 complex, we hypothesized that modulating ionic strength could improve assay performance.

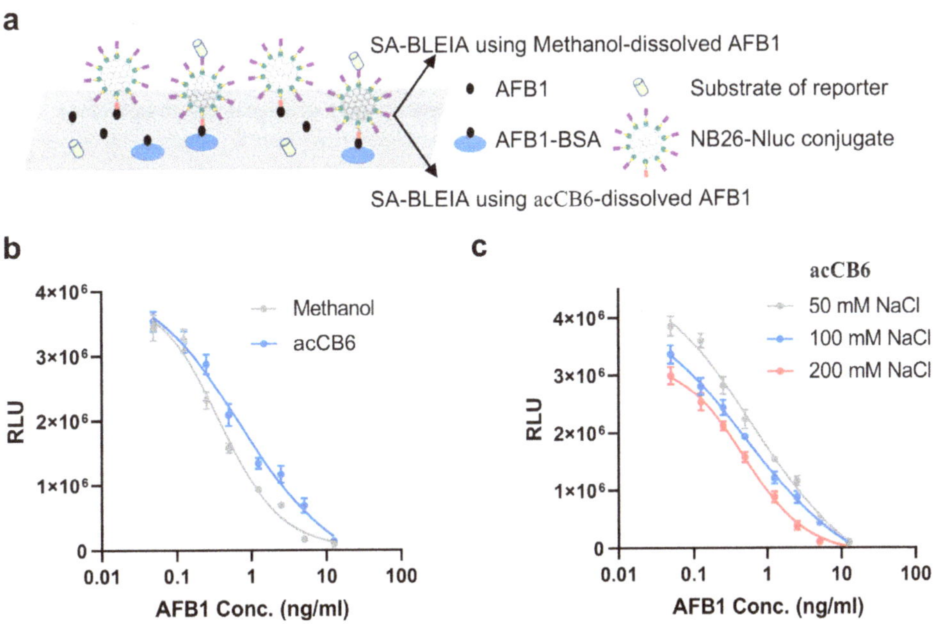

Figure 4. Performance of the scaffold-assembly-based bioluminescent enzyme immunoassay (SA-BLEIA) analysis with AFB1 standards dissolved in different solutions. (**a**) Schematic representation of the SA-BLEIA detection principle, comparing traditional methanol-dissolved AFB1 standards with acCB6-dissolved AFB1 standards. (**b**) Standard curves illustrating the detection performance of methanol-dissolved AFB1 (gray) and acCB6-dissolved AFB1 (blue) standards, highlighting differences in sensitivity. (**c**) Influence of ionic strength on SA-BLEIA performance using acCB6-dissolved AFB1 standards at varying NaCl concentrations (50, 100, and 200 mM).

Table 3. Analytical performance comparison of SA-BLEIA with methanol and acCB6 as solvents for AFB1 standards under varied buffer conditions.

Buffer Conditions	IC_{50} (ng/mL)	R^2
20% methanol	0.3730	0.9837
acCB6	0.7045	0.9814
acCB6 + 50 mM NaCl	0.6460	0.9885
acCB6 + 100 mM NaCl	0.5229	0.9926
acCB6 + 200 mM NaCl	0.4825	0.9924

To evaluate the effect of ionic strength, the SA-BLEIA system was tested under varying NaCl concentrations ranging from 50 to 200 mM. The results revealed a clear trend of

4.4. UV–Visible Absorption Spectra

Stock solutions of acCB6 (1 mM) and AFB1 (1×10^{-2} mM) were prepared in ultrapure water and subsequently diluted to a final concentration of 1 µM. An aqueous solution of the AFB1-acCB6 complex (1 µM) was prepared by combining the diluted stock solutions. The UV–visible absorption measurements were performed using a Unico UV-4802 double-beam spectrophotometer (UNICO, Shanghai, China).

4.5. Fluorescence Emission Spectra

Fluorescence emission spectra were measured using a PerkinElmer FL6500 fluorescence spectrophotometer (PerkinElmer Inc., Waltham, MA, USA). The sample preparation followed the procedure outlined in Section 4.4. After 1 h of equilibration at room temperature, spectra were recorded using an excitation wavelength of 360 nm with a slit width of 20 nm. Emission spectra were measured at 440 nm with a slit width of 10 nm.

4.6. Isothermal Titration Calorimetry (ITC) Experiments

ITC experiments were conducted at 25 °C using a MicroCal PEAQ-ITC calorimeter (Malvern Panalytical, Malvern, UK). A solution of acCB6 (0.1 mM) was titrated into an AFB1 solution (0.01 mM) prepared in ultrapure water. Data were analyzed using the manufacturer's software (MicroCal PEAQ-ITC Analysis Software 1.1.0.1262), employing a one-site binding model.

4.7. SA-BLEIA Procedure

The SA-BLEIA experiments were performed following previously described protocol [9]. High-binding white 96-well microplates were coated overnight at 4 °C with 100 µL of 5 µg/mL AFB1-BSA in PBS buffer (10 mM phosphate, pH 7.4). After incubation, the plates were blocked using 3% skim milk in PBS. Subsequently, 50 µL of 0.5 µg/mL nanobody–luciferase conjugates were pre-incubated with 50 µL of varying AFB1 (TMRM, Changzhou, China, 99% purity) concentrations prepared in PBS buffer with either 20% methanol (Sinopharm Chemical Reagent Co., Ltd., Shanghai, China) or 1.0 molar equivalents of acCB6. To evaluate salt-dependent performance, NaCl was added to the PBS buffer at final concentrations of 50, 100, or 200 mM. This mixture was added to the antigen-coated wells and incubated for 30 min at 37 °C. Plates were washed three times with PBST buffer (PBS containing 0.05% Tween-20), followed by the addition of luminescence substrates in the optimized assay buffer. Luminescence signals were recorded using a Varioskan Flash spectral scanning multimode reader (Thermo Fisher Scientific, Waltham, MA, USA). Standard curves were generated by plotting luminescence against the logarithm of AFB1 concentration and analyzed using four-parameter nonlinear regression for data analysis.

4.8. Molecular Dynamics (MD) Simulations

MD simulations were conducted using GROMACS 2020 with the GAFF force field for acCB6 and AFB1, and the modified TIP3P water model. A 150 mM NaCl concentration was included to neutralize the system [33,34]. Energy minimization was performed using the steepest descent algorithm for 50,000 steps. Equilibration was achieved through 1 ns simulations under the isothermal–isochoric (NVT) and isothermal–isobaric (NPT) ensembles [35], at 298 K, using the v-rescale thermostat and the Parrinello–Rahman barostat with a compressibility of 4.5×10^{-5} bar and a τp of 2 ps [36,37]. Production simulations were conducted for three independent 1000 ns runs with a 2 fs time step [38]. H bonds were constrained using the LINCS algorithm. The cutoff distance for electrostatic and van der Waals interactions was set to 1.2 nm. Long-range electrostatics were calculated using the particle mesh Ewald (PME) method [39]. Simulation trajectories were analyzed using GROMACS tools.

Author Contributions: Conceptualization, S.W. and S.-B.Y.; Data curation, S.W., K.F., J.N., J.X., H.M. and S.-B.Y.; Formal analysis, S.W., K.F. and S.-B.Y.; Funding acquisition, S.W., S.-B.Y. and S.Y.; Investigation, S.W., K.F., J.N., X.S. and S.-B.Y.; Methodology, S.W. and S.-B.Y.; Project administration, S.Y.; Resources, S.W., S.-B.Y. and S.Y.; Supervision, Z.-T.L. and S.Y.; Validation, S.W. and S.-B.Y.; Writing—original draft, S.W.; Writing—review & editing, S.-B.Y., Z.-T.L. and S.Y. All authors have read and agreed to the published version of the manuscript.

Funding: This research was funded by the Project of Collaborative Innovation Center of GDAAS (XT202210, XT202504), Shanghai Sailing Program (No. 22YF1458300), the Natural Science Foundation of China (No. 22307022 and 22201293), the Natural Science Foundation of Guangdong Province (2024A1515010062), and the Special Fund for Scientific Innovation Strategy-Construction of High Level Academy of Agriculture Science (R2023PY-JX024; R2021YJ-YB1004; R2023PY-JG025; 2023QZ-NK04; GDNKY-ZQQZ-K5).

Institutional Review Board Statement: Not applicable.

Informed Consent Statement: Not applicable.

Data Availability Statement: The original contributions presented in this study are included in the article. Further inquiries can be directed to the corresponding author.

Conflicts of Interest: The authors declare no conflict of interest.

References

1. Caceres, I.; Al Khoury, A.; Khoury, R.E.; Lorber, S.; Oswald, I.P.; Khoury, A.E.; Atoui, A.; Puel, O.; Bailly, J.-D. Aflatoxin Biosynthesis and Genetic Regulation: A Review. *Toxins* **2020**, *12*, 150. [CrossRef]
2. Uka, V.; Cary, J.W.; Lebar, M.D.; Puel, O.; De Saeger, S.; Diana Di Mavungu, J. Chemical repertoire and biosynthetic machinery of the Aspergillus flavus secondary metabolome: A review. *Compr. Rev. Food Sci. Food Saf.* **2020**, *19*, 2797–2842. [CrossRef]
3. Streit, E.; Schatzmayr, G.; Tassis, P.; Tzika, E.; Marin, D.; Taranu, I.; Tabuc, C.; Nicolau, A.; Aprodu, I.; Puel, O.; et al. Current situation of mycotoxin contamination and co-occurrence in animal feed—Focus on Europe. *Toxins* **2012**, *4*, 788–809. [CrossRef]
4. van Egmond, H.P.; Schothorst, R.C.; Jonker, M.A. Regulations relating to mycotoxins in food: Perspectives in a global and European context. *Anal. Bioanal. Chem.* **2007**, *389*, 147–157. [CrossRef] [PubMed]
5. Zheng, M.Z.; Richard, J.L.; Binder, J. A review of rapid methods for the analysis of mycotoxins. *Mycopathologia* **2006**, *161*, 261–273. [CrossRef]
6. Marroquín-Cardona, A.G.; Johnson, N.M.; Phillips, T.D.; Hayes, A.W. Mycotoxins in a changing global environment—A review. *Food Chem. Toxicol.* **2014**, *69*, 220–230. [CrossRef] [PubMed]
7. Anfossi, L.; Giovannoli, C.; Baggiani, C. Mycotoxin detection. *Curr. Opin. Biotechnol.* **2016**, *37*, 120–126. [CrossRef]
8. Ren, W.; Li, Z.; Xu, Y.; Wan, D.; Barnych, B.; Li, Y.; Tu, Z.; He, Q.; Fu, J.; Hammock, B.D. One-Step Ultrasensitive Bioluminescent Enzyme Immunoassay Based on Nanobody/Nanoluciferase Fusion for Detection of Aflatoxin B1 in Cereal. *J. Agric. Food Chem.* **2019**, *67*, 5221–5229. [CrossRef] [PubMed]
9. Wu, S.; Xu, J.; Chen, W.; Wang, F.; Tan, X.; Zou, X.; Zhou, W.; Huang, W.; Zheng, Y.; Wang, S.; et al. Protein nanoscaffold enables programmable nanobody-luciferase immunoassembly for sensitive and simultaneous detection of aflatoxin B1 and ochratoxin A. *J. Hazard. Mater.* **2024**, *462*, 132701. [CrossRef] [PubMed]
10. Beatty, M.A.; Hof, F. Host–guest binding in water, salty water, and biofluids: General lessons for synthetic, bio-targeted molecular recognition. *Chem. Soc. Rev.* **2021**, *50*, 4812–4832. [CrossRef]
11. Zhang, X.; Cheng, Q.; Li, L.; Shangguan, L.; Li, C.; Li, S.; Huang, F.; Zhang, J.; Wang, R. Supramolecular therapeutics to treat the side effects induced by a depolarizing neuromuscular blocking agent. *Theranostics* **2019**, *9*, 3107. [CrossRef]
12. Geng, W.-C.; Sessler, J.L.; Guo, D.-S. Supramolecular prodrugs based on host–guest interactions. *Chem. Soc. Rev.* **2020**, *49*, 2303–2315. [CrossRef] [PubMed]
13. Yin, H.; Bardelang, D.; Wang, R. Macrocycles and related hosts as supramolecular antidotes. *Trends Chem.* **2021**, *3*, 1–4. [CrossRef]
14. Yin, H.; Zhang, X.; Wei, J.; Lu, S.; Bardelang, D.; Wang, R. Recent advances in supramolecular antidotes. *Theranostics* **2021**, *11*, 1513. [CrossRef] [PubMed]
15. Li, Z.-T.; Yu, S.-B.; Liu, Y.; Tian, J.; Zhang, D.-W. Supramolecular organic frameworks: Exploring water-soluble, regular nanopores for biomedical applications. *Acc. Chem. Res.* **2022**, *55*, 2316–2325. [CrossRef] [PubMed]

16. Yu, S.; Zhou, W.; Tian, J.; Ma, D.; Zhang, D.; Li, Z. Host-guest chemistry for the design of drug antagonists. *Sci. Sin. Chim.* **2023**, *53*, 2345–2356. [CrossRef]
17. Liu, Y.-Y.; Yu, X.-Y.; Pan, Y.-C.; Yin, H.; Chao, S.; Li, Y.; Ma, H.; Zuo, M.; Teng, K.-X.; Hou, J.-L.; et al. Supramolecular systems for bioapplications: Recent research progress in China. *Sci. China Chem.* **2024**, *67*, 1397–1441. [CrossRef]
18. Zhang, B.; Isaacs, L. Acyclic cucurbit [n] uril-type molecular containers: Influence of aromatic walls on their function as solubilizing excipients for insoluble drugs. *J. Med. Chem.* **2014**, *57*, 9554–9563. [CrossRef] [PubMed]
19. Feng, K.; Liu, Y.-Y.; Zong, Y.; Lei, Z.; Wu, Y.; Yang, J.; Lin, F.; Qi, Q.-Y.; Li, Q.; Zhuang, S.-Y.; et al. Structure–Activity Relationship Studies Leading to the Discovery of Highly Water-Soluble and Biocompatible Acyclic Cucurbit [n] uril FY-3451 as a Universal Antagonist That Rapidly Reverses Neuromuscular Blocking Agents In Vivo. *J. Med. Chem.* **2024**, *67*, 17905–17918. [CrossRef] [PubMed]
20. Ma, D.; Zavalij, P.Y.; Isaacs, L. Acyclic cucurbit [n] uril congeners are high affinity hosts. *J. Org. Chem.* **2010**, *75*, 4786–4795. [CrossRef]
21. Liu, H.; Guo, Y.-J. Acyclic cucurbiturils and their applications. *J. Incl. Phenom. Macrocycl. Chem.* **2022**, *102*, 723–733. [CrossRef]
22. IARC Working Group on the Evaluation of Carcinogenic Risks to Humans. Some Traditional Herbal Medicines, Some Mycotoxins, Naphthalene and Styrene. *IARC Monogr. Eval. Carcinog. Risks Hum.* **2002**, *82*, 1–556.
23. Armstrong, L.; Chang, S.L.; Clements, N.; Hirani, Z.; Kimberly, L.B.; Odoi-Adams, K.; Suating, P.; Taylor, H.F.; Trauth, S.A.; Urbach, A.R. Molecular recognition of peptides and proteins by cucurbit [n] urils: Systems and applications. *Chem. Soc. Rev.* **2024**, *53*, 11519–11556. [CrossRef] [PubMed]
24. Burrows, J.; Sain, P.; Saunders, G.C. A Variable Temperature Study of the π–π Stacking Interaction in the Co-Crystal Naphthalene-Octafluoronaphthalene. *Open Chem. J.* **2019**, *6*, 66–73. [CrossRef]
25. Nishijima, A.; Osugi, Y.; Uemura, T. Fabrication of Self-Expanding Metal–Organic Cages Using a Ring-Openable Ligand. *Angew. Chem.* **2024**, *136*, e202404155. [CrossRef]
26. Zebaze Ndendjio, S.A.; Isaacs, L. Molecular recognition properties of acyclic cucurbiturils toward amino acids, peptides, and a protein. *Supramol. Chem.* **2019**, *31*, 432–441. [CrossRef]
27. Graña-Suárez, L.; Verboom, W.; Egberink, R.J.; Sarkar, S.; Mahalingam, V.; Huskens, J. Host–Guest and Electrostatic Interactions in Supramolecular Nanoparticle Clusters. *Eur. J. Org. Chem.* **2016**, *2016*, 5511–5518. [CrossRef]
28. Jordan, J.H.; Ashbaugh, H.S.; Mague, J.T.; Gibb, B.C. Buffer and salt effects in aqueous host–guest systems: Screening, competitive binding, or both? *J. Am. Chem. Soc.* **2021**, *143*, 18605–18616. [CrossRef]
29. Sun, Z.; Wang, X.; Tang, Z.; Chen, Q.; Liu, X. Development of a biotin-streptavidin-amplified nanobody-based ELISA for ochratoxin A in cereal. *Ecotoxicol. Environ. Saf.* **2019**, *171*, 382–388. [CrossRef] [PubMed]
30. Muheyati, M.; Wu, G.; Li, Y.; Pan, Z.; Chen, Y. Supramolecular nanotherapeutics based on cucurbiturils. *J. Nanobiotechnol.* **2024**, *22*, 790. [CrossRef]
31. Wang, Y.; Wu, Y.; Gesang, D.; Dong, Z.; Qin, Z.; Li, Q.; Li, J.; Zhou, Q.; Shi, G. Non-organic solvent extraction of capsaicinoids from oil combined with fluorescent lateral flow immunoassay strips for on-site identification of illegally recycled waste cooking oil. *Sens. Diagn.* **2025**, *4*, 147–158. [CrossRef]
32. Niu, J.; Yu, J.; Wu, X.; Zhang, Y.-M.; Chen, Y.; Yu, Z.; Liu, Y. Host–guest binding between cucurbit [8] uril and amphiphilic peptides achieved tunable supramolecular aggregates for cancer diagnosis. *Chem. Sci.* **2024**, *15*, 13779–13787. [CrossRef]
33. Abraham, M.J.; Murtola, T.; Schulz, R.; Páll, S.; Smith, J.C.; Hess, B.; Lindahl, E. GROMACS: High performance molecular simulations through multi-level parallelism from laptops to supercomputers. *SoftwareX* **2015**, *1*, 19–25. [CrossRef]
34. Wang, J.; Wolf, R.M.; Caldwell, J.W.; Kollman, P.A.; Case, D.A. Development and testing of a general amber force field. *J. Comput. Chem.* **2004**, *25*, 1157–1174. [CrossRef] [PubMed]
35. Deift, P.; Zhou, X. A steepest descent method for oscillatory Riemann--Hilbert problems. Asymptotics for the MKdV equation. *Ann. Math.* **1993**, *137*, 295–368. [CrossRef]
36. Parrinello, M.; Rahman, A. Crystal structure and pair potentials: A molecular-dynamics study. *Phys. Rev. Lett.* **1980**, *45*, 1196. [CrossRef]
37. Hoover, W.G. Canonical dynamics: Equilibrium phase-space distributions. *Phys. Rev. A* **1985**, *31*, 1695–1697. [CrossRef] [PubMed]
38. Hess, B.; Bekker, H.; Berendsen, H.J.; Fraaije, J.G. LINCS: A linear constraint solver for molecular simulations. *J. Comput. Chem.* **1997**, *18*, 1463–1472. [CrossRef]
39. Essmann, U.; Perera, L.; Berkowitz, M.L.; Darden, T.; Lee, H.; Pedersen, L.G. A smooth particle mesh Ewald method. *J. Chem. Phys.* **1995**, *103*, 8577–8593. [CrossRef]

Disclaimer/Publisher's Note: The statements, opinions and data contained in all publications are solely those of the individual author(s) and contributor(s) and not of MDPI and/or the editor(s). MDPI and/or the editor(s) disclaim responsibility for any injury to people or property resulting from any ideas, methods, instructions or products referred to in the content.

toxins

MDPI

Article

Susceptibility and Mechanism of Aflatoxin Contamination of *Ziziphus jujuba* var. *spinosa*

Abdelrahman Elamin and Shohei Sakuda *

Department of Biosciences, Teikyo University, 1-1 Toyosatodai, Utsunomiya 320-8551, Tochigi, Japan; a.elamin@nasu.bio.teikyo-u.ac.jp
* Correspondence: sakuda@nasu.bio.teikyo-u.ac.jp

Abstract: The susceptibility and mechanism of aflatoxin (AF) contamination in *Ziziphus jujuba* var. *spinosa*, whose seeds are important for medicinal use, were evaluated in this study. First, the susceptibility of intact fruits, classified into four maturity groups, to AF accumulation was assessed through artificial contamination with an aflatoxigenic *Aspergillus flavus* strain. AF analysis revealed that mid-mature fruits were highly susceptible to AF contamination. Next, AF accumulation in seed parts was examined by artificially inoculating *A. flavus* on intact fruits, showing AF presence in seeds after 30 days of incubation. The susceptibility of jujube kernels to AF accumulation in seed parts was then studied. The artificial inoculation of *A. flavus* on kernels, classified into three groups based on the pedicel condition, showed no correlation between AF contamination and the pedicel condition, with large fluctuations within each group. Finally, the effect of the hilar region morphology on AF contamination in seeds was investigated. The microscopic investigation of artificially contaminated seeds and AF quantification revealed that variations in AF concentration were linked to differences in the hilar region morphology.

Keywords: jujube seed; food supplement; aflatoxins; hilar region; HPLC; SEM

Key Contribution: This study elucidates the susceptibility of jujube fruits and kernels to aflatoxin contamination. It highlights how variations in the hilar region morphology influence fungal mycelial penetration and aflatoxin accumulation in seeds.

1. Introduction

Jujube fruit is widely used as food and a dietary supplement due to its rich phytochemical content, including triterpene acids and their saponins, unsaturated fatty acids, flavonoid C-glycosides, alkaloids, and indole derivatives, which promote a healthy diet [1–3]. Among jujube varieties, *Ziziphus jujuba* var. *spinosa* is notable for its medicinal seed, known as "Sansonin" in Japanese, which exhibits anti-anxiety and antidepressant properties [4].

Mycotoxins are toxic secondary metabolites produced by various fungal species, including *Aspergillus*, *Penicillium*, *Fusarium*, *Claviceps*, and *Alternaria*. With around 400 known compounds, mycotoxins are widespread in agricultural commodities and medicinal herbs across diverse regions [5]. Among those posing the greatest threat to medicinal herbs are aflatoxins (AFs), ochratoxins (OTs), fumonisins, zearalenone, and deoxynivalenol [6].

AFs are the most commonly detected mycotoxins in crude medicinal plants. Primarily produced by *Aspergillus flavus* and *Aspergillus parasiticus*, AFs include four major types: aflatoxin B_1 (AFB_1), aflatoxin B_2 (AFB_2), aflatoxin G_1 (AFG_1), and aflatoxin G_2 (AFG_2) [7]. Recognized as Group 1 carcinogens by the International Agency for Research on Cancer (IARC), AFs are highly carcinogenic to humans [8].

Several studies have reported fungal infection and AF contamination in jujube fruits [9–12]. Aflatoxigenic *A. flavus* strains have been isolated from fruits in Iraq, pre-harvest fruits in India, and AF-contaminated fruits in Zambian markets. Ripe fruits in Bangkok showed AF contamination at 2.5–6.1 ppb. Due to the lack of basic studies on AF contamination prevention in jujube fruits, we conducted artificial inoculation experiments to assess the susceptibility of *Z. jujuba* var. *spinosa* to AF contamination, which we preliminarily reported in 2021 [4]. Our experiments revealed that mid-mature fruits were highly susceptible to AF contamination when *A. flavus* was inoculated on artificially wounded fruits. After cultivation, pooled fruits at the same maturity stage were analyzed for AF content. To confirm the susceptibility findings, additional experiments using nonwounded fruits and AF analysis in individual fruits are necessary.

Because jujube seeds are used in medicine, understanding AF contamination mechanisms is crucial for prevention. Our previous report speculated, based on microscopic observations, that fungal mycelial penetration occurs through the seed hilar region. However, the relationship between the hilar region morphology and AF contamination levels remains unclear.

Therefore, this study aims to evaluate (1) the susceptibility of intact jujube fruits at different maturity stages to AF contamination and the possibility of AF accumulation in seed parts; (2) the effect of pericarp layers (seed-surrounding layers) on AF accumulation in jujube seeds; and (3) the role of hilar region morphology in fungal mycelial penetration and AF accumulation in jujube seeds.

2. Results

2.1. Susceptibility of Intact Jujube Fruits at Different Stages of Maturity to AF Contamination

The HPLC analysis (Table 1) of the AFB_1 and AFB_2 levels in whole fruits revealed high AF accumulation in GBG (mid-mature) fruits after 15 and 30 days of incubation and in BG (mid-mature) fruits after 30 days. In contrast, GG (immature) and DBG (mature) fruits showed little or no AF contamination. The high susceptibility of mid-mature fruits to AF contamination was consistent with our previous experiments using artificially wounded fruits and pooled samples. However, individual fruit analysis revealed large variations in AF concentrations within the same maturity group. For instance, the AFB_1 concentrations in GBG fruits after 15 days of incubation ranged from 0.7 to 3499 μg/kg.

Table 1. AF concentrations of intact fruits exposed to *A. flavus* for 15 and 30 days.

Group	Rep.	15 Days of Incubation		Rep.	30 Days of Incubation	
		AFB_1 (μg/kg)	AFB_2 (μg/kg)		AFB_1 (μg/kg)	AFB_2 (μg/kg)
GG	1	<0.25 (b)	ND (a)	1	ND	ND
	2	1.0	ND	2	ND	ND
	3	ND	ND	3	ND	ND
	4	ND	ND	4	ND	ND
	5	<0.25	ND	5	<0.25	ND
GBG	1	14.6	<0.25	1	31.4	0.8
	2	2.3	ND	2	11.7	<0.25
	3	3499	87.7	3	<0.25	ND
	4	6.7	<0.25	4	<0.25	ND
	5	0.7	ND	5	0.3	ND

Table 1. *Cont.*

Group	Rep.	15 Days of Incubation		Rep.	30 Days of Incubation	
		AFB$_1$ (μg/kg)	**AFB$_2$ (μg/kg)**		**AFB$_1$ (μg/kg)**	**AFB$_2$ (μg/kg)**
BG	1	<0.25	ND	1	ND	ND
	2	<0.25	ND	2	ND	ND
	3	<0.25	ND	3	0.5	ND
	4	<0.25	ND	4	52.0	1.0
	5	ND	ND	5	0.5	ND
DBG	1	ND	ND	1	ND	ND
	2	ND	ND	2	ND	ND
	3	ND	ND	3	ND	ND
	4	ND	ND	4	ND	ND
	5	ND	ND	5	ND	ND

(a) below the LOD (0.1 μg/kg). (b) below the LOQ (0.25 μg/kg).

2.2. AF Accumulation in the Seed Parts of Mid-Mature Fruits

To confirm the AF accumulation in seed parts, *A. flavus* was artificially inoculated on intact GBG fruits using the same method as in Section 2.1. After 30 days of incubation, the AF levels in the combined exocarp and mesocarp, as well as in seed parts, were analyzed (Table 2). Among 12 fruits, AF accumulation was detected in the combined parts of 7 fruits and the seeds of 3 fruits, indicating that AF can accumulate in seeds after 30 days. In two replicates (No. 2 and 11), the AF concentrations were higher in seeds than in the combined parts. Large fluctuations in the AF levels among fruits were reproduced.

Table 2. AF concentrations of the two parts of jujube fruits when the intact fruits were exposed to *A. flavus* for 30 days and then divided.

Rep.	Part	AFB$_1$ (μg/kg)	AFB$_2$ (μg/kg)
1	exocarp + mesocarp	1819.0	42.1
	seed	1.8	ND (a)
2	exocarp + mesocarp	1.1	ND
	seed	1.8	ND
3	exocarp + mesocarp	1.1	ND
	seed	ND	ND
4	exocarp + mesocarp	ND	ND
	seed	<0.25 (b)	ND
5	exocarp + mesocarp	<0.25	ND
	seed	ND	ND
6	exocarp + mesocarp	0.66	ND
	seed	ND	ND
7	exocarp + mesocarp	3.9	ND
	seed	ND	ND
8	exocarp + mesocarp	<0.25	ND
	seed	ND	ND

Table 2. *Cont.*

Rep.	Part	AFB$_1$ (μg/kg)	AFB$_2$ (μg/kg)
9	exocarp + mesocarp	<0.25	ND
	seed	ND	ND
10	exocarp + mesocarp	0.4	ND
	seed	ND	ND
11	exocarp + mesocarp	<0.25	ND
	seed	0.3	ND
12	exocarp + mesocarp	1.4	ND
	seed	ND	ND

[a] below the LOD (0.1 μg/kg). [b] below the LOQ (0.25 μg/kg).

2.3. Susceptibility of Jujube Kernels to AF Accumulation in Their Seed Parts

Jujube seeds are enclosed by three layers: the exocarp, mesocarp, and endocarp [13]. The endocarp plays a crucial role in protecting the seed [14]. To examine its effect on the susceptibility of seeds to AF contamination, GBG fruits were divided into three subgroups: GBG-A (fruits with pedicels), GBG-B (fruits without pedicels), and GBG-C (fruits without pedicels that were artificially damaged at the pedicel contact area). The exocarp and mesocarp were removed to prepare kernels. Five kernels per group were inoculated with *A. flavus* spores at the pedicel area and incubated for 5 or 10 days. The AF concentrations in the endocarp and seed of each kernel were analyzed (Table 3). AF contamination in the seed was observed in two GBG-B and one GBG-C kernels after 5 days and in two GBG-A kernels and one GBG-C kernel after 10 days. This indicates that AF can accumulate in seeds within 5–10 days, but that the pedicel area conditions did not affect the AF levels. After 10 days, only two seeds, Rep. 4 of GBG-A and Rep. 2 of GBG-C, exhibited exceedingly high AF concentrations, highlighting large fluctuations in the AF levels among seeds.

Table 3. AF concentrations of endocarp and seed part of kernels after inoculation with *A. flavus* for 5 and 10 days.

Incubation Period	Rep.	Part	GBG-A		GBG-B		GBG-C	
			AFB$_1$ (μg/kg)	AFB$_2$ (μg/kg)	AFB$_1$ (μg/kg)	AFB$_2$ (μg/kg)	AFB$_1$ (μg/kg)	AFB$_2$ (μg/kg)
5 days	1	Endocarp	<0.25 [a]	ND [b]	2.3	0.3	1.7	ND
		Seed	ND	ND	14.1	ND	ND	ND
	2	Endocarp	1.3	<0.25	1.8	<0.25	1.8	ND
		Seed	ND	ND	ND	ND	ND	ND
	3	Endocarp	<0.25	ND	1.1	ND	1.3	ND
		Seed	ND	ND	ND	ND	1.9	ND
	4	Endocarp	0.4	ND	0.9	ND	1.7	ND
		Seed	ND	ND	ND	ND	ND	ND
	5	Endocarp	4.8	0.8	1.1	ND	1.9	ND
		Seed	ND	ND	ND	ND	ND	ND

Table 3. *Cont.*

Incubation Period	Rep.	Part	GBG-A		GBG-B		GBG-C	
			AFB$_1$ (μg/kg)	AFB$_2$ (μg/kg)	AFB$_1$ (μg/kg)	AFB$_2$ (μg/kg)	AFB$_1$ (μg/kg)	AFB$_2$ (μg/kg)
10 days	1	Endocarp	<0.25	ND	2.3	ND	2.2	<0.25
		Seed	ND	ND	ND	ND	ND	ND
	2	Endocarp	1.4	ND	2.9	ND	1.6	ND
		Seed	ND	ND	ND	ND	990	23.9
	3	Endocarp	0.7	ND	2.1	ND	2.7	ND
		Seed	ND	ND	ND	ND	ND	ND
	4	Endocarp	1.7	ND	5.3	ND	2.1	ND
		Seed	4605	46	ND	ND	ND	ND
	5	Endocarp	1.8	ND	2.4	ND	2.3	<0.25
		Seed	14.2	ND	ND	ND	ND	ND

[(a)] below the LOQ (0.25 μg/kg). [(b)] below the LOD (0.1 μg/kg).

When GBG-B kernels were inoculated by dipping them in *A. flavus* spore suspension and incubated for 30 days, AF accumulation in seeds was observed in 4 out of 10 kernels (Table 4), further confirming large variations in the AF concentrations among seeds.

Table 4. AF concentrations of the kernels' seeds when the jujube kernels were dipped in the spore suspension of *A. flavus* and incubated for 30 days.

Rep.	AFB$_1$ (μg/kg)	AFB$_2$ (μg/kg)
1	1.0	ND [(a)]
2	0.7	ND
3	ND	ND
4	<0.25	ND
5	ND	ND
6	ND	ND
7	ND	ND
8	ND	ND
9	21,572	535
10	0.7	ND

[(a)] below the LOD (0.1 μg/kg).

2.4. Relationship of the Hilar Region Morphology with the AF Contamination Level in the Seed

Since the seed hilar region is the primary entry point for water and can become permeable under environmental stresses such as microbial attack [4], fluctuations in the AF concentrations in seeds may be linked to the response of the hilar region to fungal mycelial stress. To investigate this, a microscopic study was conducted to assess differences in the hilar region morphology and their effect on fungal penetration and AF accumulation.

Seeds from GBG fruits were placed on modified CZA, with or without *A. flavus* spores, and incubated at 25 °C for 5 days. Based on a visual assessment of SEM images, three classifications of hilar fissure morphology in seeds incubated without spores were revealed (Figure 1): nearly closed (Figure 1A–C), semi-closed (Figure 1D–F), and open (Figure 1G–J). A similar classification was observed in seeds incubated with spores (Figure 2): nearly closed (Figure 2A–F), semi-closed (Figure 2G–K), and open (Figure 2L). The AF concentrations in the seeds (Table 5) appeared to correlate with the hilar fissure classification. The accumulation of fungal mycelia, which formed infection pads on the hilar fissure (Figure 2), increased with the width of the fissure opening.

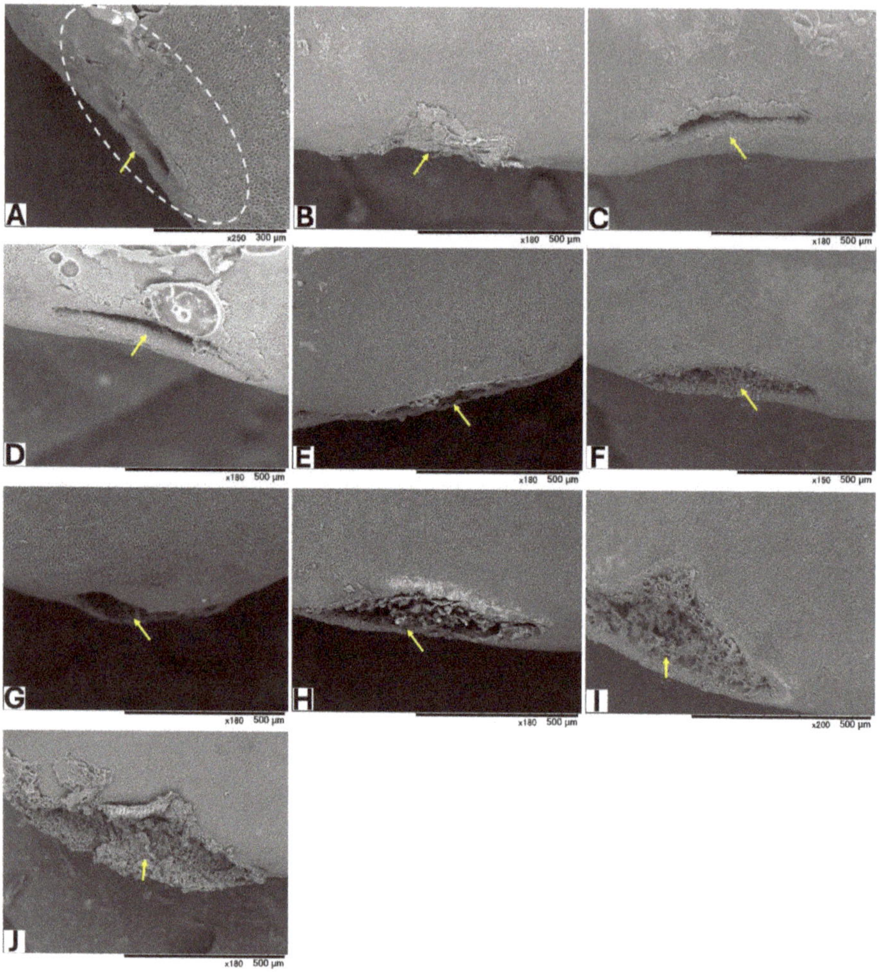

Figure 1. SEM images of seeds without inoculation of *A. flavus*. The white dotted circle denotes the HR (hilar region). Yellow arrows show HF (hilar fissure).

Table 5. AF concentrations of the seeds inoculated with *A. flavus*.

Seed (Figure No.)	HF Status	AFB_1 (μg/kg)	AFB_2 (μg/kg)
2A	nearly closed	3.4	ND [(a)]
2B		4.9	ND
2C		7.8	ND
2D		2.8	ND
2E		3.1	ND
2F		8.0	ND

Table 5. *Cont.*

Seed (Figure No.)	HF Status	AFB_1 (μg/kg)	AFB_2 (μg/kg)
2G	Semi-closed	24.7	ND
2H		111.2	0.8
2I		8.7	ND
2J		9.8	ND
2K		17.9	ND
2L	opened	118,456	1754

[(a)] below the LOD (0.1 μg/kg).

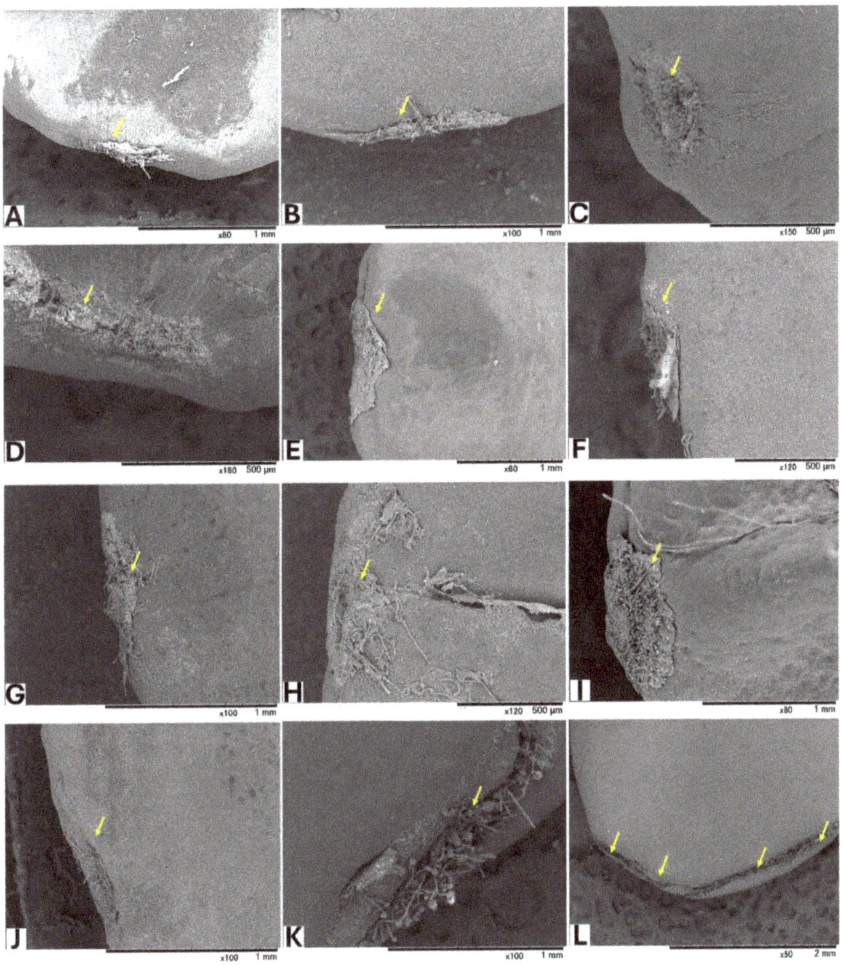

Figure 2. SEM images of seeds incubated with *A. flavus*. Yellow arrows show HF of the seeds after removing the accumulated fungal mycelia.

3. Discussion

Jujube fruit is one of the most widely used medicinal herbs globally, and AF contamination [11,12] has raised significant concerns in the herbal medicine industry. Studies have explored the relationship between fruit maturation, nutrient concentrations, and AF levels, showing that unsaturated fatty acids, soluble sugars, and asparagine may affect susceptibility to AF contamination [4,15–18].

Our previous study [4] evaluated AF susceptibility in whole, wounded jujube fruits and their separate parts at different maturity stages, concluding that mid-mature wounded fruits and their seed parts are the most susceptible. This study aimed to determine whether AFs can also contaminate intact fruits and kernels and accumulate in seeds. Intact fruits protect their seeds through a pericarp composed of three layers: the exocarp, mesocarp, and endocarp [13]. The exocarp consists mainly of cuticle and epidermal cells, whose shape, size, and arrangement vary according to the stage of maturation, changing from compact to loose as the fruit matures [19]. The epidermal cell thickness decreases with maturation, while the pericarp cellulose content declines as lignin and hemicellulose levels increase [18]. The endocarp hardens through secondary cell wall formation and lignification, crucial for resistance to biotic stress [19]. Understanding fungal penetration through these layers to accumulate aflatoxins (AF) in jujube seeds is essential [20].

In this study, intact fruits at different maturity stages (Figure S1) were inoculated with *A. flavus* spores and incubated for 15 and 30 days. Consistent with previous findings [4], mid-mature fruits were the most susceptible to AF contamination, while mature fruits were least susceptible. However, mid-mature (GBG) replicates exhibited fluctuating AF concentrations, suggesting a variation in *A. flavus* mycelial penetration [21,22].

To confirm the fungal penetration and AF accumulation in seeds of intact fruits, mid-mature fruits were inoculated with *A. flavus* and incubated for 30 days. Each replicate was divided into seed and combined exocarp-mesocarp parts. AF analysis confirmed AF accumulation in seeds and reproduced fluctuations in AF levels.

Since the endocarp protects the seed and the pedicel is the only connection between the seed and the external environment, mid-mature (GBG) fruits were divided into three subgroups to examine mycelial penetration (Figure S2). The exocarp and mesocarp were removed, leaving only the kernels (endocarp with seeds). The pedicels were left intact in GBG-A, removed in GBG-B, and artificially damaged in GBG-C. After incubation, AF analysis unexpectedly showed the highest AF accumulation in a GBG-A replicate, suggesting that subgroup preparation did not influence the AF levels. Fluctuations in the AF concentrations were observed, consistent with previous fruit experiments. A separate experiment, in which mid-mature kernels were dipped in an *A. flavus* spore suspension and incubated for 30 days, reproduced these fluctuations.

Microscopic observations of control and artificially contaminated GBG seeds revealed differences in the hilar region (HR) and hilar fissure (HF) morphology. AF concentrations correlated with the HF width, which naturally varies among seeds. This aligns with studies on jujube seed structure and physical characteristics. *Z. jujuba* seeds are water-impermeable and exhibit physical dormancy [23]. The seed coat (SC) has a thick, hydrophobic palisade cell layer, making it impermeable. However, the HR differs structurally from the SC [24,25] and can become permeable under fungal attack [4]. Fungal penetration in intact plants [26] involves hyphal accumulation forming infection pads that modify the morphology and generate pressure to aid penetration. Variations in the seed structure and infection pad stress of jujube may explain the differences in the HF response and, consequently, the variability in AF accumulation.

4. Conclusions

Whole mid-mature jujube fruits and their seeds are highly susceptible to AF contamination, making this stage more vulnerable than others. Despite protective barriers (exocarp, mesocarp, and endocarp), aflatoxigenic fungi can penetrate intact fruit and kernels. Once inside, the fungi colonize the seed, leading to significant AF accumulation. The AF levels in seeds are influenced by variations in the hilar region's morphology, which differ among seeds of the same maturity. This highlights the complex relationship between seed structure and fungal pathogenicity, suggesting that morphological variations in the hilar region significantly impact susceptibility to fungal invasion and AF accumulation.

5. Materials and Methods

5.1. Samples Preparation

The fruits of *Z. jujuba* var. *spinosa* were collected from multiple trees in a field in Ibaraki, Japan, in October 2022–2024. Intact fruits, kernels and seeds were evaluated in five experiments to determine the susceptibility of the seeds of jujube fruits to AF contamination and its mechanism.

5.1.1. Intact Jujube Fruits and Jujube Kernels

Intact jujube fruits at different maturity stages were selected and classified into four color-based groups: green (GG), greenish-brown (GBG), brown (BG), and dark brown (DBG). Because the selected fruits inconsistently included pedicels, any attached pedicels were removed. Each fruit group was washed with ethanol:water (1:1 *v*/*v*), placed in autoclaved beakers (Figure S1), and stored in a biosafety cabinet overnight to allow ethanol evaporation.

Subsequently, the fruits in each group (n = 10) were divided into two microplates (each with 12 wells). Two milliliters of sterilized distilled water were added to each of the six wells in the front and back rows to maintain moisture and promote fungal growth. The contact point of the pedicel and each fruit was inoculated with 5 μL of an *A. flavus* IFM 47798 spore suspension (4.6×10^6/mL). The microplates were wrapped with parafilm, with two pieces of autoclaved sponge placed between the cover and body to allow humid air circulation. They were incubated at 25 °C for 15 and 30 days (Figures S3 and S4). After incubation, the fruits were washed with ethanol:water (1:1 *v*/*v*), desiccated in an oven at 50 °C, ground, weighed, and analyzed by HPLC-FLD to determine AF contents.

To confirm the ability of *A. flavus* mycelia to penetrate the seeds of intact jujube fruits and accumulate AFs, GBG fruits (mid-mature stage) that showed high susceptibility to AF contamination in Section 2.1 were selected. The fruits (n = 12) were collected, sterilized, inoculated as described above, and incubated for 30 days. They were then divided into combined exocarp and mesocarp parts and seed parts, desiccated at 50 °C, ground, weighed, and analyzed by HPLC-FLD to determine the AF contents.

To investigate the fluctuation in AF levels in the seeds of intact fruits, GBG fruits (mid-mature stage) were divided into three subgroups (Figure S2): GBG-A (fruits with pedicels), GBG-B (fruits without pedicels), and GBG-C (fruits without pedicels that were artificially damaged at the pedicel contact area). The exocarp and mesocarp were removed from all fruits, and the kernels were sterilized and treated to evaporate ethanol as described above. The kernels of each subgroup (n = 10) were divided into two microplates (each with 12 wells), prepared and inoculated as with intact fruits, and incubated at 25 °C for 5 and 10 days (Figures S5 and S6). The kernels were washed with ethanol:water (1:1 *v*/*v*), divided into endocarp and seed parts, desiccated separately at 50 °C, ground, weighed, and analyzed by HPLC-FLD to determine the AF contents.

As Section 2.3 showed that protective barriers do not directly influence AF level fluctuations in kernel seeds, new experiments were performed to confirm this. The GBG kernels (without pedicels) (n = 10) were sterilized as described above and dipped in an *A. flavus* IFM 47798 spore suspension (4.6×10^6/mL) for 15 min. The kernels were placed in microplates, prepared as previously described, and incubated for 30 days. After incubation, the kernels were sterilized, divided into endocarp and seed parts, desiccated separately at 50 °C, ground, weighed, and analyzed by HPLC-FLD to determine the AF contents.

5.1.2. Jujube Seeds

Seeds of GBG fruits were collected (*n* = 22). Sucrose-free Czapek–Dox agar (CZA) was prepared as previously described (Elamin et al., 2018) [27]. A 100 µL spore suspension (4.6×10^6/mL) of *A. flavus* IFM 47798 was spread on the surface of the medium in a Petri dish (90 mm × 15 mm). The seeds were washed with ethanol:water (1:1 *v*/*v*), placed in sterilized beakers, and stored in a biosafety cabinet overnight for desiccation. Ten seeds were spread on the surface of uninoculated agar and incubated for 5 days at 25 °C. The other 12 seeds were placed on *A. flavus*-inoculated agar and incubated under the same conditions. After incubation, the seeds were pooled, washed with ethanol:water (1:1 *v*/*v*), desiccated in an oven at 50 °C overnight, examined by SEM, and analyzed by HPLC-FLD to determine the AF contents.

5.2. AF Quantification Using HPLC

Each whole intact fruit from the GG, GBG, BG, and DBG groups (*n* = 40); the combined exocarp and mesocarp part and seed of each GBG intact fruit (*n* = 24); the endocarp and seed of each GBG kernel (*n* = 60); and the seeds from the GBG fruits used in the microscopic study (*n* = 10) were ground (Wonder Crusher WC-3; Osaka Chemical Co., Ltd., Osaka, Japan), weighed, and mixed with 4 mL of acetonitrile:water:methanol (6:4:1, *v*/*v*/*v*). The mixture was vortexed for 5 min at room temperature. After centrifugation (4770× *g*, 10 min, 4 °C), 0.4 mL of the supernatant was diluted to 10 mL with phosphate-buffered saline (PBS) containing 0.01% Tween 20 (Kanto Chemical Co., Inc., Tokyo, Japan). The mixture was filtered through a glass-fiber filter (GA-100; Advantec Toyo Kaisha, Ltd., Tokyo, Japan) and transferred to an immunoaffinity column (Aflaking, Horiba, Ltd., Kyoto, Japan). The column was washed twice with 3 mL of PBS and 3 mL of water, then eluted with 3 mL of acetonitrile. The eluate was dried under N_2 gas, mixed with 0.1 mL of water:acetonitrile (9:1, *v*/*v*) and 0.1 mL of trifluoroacetic acid, and vortexed for 15 min. After adding 0.3 mL of water:acetonitrile (9:1, *v*/*v*), the reaction mixture was filtered through a 0.2 µm syringe filter (Minisart® RC 4; Sartorius Stedim Lab. Ltd., Stonehouse, UK) and analyzed by HPLC-FLD (Capcell Pak C_{18} UG 120 column, 250 × 4.6 mm inner diameter; Osaka Soda Co., Ltd., Osaka, Japan) using the same method as in our previous study [22]. The AF mixture standard solution (FUJIFILM Wako Chemicals, Osaka, Japan) was used to prepare calibration solutions for HPLC-FLD determination and recovery experiments. The retention times, limits of detection, and limits of quantification for AFB_1 and AFB_2 were 7.15 and 12.60 min, 0.1 and 0.1 µg/kg, and 0.25 and 0.25 µg/kg, respectively. The linear range for AFB_2 and AFB_2 was 0.078 to 25 µg/kg, with a correlation coefficient (R^2) greater than 0.9946. The samples that were outside the calibration range were diluted and reanalyzed to ensure accurate measurement.

5.3. SEM

SEM was performed using a TM3030 tabletop microscope (Hitachi, Ltd., Tokyo, Japan) under high vacuum. Images were captured in backscattered electron mode at 15 kV with a 5.00 mm working distance. SEM was used to observe the hilar region's shape in the control and artificially contaminated seeds and to assess its permeability to fungal mycelia in artificially contaminated GBG jujube seeds.

Supplementary Materials: The following supporting information can be downloaded at: https://www.mdpi.com/article/10.3390/toxins17030113/s1, Figure S1. The four groups of jujube fruits. Figure S2. Artificially prepared samples of the three subgroups. Figure S3. The jujube fruits were inoculated with the spores of *A. flavus* and incubated for 15 days at 25 °C. Figure S4. The jujube fruits were inoculated with the spores of *A. flavus* and incubated for 30 days at 25 °C. Figure S5. The sample of subgroups GBGA–C before and after 5 days of incubation. Figure S6. The sample of subgroups GBGA–C before and after 10 days of incubation.

Author Contributions: Conceptualization, A.E. and S.S.; methodology, A.E.; software, A.E. and S.S.; validation, A.E. and S.S.; formal analysis, A.E.; investigation, A.E.; data curation, A.E.; writing—original draft preparation, A.E.; writing—review and editing, A.E. and S.S.; visualization, A.E.; supervision, S.S.; project administration, S.S.; funding acquisition, S.S. All authors have read and agreed to the published version of the manuscript.

Funding: This study was funded by Tsumura & Co., Tokyo, Japan.

Institutional Review Board Statement: Not applicable.

Informed Consent Statement: Not applicable.

Data Availability Statement: The original contributions presented in this study are included in this article and Supplementary Materials. Further inquiries can be directed to the corresponding author.

Acknowledgments: We thank Tsumura & Co. for the kind gift of jujube fruits.

Conflicts of Interest: The authors declare no conflicts of interest.

References

1. Agrawal, P.; Singh, T.; Pathak, D.; Chopra, H. An updated review of *Ziziphus jujube:* Major focus on its phytochemicals and pharmacological properties. *Pharmacol. Res. Mod. Chin. Med.* **2023**, *8*, 100297. [CrossRef]
2. Yang, B.; Yang, H.; Chen, F.; Huad, Y.; Jiang, Y. Phytochemical analyses of *Ziziphus jujuba* Mill. var. *spinosa* seed by ultrahigh performance liquid chromatography-tandem mass spectrometry and gas chromatography-mass spectrometry. *Analyst* **2013**, *138*, 6881–6888. [CrossRef] [PubMed]
3. Guo, S.; Duan, J.-A.; Li, Y.; Wang, R.; Yan, H.; Qian, D.; Tang, Y.; Su, S. Comparison of the Bioactive Components in Two Seeds of Ziziphus Species by Different Analytical Approaches Combined with Chemometrics. *Front. Pharmacol.* **2017**, *8*, 609. [CrossRef] [PubMed]
4. Elamin, A.; Sakuda, S. Evaluation of the susceptibility of *Ziziphus jujuba* var. *spinosa* fruit to aflatoxin contamination and infection of aflatoxigenic fungus based on ripening stages and fruit parts. *JSM Mycotoxins* **2021**, *71*, 63–67. [CrossRef]
5. Tulayakul, P.; Sugita-Konishi, Y. Mycotoxin contamination in foodstuffs and feeds-health concerns in Thailand. *Jpn. J. Vet. Res.* **2017**, *65*, 173–183.
6. Ałtyn, I.; Twarużek, M. Review: Mycotoxin contamination concerns of herbs and medicinal plants. *Toxins* **2020**, *12*, 182. [CrossRef]
7. Namjoo, M.; Salamat, F.; Rajabli, N.; Hajihoseeini, R.; Niknejad, F.; Kohsar, F.; Joshaghani, H. Quantitative determination of aflatoxin by high performance liquid chromatography in wheat silos in Golestan province, north of Iran. *Iran J. Public Health* **2016**, *45*, 905–910.
8. Alshannaq, A.F.; Yu, J.-H. A liquid chromatographic method for rapid and sensitive analysis of aflatoxins in laboratory fungal cultures. *Toxins* **2020**, *12*, 93. [CrossRef]
9. Singh, Y.P.; Sumbali, G. Natural incidence of toxigenic *Aspergillus flavus* strain on the surface of pre-harvest jujube fruits. *Indian Phytopathol.* **2000**, *53*, 404–406.
10. Al-Meamar, T.S.; Al-Jassani, M.J.; Hamad, N.S. Aflatoxins and aflatoxigenic fungi contamination of dried fruits in Iraqi market. *J. Glob. Pharma Technol.* **2017**, *9*, 299–308.
11. Rangsipanuratn, W.; Kammarnjassadakul, P.; Janwithayanuchit, I.; Paungmoung, P.; Ngamurulert, S.; Sriprapun, M.; Yangen, S.; Soottitantawat, V.; Sandee, A. Detection of microbes, aflatoxin and toxic heavy metals in Chinese medicinal herbs commonly consumed in Thailand. *Pharm. Sci. Asia* **2017**, *44*, 162–171. [CrossRef]
12. Kachapulula, P.W.; Bandyopadhyay, R.; Cotty, P.J. Aflatoxin contamination of non-cultivated fruits in Zambia. *Front. Microbiol.* **2019**, *10*, 1840. [CrossRef]
13. Sapkota, S.; Sapkota, S.; Wang, S.; Liu, Z. Phenological study of Chinese jujube trees using biologische Bundesanstalt, Bundessortenamt and Chemische Industrie (BBCH) scale. *J. Hortic. Sci. Res.* **2020**, *3*, 68–73.
14. Khan, M.K.U.; Muhammad, N.; Jia, Z.; Peng, J.; Liu, M. Mechanism of stone (hardened endocarp) formation in fruits: An attempt toward pitless fruits, and its advantages and disadvantages. *Genes* **2022**, *13*, 2123. [CrossRef] [PubMed]
15. Reche, J.; Almansa, M.S.; Hernández, F.; Amorós, A.; Legua, P. Physicochemical and antioxidant capacity of jujube (*Ziziphus jujuba* Mill.) at different maturation stages. *Agronomy* **2021**, *11*, 132. [CrossRef]

16. Song, J.; Bi, J.; Chen, Q.; Wu, X.; Lyu, Y.; Meng, X. Assessment of sugar content, fatty acids, free amino acids, and volatile profiles in jujube fruits at different ripening stages. *Food Chem.* **2019**, *270*, 344–352. [CrossRef] [PubMed]
17. Guil-Guerrero, J.L.; Delgado, A.D.; González, M.C.M.; Isasa, M.E.T. Fatty acids and carotenes in some ber (*Ziziphus jujuba* Mill) varieties. *Plant Foods Hum. Nutr.* **2004**, *59*, 23–27. [CrossRef]
18. Zhang, Q.; Wang, L.; Wang, Z.; Liu, Z.; Zhao, Z.; Zhou, G.; Liu, M.; Liu, P. Variations of the nutritional composition of jujube fruit (*Ziziphus jujuba* Mill.) during maturation stages. *Int. J. Food Prop.* **2020**, *23*, 1066–1081. [CrossRef]
19. Zhang, Q.; Wang, L.; Wang, Z.; Zhang, R.; Liu, P.; Liu, M.; Liu, Z.; Zhao, Z.; Wang, L.; Chen, X.; et al. The regulation of cell wall lignification and lignin biosynthesis during pigmentation of winter jujube. *Hortic. Res.* **2021**, *8*, 238. [CrossRef] [PubMed]
20. Fouda, T.; Abdelsalam, A.; Swilam, A.; El-Didamony, M. Determination of physical properties of some seeds. *Sci. Pap. Ser. Manag. Econ. Eng. Agric. Rural Dev.* **2022**, *22*, 223–238.
21. Elamin, A.; Enomoto, H.; Watanabe, M.; Sakuda, S. The Mechanism of ochratoxin contamination of artificially inoculated licorice roots. *Toxins* **2023**, *15*, 219. [CrossRef] [PubMed]
22. Elamin, A.; Sultana, S.; Sakuda, S. Evaluation of the susceptibility of lotus seeds (*Nelumbo nucifera* Gaertn.) to *Aspergillus flavus* infection and aflatoxin contamination. *Toxins* **2024**, *16*, 29. [CrossRef]
23. Baskin, C.C. Breaking physical dormancy in seeds—Focussing on the lens. *New Phytol.* **2003**, *158*, 229–232. [CrossRef]
24. Jayasuriya, K.M.G.G.; Baskin, J.M.; Geneve, R.L.; Baskin, C.C. Morphology and anatomy of physical dormancy in Ipomoea lacunosa: Identification of the water gap in seeds of Convolvulaceae (Solanales). *Ann. Bot.* **2007**, *100*, 13–22. [CrossRef] [PubMed]
25. Pinto, T.T.; Geisler, G.E.; Santos, M.; Paulilo, M.T.S. Germination of physically dormant seeds of *Colubrina glandulosa* Perkins (Rhamnaceae). *Braz. J. Bot.* **2020**, *43*, 91–97. [CrossRef]
26. Mendgen, K.; Hahn, M.; Deising, H. Morphogenesis and mechanisms of penetration by plant pathogenic fungi. *Annu. Rev. Phytopathol.* **1996**, *34*, 364–386. [CrossRef]
27. Elamin, A.; Takatori, K.; Matsuda, Y.; Tsukada, M.; Kirino, F. Fungal biodeterioration of artificial aged linen textile: Evaluation by microscopic, spectroscopic and viscometric methods. *Mediterr. Archaeol. Archaeom.* **2018**, *18*, 103–120.

Disclaimer/Publisher's Note: The statements, opinions and data contained in all publications are solely those of the individual author(s) and contributor(s) and not of MDPI and/or the editor(s). MDPI and/or the editor(s) disclaim responsibility for any injury to people or property resulting from any ideas, methods, instructions or products referred to in the content.

toxins

MDPI

Review

A Review of Decontamination of *Aspergillus* spp. and Aflatoxin Control for Grains and Nuts with Atmospheric Cold Plasma

Miral Javed, Wei Cao, Linyi Tang and Kevin M. Keener *

School of Engineering, University of Guelph, 50 Stone Road East, Guelph, ON N1G 2W1, Canada; mjaved11@uoguelph.ca (M.J.); wcao02@uoguelph.ca (W.C.); ltang07@uoguelph.ca (L.T.)
* Correspondence: kkeener@uoguelph.ca

Abstract: *Aspergillus* spp. and their produced aflatoxins are responsible for contaminating 25–30% of the global food supply, including many grains, and nuts which when consumed are detrimental to human and animal health. Despite regulatory frameworks, *Aspergillus* spp. and aflatoxin contamination is still a global challenge, especially in cereal-based matrices and their derived by-products. The methods for reducing *Aspergillus* spp. and aflatoxin contamination involve various approaches, including physical, chemical, and biological control strategies. Recently, a novel technology, atmospheric cold plasma (ACP), has emerged which can reduce mold populations and also degrade these toxins. ACP is a non-thermal technology that operates at room temperature and atmospheric pressure. It can reduce mold and toxins from grains and seeds without affecting food quality or leaving any chemical residue. ACP is the conversion of a gas, such as air, into a reactive gas. Specifically, an electrical charge is applied to the "working" gas (air) leading to the breakdown of diatomic oxygen, diatomic nitrogen, and water vapor into a mixture of radicals (e.g., atomic oxygen, atomic nitrogen, atomic hydrogen, hydroxyls), metastable species, and ions (e.g., nitrate, nitrite, peroxynitrate). In a cold plasma process, approximately 5% or less of the working gas is ionized. However, cold plasma treatment can generate over 1000 ppm of reactive gas species (RGS). The final result is a range of bactericidal and fungicidal molecules such as ozone, peroxides, nitrates, and many others. This review provides an overview of the mechanisms and chemistry of ACP and its application in inactivating *Aspergillus* spp. and degrading aflatoxins, serving as a novel treatment to enhance the safety and quality of grains and nuts. The final section of the review discusses the commercialization status of ACP treatment.

Keywords: cold plasma; molds; mycotoxins; inactivation; detoxification; food safety

Key Contribution: This review offers a comprehensive overview of atmospheric cold plasma (ACP) and its applications in the inactivation of *Aspergillus* and the degradation of aflatoxin, highlighting its potential as an innovative treatment to improve the safety and quality of grains and nuts. Additionally, it provides insights into the current commercialization status of ACP technology.

1. Introduction

Grains and nuts such as corn, rice, wheat, oats, barley, peanuts, walnuts, hazelnuts, cashew nuts, etc., are prone to aflatoxin contamination along the agricultural chain from the pre-harvest stage through the harvesting stage and post-harvesting stage. Aflatoxin contamination of these common produces has detrimental consequences on the output of the agricultural sector generally while constraining the attainability of the four food security pillars; availability, access, utilization, and stability [1]. Fungal diseases in crops have a major impact on food security, and there are direct measurable economic consequences [2]. Fungi have the ability to grow on all kinds of grains and nuts, where they cause food spoilage such as toxin production, discoloration, off-flavor development, rotting, and formation of pathogenic and allergenic propagules [3]. Maize farmers in the United States face an annual loss of $160 million due to aflatoxin-related problems. In developing countries, particularly in sub-Saharan Africa, the losses are even greater, totaling $450 million, which accounts for 38% of the global agricultural losses caused by aflatoxin [4].

Aspergillus spp. is primarily responsible for aflatoxin production, which can occur at any stage of the crop production chain, influenced by environmental factors such as temperature, humidity, and rainfall, as well as farm management practices like cropping, harvesting, and storage conditions [5]. For example, aflatoxin contamination in maize across China varies by region, with incidence rates ranging from below 40% to over 90% [6]. The average aflatoxin contamination levels range from 1.15 to 107.93 μg/kg. In a separate survey conducted in France, 6% of 114 maize field samples and 15% of 81 maize silo samples tested positive for aflatoxins [7].

Plant stress factors, including drought, injury caused by insects or birds on the host, and inadequate fertilization, are known to accelerate aflatoxin production by *Aspergillus* spp. [8]. The contamination of crops by *Aspergillus* spp. is also influenced by handling and storage conditions at the post-harvest stage [9]. Maize, groundnuts, and tree nuts are common hosts infected by *Aspergillus* spp. at the pre-harvest stage, while wheat, sorghum, and rice are more susceptible at the post-harvest stage [10]. AFB_1, AFB_2, AFG_1, and AFG_2 are the most common aflatoxins [11]. Agricultural products from Sub-Saharan African countries, Gambia, Uganda, Kenya, and Tanzania and Southeast Asian countries such as China, Thailand, Vietnam, and Indonesia have been associated with the most prevalent occurrence of aflatoxins, which has corresponded to the highest rates (40% for Africa and 27% for Southeast Asia) of hepatocellular carcinoma and aflatoxicosis outbreaks [12]. Aflatoxin contamination in grains and nuts is a worldwide concern due to its toxicological effects on humans and animals [13].

To mitigate aflatoxin contamination, current methods of *Aspergillus* reduction and toxin decontamination in grains and nuts include processes with lengthy treatments, high temperatures, or harsh chemicals. Many of these treatments negatively impact the food causing color loss, texture changes, flavor changes, and lower nutritional content [14]. Aflatoxin is known for its high heat tolerance, remaining stable at temperatures reaching 350 °C; thus, heat-based methods of food processing are insufficient to eliminate the toxicity [15]. Extrusion has degraded 77.6% AFB_1 from contaminated peanut meals (35.8 ppb); however, the extrusion process at high temperature (150 °C) turned the peanuts into "peanut butter" [16]. A 65.9% reduction in AFB_1 from the artificially contaminated peanuts (200 ppb) was observed after exposure to a concentration of 6.0 mg/L ozone for 30 min [17]. More than 95% AFB_1 was reduced from artificially contaminated peanuts (300 ppb) with a combination of microwave and gamma irradiation; however, the process impacted peanut quality such as moisture content, color, hardness, and peroxide value [18].

Atmospheric cold plasma (ACP) is an emerging non-thermal food processing technology, which can deliver beneficial results in a short treatment time and with minimal inputs of air and electricity, without heating effects on the quality of food products, and leaves no residues compared to conventional decontamination technologies [19]. The application of ACP can inactivate *A. flavus* and reduce aflatoxin from contaminated commodities, such as pistachio [20–22], hazelnuts [23–27], and maize [28]. A hazelnut study has shown that *A.*

flavus and *A. parasiticus* were reduced by 4.50 log CFU/g and 4.19 log CFU/g, respectively, after 5-minute treatments at 655 Watts with dry air as the plasma-forming gas [23].

In this review, cold plasma technology for the inactivation of *Aspergillus* spp. and degradation of aflatoxins and their mechanism of decontamination have been elaborated. This review has also focused on the application of cold plasma in grains and nuts. Furthermore, the commercialization of ACP treatment has been discussed.

2. Atmospheric Cold Plasma

In recent years, consumers have increased demands for convenient, nutritious, safe, and long shelf-life foods. Existing traditional thermal food processing technologies make it difficult to meet these requirements; therefore, innovative non-thermal food processing and preservation technologies are expected [29]. ACP as a simple, fast, and cost-effective technology with minimized food quality impacts, has been drawing more attention. Current cold plasma research for the food industry focuses on its applications for food decontamination and contaminant degradation [30,31].

Plasma is an ionized gas, generated by adding energy via light, heat, or electricity to a gas. During the process of plasma generation, electrons are stripped away from their respective atoms resulting in ionization. As more energy is added, more "plasma" is generated. Plasma is defined as the fourth fundamental state of matter. When the gas is fully ionized, the created plasma contains light, photons, ions, electrons, free radicals, reactive species, and electrically charged particles [32]. ACP (also known as non-thermal or low-temperature plasmas) are partially ionized gas where approximately 5% or less of the gas exists as plasma [33]. Depending on the system applied, the mechanisms of cold plasma may involve a combination of electric field, ultraviolet light, and reactive gas species. The RGS generated may include ions (H^+, H_3O^+, O^+, H^-, O^-, OH^-, N_2^+, etc.), molecular species (N_2, O_2, O_3, H_2O_2, etc.), and reactive radicals (O•, H•, OH•, NO•, etc.) [21]. RGS with varying stability and half-lives are produced by ACP. Table 1 provides examples of some reactive species along with their half-lives. Species with a half-life of less than 0.5 s are classified as short half-life species [34]. ACP can be generated in air and other gases at room temperature and atmospheric pressure with low energy consumption.

Table 1. Examples of reactive oxygen and nitrogen species produced during ACP treatment and their half-life.

Reactive Species	Half-Life	References
$ONOO^{\bullet-}$ (peroxynitrite)	5–20 milliseconds	[34]
$O_2^{\bullet-}$ (superoxide)	~1 microseconds	[34]
H_2O_2 (hydrogen peroxide)	seconds to minutes	[34]
O_3 (ozone)	99 min under refrigeration	[35]
NO_3 (nitrate)	5–8 h (human blood)	[36]
NO_2 (nitrogen dioxide)	13 min (when $O_3 < 1.0$ ppm)	[37]
NO (nitric oxide)	15 s	[38]
N_2 ($C^3\pi_u$)	38 nanoseconds	[39]
N_2 ($B^3\pi_g$)	13 nanoseconds	[39]
OH• (hydroxyl radical)	~10^{-9} s	[40]

2.1. Dielectric Barrier Discharge (DBD)

One of the most popular techniques for producing ACP is DBD. Here, a dielectric material is applied to one or both electrodes arranged in a parallel plate configuration to create a plasma discharge between them (Figure 1a). In DBD-ACP, the current discharge (i.e., electrical arcs) is limited by the dielectric barrier layer, such as alumina, glass, silica glass, ceramic materials, and thin enamel or polymer layers, which cover one or both electrodes, or it can be suspended between two electrodes; hence, a spark or arc discharge can be avoided. This allows the electric field to interact with the working gas while preventing excess heating from sparks or arcs. The RGS generated is dependent on the equipment setup and process parameters employed such as voltage, working gas, electrode shape, electrode gap, gas volume, food type, packaging materials, and the exposure mode (direct or indirect). The versatility of geometrical configurations and miniaturization, operational parameters, simple design, cost, safety, and characteristics of the power supply are only a few of the many benefits that DBD offers [41]. Dielectric barriers serve as a stabilizing material, preventing arc transitions and aiding in the production of a large number of micro-discharges for homogenous treatments [41]. A DBD-ACP treatment was applied to pistachios contaminated with *A. flavus*. The treatment utilized air as the working gas, with an electric power consumption of 2.49 W/cm^3 and a voltage of 10.7 kV. The pistachios were exposed to the plasma for 18 min, resulting in *A. flavus* from the samples being reduced to an undetectable level. However, no quality evaluations were conducted to assess the impact of the treatment on the pistachios' sensory or nutritional properties [22]. Another study used the DBD-ACP treatment on corn kernels to remove AFB_1 from the contaminated surface. The ACP was operated at two different discharge power levels: low (0.18 W/cm) and high (0.31 W/cm). Treatment durations varied, including 30, 60, 120, 240, and 480 s, with air as the working gas. The treatment successfully achieved 100% removal of AFB_1 from the corn kernels [42]. Therefore, the DBD system is simple and could be easily scaled up for high production volume.

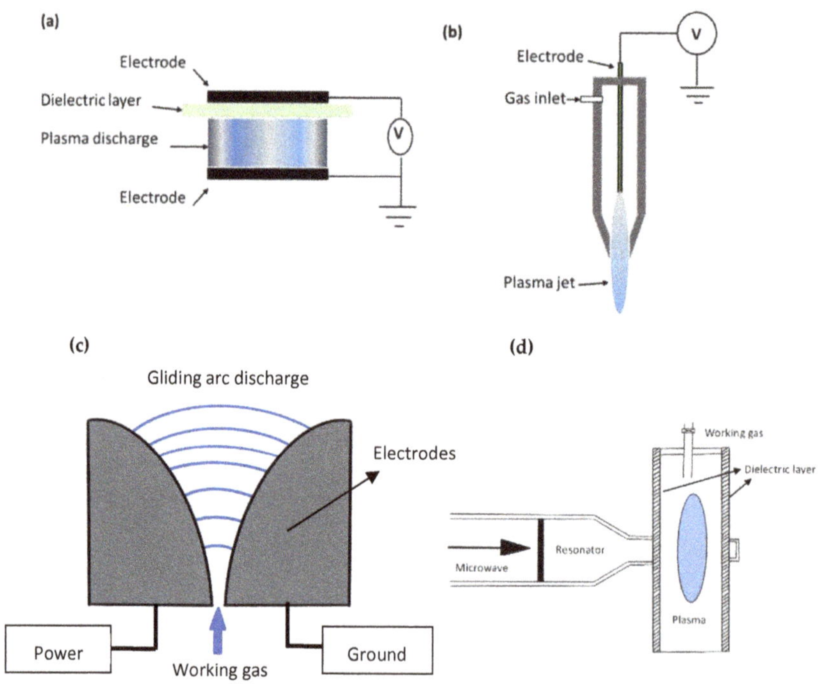

Figure 1. Schematic diagram of dielectric barrier cold plasma (**a**), plasma jet system (**b**) [31], gliding arc discharge (**c**) [43], and microwave discharge (**d**) [44].

2.2. *Atmospheric Pressure Plasma Jets*

The plasma jet can realize the operation of gas discharge in a non-sealed electrode arrangement and the projection of the discharge plasma species into an open environment [45]. The distinctive feature of the atmospheric pressure plasma jet configuration is its ability to launch stable plasma species into a separate environment, where the electric field can be very low [46]. This system usually consists of two concentric rings or coaxial electrodes with a gap of 0.5–6 mm, in which the gas flows between the pair of electrodes at a range of several liters per minute [45]. The outer electrode is grounded, and the central electrode is excited by radiofrequency, typically 13.56 MHz, accelerating the free electrons that collide with gas molecules forming various reactive species [47]. The gas flowing at a high flow rate, usually a noble gas or a mixture with a reactive gas, pushes the plasma formed outside the electrode region, projected as a jet, and discharges plasma species into the open environment [48]. The plasma jet produces a stable, homogenous, and uniform discharge at atmospheric pressure. A plasma jet treatment was applied to 5 g of hazelnuts contaminated with 3 μg/kg of AFB_1. The treatment was operated with dry air (purity 99.99%) at 3000 L/h of gas flow rate, 25 kHz frequency, 655 W of power, and a 7 cm distance between the plasma jet and the sample. The hazelnuts were treated for 1.7 min at room temperature, resulting in a 72% reduction in AFB_1 levels. However, there were no data provided on the plasma jet or hazelnut surface temperature and whether some or all of the reported reductions came from thermal effects. Importantly, the treatment did not compromise the sensory attributes of the hazelnuts, as assessed by a sensory panel, indicating that the plasma jet technology can effectively reduce harmful contaminants like AFB_1 without negatively impacting the food's quality or taste [25,26,49]. Another study was performed using atmospheric pressure plasma jet treatment on 10 g of maize contaminated with *A. flavus* and *A. parasiticus* [50]. The plasma reactor was made from glass with a 49 mm diameter, 147 mm length, and an L/D ratio of 3. A stainless steel stand was designed to hold the reactor on top of the plasma jet (a nozzle with an inner diameter of 4 mm). The generated plasma expanded to the outside of the nozzle with a length of 20 mm. The treatment used dry air as the working gas, with a flow rate of 3000 L/h, a voltage of 10 kV, and a power output of 655 Watts at room temperature. The maize was exposed to the plasma jet for 5 min, resulting in a significant reduction in microbial contamination. Specifically, *A. flavus* was reduced by 5.48 log CFU/g, while *A. parasiticus* showed a 5.20 log CFU/g reduction. However, there were no data provided on the surface temperature and whether some or all of the reductions came from thermal effects. The quality evaluations were conducted to assess the impact of the treatment on the maize's sensory or nutritional properties [23,50]. However, this plasma is limited when aiming at a large area, but the arrangement of several jets must be applied in this case.

2.3. Gliding Arc Discharge

Generating a gliding arc discharge requires a high-velocity gas flowing between two or more divergent electrodes (electrode gap of 1–5 mm) that are connected to a high-voltage power supply (1 kV to 10 kV or higher, and 1–10 kHz frequency) [51]. The gas, such as humid air, pushes the formed arc plasma in the same direction along the axis of the reactor resulting in a series of sliding discharges that stretch across the inter-electrode region. When the gas flow rate between the electrodes increases, it extends the length of the arc along the electrodes and requires larger power to maintain the arc. The plasma gas temperature generated from the gliding arc is several hundred degrees Celsius [52]. During gliding arc discharge plasma, a variety of reactive species are generated due to the interaction of the electrical discharge with the working gas, which is typically air or a mixture of gases. Key reactive oxygen species (ROS) produced include ozone (O_3), singlet oxygen ($^1O_2$), and oxygen atoms (O), all of which arise from the dissociation of oxygen molecules and subsequent recombination or energy transfer [53]. Additionally, peroxy radicals such as HO_2 and $O_2\bullet$ are formed, further enhancing the reactivity. In terms of reactive nitrogen species (RNS), nitric oxide (NO) and nitrogen dioxide (NO_2) are generated through reactions between nitrogen and oxygen in the plasma, while nitrous oxide (N_2O) and peroxynitrite ($ONOO^-$) can also form under certain conditions. Along with these, highly reactive radicals like hydroxyl radicals (OH•), superoxide radicals (O_2•$^-$), and hydrogen peroxide (H_2O_2) are produced, particularly when water vapor is present in the plasma. Excited species, including excited oxygen (O_2*) and nitrogen molecules (N_2*), also play a role in the generation of these reactive species [51,53]. These species contribute to the antifungal effects of gliding arc plasma, as they can damage microbial cell membranes, proteins, and nucleic acids, leading to the inhibition or destruction of *Aspergillus* spp. A gliding arc cold plasma system consists of three parts: the plasma reactor, the mixing of dry air with wet air, and the voltage measurement and recording system, which was used to inactivate *A. flavus* and *A. parasiticus on the wheat* [53]. The plasma reactor includes two stainless steel tubular electrodes with an external diameter of 6 mm and a radius of curvature of 180 mm. These electrodes were positioned at an upper distance of 2 mm and a lower distance of 80 mm from each other, enclosed within a Teflon and glass frame. The reactor bed, where wheat samples were situated, was at a distance of 70 mm from the discharge opening. The treatment was conducted at a frequency of 20 kHz and using dry air (21% oxygen and 79% nitrogen) as the working gas, with an added humidity of 1000 ppm water vapor concentration inside the reactor. The reactor was stated to operate at room temperature; however, there were no data provided on the plasma jet or wheat surface temperature and whether some or all of the reported reductions came from thermal effects. The plasma treatment resulted in significant reductions in fungal growth, with *A. flavus* showing a 78.74% reduction and *A. parasiticus* a 68.57% reduction. Notably, the plasma treatment inhibited fungal growth by extending the lag phase and reducing the overall growth rate of the fungi, highlighting its potential as a viable alternative to fungal contamination [53]. However, the lower the gas flow rate, the greater the NOx concentration produced in the water [54].

2.4. Microwave Discharge Plasma

Microwave discharges are the electrical discharges generated by electromagnetic waves with frequencies exceeding 300 MHz. In microwave plasma generators, electromagnetic waves typically at 2.45 GHz are emitted by a magnetron and used to produce microwave discharges, with a power of 500 W to 2 kW, a voltage (Magnetron) of 2 kV to 4 kV, a gas temperature of 300 K to 1000 K for non-thermal plasma, a gas flow rate of 1–10 L/min, and a pressure of 0.1 Torr to atmospheric pressure [44,55]. The microwave electric field accelerates the electrons of gas molecules and forms cold plasma as a result, without electrodes. This system can produce plasma at low atmospheric pressures. Some important advantages of using microwaves for cold plasma production include increased electron density, high efficiency in generating reactive species, and contamination-free processes [56]. At the plasma–water interphase, surface-wave microwaves generate plasma discharge. Varying the initial gas mixture composition (Ar, $Ar/N_2/O_2$ binary, and ternary mixture) and the distance between the edge of the plasma-discharge tube and the water surface altered the electron and species densities at the interaction between the plasma and water. H_2O_2 was undetected in plasma-activated water (PAW) via the microwave plasma system; however, $ONOO^-$ was generated as an isomer of NO^{3-}. A study was conducted using the afterglow of a low-pressure $Ar/N_2/O_2$ (N_2-20%O_2, N_2-10%O_2, N_2-2%O_2, Ar-20%O_2, Ar-40%O_2, and Ar-20%O_2+N_2-2%O_2) microwave discharge at 200–800 Pa pressure, rich in oxygen and nitrogen atoms, O_2 and NO molecules, and UV radiation. The study explored the impact of ROS and RNS on barley and corn seed germination, as well as their surface disinfection from two types of *Fusaria*, *Fusarium graminearum*, and *Fusarium verticillioides*. The results showed that the germination and vigor of non-infected seeds were not significantly affected when barley was treated for 120 s at 200 Pa (only 2% of the pressure of one atmosphere) and corn for 240 s at 400 Pa. Afterglow treatments, independently of the mixture composition, do not significantly influence the germination and the vigor of seeds, although they show a slightly decreasing trend for longer treatment times [56]. However, the seeds were effectively decontaminated from the germination inhibitors, *Fusarium graminearum*, and *Fusarium verticillioides*, when barley was treated for 3 min at 400 Pa with an 80% Ar-20%O_2 discharge afterglow, and corn for 4 min at 800 Pa in the same gas mixture, followed by a 2-min treatment with an N_2-2%O_2 afterglow. The temperature of the gas was lower than 40 °C [56]. These treatments also improved the germination of the infected seeds from 35% and 50% to above 80%, respectively. Additionally, it was found that high NO content in the mixtures and surface heating of the seeds due to the recombination of O and N atoms inhibited barley germination. These findings highlight the importance of carefully optimizing plasma species to effectively decontaminate.

3. Application of ACP Against *Aspergillus flavus* and Aflatoxins in Nuts and Grains

3.1. Nuts

In recent years, there has been increased interest in the potential applications of ACP for microbial decontamination and mycotoxin treatments, including *A. flavus* inactivation and aflatoxin degradation from contaminated commodities, such as pistachios [20,21], hazelnuts [24,27], and peanuts [57]. Promising results of *Aspergillus* inactivation and aflatoxin degradation have been reported. Table 2 summarizes the recent research using cold plasma to inactivate *Aspergillus* spp. in nuts. The plasma jet has successfully reduced 5.14 log CFU/g from pistachios with a 15 min treatment [20], and 4.64 log CFU/g from hazelnuts after a 1.7 min treatment [25]. The temperature of the treated samples was measured at 20 kV due to the higher temperature of the plasma jet at this voltage and recorded at 27 °C. The quality properties of pistachios such as moisture content, peroxide value, free fatty acid, and total polyphenol were measured after plasma treatment and compared with untreated samples, which indicated no significant differences. Meanwhile, a sensory evaluation was performed to determine the acceptance of the color, smell, and taste of plasma-treated pistachio samples, which received satisfactory scores [21]. No quality experiments were conducted; instead, they performed a sensory evaluation and reported satisfactory scores for texture, overall color, odor, and overall acceptance of plasma-treated hazelnuts. A plasma jet was also applied to inactivate both *A. flavus* and *A. parasiticus* from hazelnuts [23] with the same treatment conditions. After 655 W and 5 min dry-air plasma treatments, 4.50 log CFU/g of *A. flavus* and 4.19 log CFU/g of *A. parasiticus* were decontaminated from hazelnuts. However, quality parameters of plasma-treated hazelnuts were not conducted.

Table 2. Summary of literature on cold plasma for inactivation of *Aspergillus* spp. in nuts.

Plasma Source	Sample Characteristics	Organism(s)	Treatment Conditions	Significant Findings	References
Plasma jet	10 pieces pistachio kernels	*A. flavus*	• Working gas: 62% argon/38% air • Voltage: 20 kV • Treatment time: 15 min • Flow rate: 3 L/min gas	• ~5.14 log CFU/g reduction • No effect on quality characteristics of pistachio kernels • The temperature of treated pistachio was 27 °C	[20]

Table 2. *Cont.*

Plasma Source	Sample Characteristics	Organism(s)	Treatment Conditions	Significant Findings	References
Plasma jet	10 g hazelnuts	*A. flavus* and *A. parasiticus*	• Working gas: air • Flow rate: 3000 L/h • Power: 655 W • Frequency: 25 kHz • Treatment time: 1.7 min	• 5.4 and 5.5 log CFU/g of reductions in *A. flavus* and *A. parasiticus*, respectively • No quality evaluations conducted • Neither the temperature of the plasma jet or the treated hazelnuts was reported.	[23]
Plasma jet	10 g hazelnuts	*A. flavus* and *A. parasiticus*	• Working gas: dry air • Flow rate: 3000 L/h • Power: 655 W • Frequency: 25 kHz • Treatment time: 5 min	• 4.50 and 4.19 log CFU/g reduction for *A. flavus* and for *A. parasiticus*, respectively • Neither the temperature of the plasma jet or the treated hazelnuts was reported No quality evaluations conducted	[23]
Plasma jet	15 g peanuts	*A. flavus and A. niger*	• Working gas: Argon • Power: 200 W • Rotary speed: 1000 rpm • Flow rate: 30 L/min • Treatment time: 3.5 min	• Reduced to undetectable level from the initial inoculation level of 6.39 log CFU/g for *A. flavus* and 5.83 log CFU/g for *A. niger*, respectively • Neither the temperature of the plasma jet or the treated peanuts was reported No effect on quality characteristics	[58]
Plasma jet	1 g peeled peanuts	*A. flavus*	• Working gas: 9 L/min Argon mixed with 90 mL/min O_2 • Power: 20 W • Treatment time: 10 min	• 100% microbial growth inhibition (initial inoculation: 50 µL of 5×10^9 conidia/mL) • Neither the temperature of the plasma jet or the treated peanuts was reported No quality evaluations conducted	[59]
DBD	4 g raw pistachio	*A. flavus*	• Working gas: air • Power: 130 W • Voltage: 15 kV • Treatment time: 3 min	• 4.0 log CFU/g reduction • Pistachio kernels became darker • Antioxidant activity of pistachio nuts slightly increased • Neither the temperature of the plasma gas or the treated pistachio nuts was reported	[21]
DBD	Pistachio	*A. flavus*	• Working gas: air • Electric power consumption: 2.49 W/cm^3 • Voltage: 10.7 kV • Treatment time: 18 min	• *A. flavus* was completely removed after treatment • Neither the temperature of the plasma gas or the treated pistachio nuts was reported • No quality evaluations conducted	[22]
DBD	35 g hazelnuts	*A. flavus*	• Working gas: 1. Ambient air 2. Synthetic air without water vapor 3. Synthetic air with water vapor • Flow rate: 3 L/min • Power: 400 W • Treatment time: 180 s	• *A. flavus* reduction: Gas 1: 2.95 log CFU/g Gas 2: 2.56 log CFU/g Gas 3: below detection limit • Neither the temperature of the plasma gas or the treated hazelnuts was reported • No quality evaluations conducted	[24]

Table 2. *Cont.*

Plasma Source	Sample Characteristics	Organism(s)	Treatment Conditions	Significant Findings	References
DBD	10 g raw groundnuts	*A. flavus* and *A. parasiticus*	• Working gas: air • Power: 60 W • Voltage: 1950 V • Treatment time: 24 min	• 99.9% and 99.5% inactivation of *A. paraticus* and *A. flavus*, respectively • Complete disintegration of fungal spore membrane due to electroporation and etching caused by the reactive species of plasma • Neither the temperature of the plasma gas or the treated groundnuts was reported • No quality evaluations conducted	[60]
DBD	10 g raw peanut kernels	*A. flavus*	• Working gas: 80% humidity air • Voltage: 90 kV • Treatment time: 10 min • Post-treatment storage: 24 h	• 99.9% reduction (initial level of 3.50 log spore per sample) of *A. flavus* • The temperature of the gas and treated peanut was lower than 42 °C • No significant effect on quality	[57]

DBD was widely applied to treat different food commodities. A study reported that 4.0 log CFU/g *A. flavus* from pistachio was reduced using DBD with 130 W and 3 min [21]. However, after 3-minute plasma treatment, the instrumental color value a* (redness) increased considerably and b* (yellowness) decreased slightly, and the overall color changes resulted in a darker pistachio nuts color. But the antioxidant activity of pistachio nuts slightly increased after treatment. Another research conducted by researchers [21] also applied DBD to inactivate *A. flavus* spores from artificially contaminated pistachio surfaces and indicated *A. flavus* was completely inactivated after an 18 min treatment. Meanwhile, the temperature, pH, moisture content, and color of pistachio samples remained unchanged after plasma treatment. Moreover, we applied DBD to inactivate *A. flavus* from raw hazelnuts using different working gases [24]. After 400 W and 3 min treatments, 2.95 log CFU/g and 2.56 log CFU/g were reduced using ambient air (21% O_2 + 78% N_2) and synthetic air (20%O_2 + 80%N_2) without water vapor, respectively. Contrarily, *A. flavus* spores were reduced to below the detection limit (1.0 log CFU/g) from an initial population of 4.6 CFU/g when using synthetic air with water vapor to generate plasma. Although their treatments effectively inactivated *A. flavus* from hazelnuts, there were no quality experiments conducted after plasma treatments (Table 2). A treatment of 5 min resulted in a 2.20 log spores/sample reduction of *A. flavus* spores on raw peanut kernels. *A. flavus* was almost completely inactivated (99.9%) by DBD treatment for 10 min with 80% RH in the air and 24 h post-treatment storage without adversely affecting peanut quality [57]. Scanning electron microscopy (SEM) was applied to investigate the impact of atmospheric pressure plasma treatment on the morphology of *A. flavus* [20,25]. They all concluded that active plasma species disintegrated the cell wall and cell contents such as cytoplasmic structures were released from plasma-treated *A. flavus* spores. Cold plasma's antimicrobial effect is a result of the action of charged particles and reactive species generated in the plasma, which can cause damage to the cell membrane, leading to further penetration of reactive species into the cell, DNA damage, and the disruption of chemical bonds [29]. In a light microscopic analysis of *A. flavus* cells treated with a radiofrequency plasma jet, the Study reported that the conidiophores and vesicles were observed to be broken after plasma treatment, resulting in cell leakage and loss of viability [50].

Table 3 illustrates the recent work with cold plasma on the degradation of aflatoxins in nuts. A study explored the AFB_1 degradation effect of a plasma jet on pistachio kernels [26]. The authors observed that up to 80% AFB_1 was reduced after 15 min of atmospheric plasma treatment. Pistachio quality and sensory evaluation were also assessed after plasma treatment and indicated no significant differences between plasma-treated samples and controls. In a study, a plasma jet was applied to 5 g of hazelnuts containing 3 μg/kg of AFB_1 [26]. The treatment parameters included a power of 655 W, air as the working gas, a flow rate of 3000 L/h, and a treatment duration of 1.7 min. The results showed a 72% reduction in AFB_1 levels. Additionally, the sensory attributes of the hazelnuts were maintained, as evaluated by a sensory panel, indicating that the plasma treatment did not negatively affect the quality of the food. A DBD system was applied with N_2 to generate plasma to degrade AFB_1 from hazelnuts [27] resulting in total aflatoxins and AFB_1 reductions of over 70% after 12 min of treatment. The authors reported that AFB_1 and AFG_1 were more sensitive to plasma treatments compared to AFB_2 and AFG_2, respectively. AFB_1 was more sensitive to plasma treatments compared to AFG_1. However, there was no quality analysis of plasma-treated hazelnuts. A 71.3% AFB_1 reduction in the raw peanut kernels was achieved with a DBD treatment for 2 min and 80% humid air as the working gas without post-treatment. The reduction of AFB_1 toxin also significantly increased with increasing treatment time, higher RH and post-treatment storage. There were no statistical differences in the color, texture, and peroxide value between treated and untreated peanuts [57]. A DBD system was used to treat AFB_1 contaminated pistachios. According to the study, the maximum AFB_1 reduction rate (52.42%) was observed after 130 W and 180 s [21]. After applying ACP to treat pure AFB_1 in 40% RH air for 5 min, six main degradation products were identified ($C_{16}H_{17}O_6$, $C_{17}H_{15}O_7$, $C_{14}H_{13}O_5$, $C_{14}H_{11}O_6$, $C_{17}H_{13}O_7$, $C_{19}H_{19}O_8$) and it was stated that AFB_1 was degraded by cold plasma by targeting the furan ring, lactone ring, cyclopentanone, and the methoxyl group [57,61].

Table 3. Application of atmospheric cold plasma for aflatoxin degradation in nuts.

Plasma Source	Sample Characteristics	Treatment Conditions	Detection Technique	Significant Findings	References
Plasma jet	5 g hazelnut (3 μg/kg AFB_1)	• Working gas: air • Flow rate: 3000 L/h of Air • Power: 655 W • Frequency: 25 kHz • Treatment time: 1.7 min	HPLC	• 72% AFB1 reduction • Maintained the sensory attributes of food evaluated by a sensory panel • Neither the temperature of the plasma gas or the treated hazelnuts was reported	[26]
Plasma jet	10 pistachio kernels (20 μg/kg aflatoxins)	• Voltage: 20 kV • Working gas: 62%argon-38%air • Flow rate: 3 L/min gas • Treatment time: 15 min	Not avaliable	• 80% AFB1, 76% AFB2, 42% AFG1, and 51% AFG2 reduction • No effect on quality characteristics of pistachio kernels • After treatment the sample temperature was 27 °C	[20]
Atmospheric pressure plasma jet (APPJ)	15 g raw peanuts	• Working gas: air • Gas flow rate: 107 L/min • Voltage: 4.4 kV • Frequency: 70–90 kHz • Treatment time: 2 min	ELISA	• 23% total aflatoxins reduction (from 62.3 to 48.2 ppb) • Less bitterness than controls • No changes in the quality of peanut oil	[62]
DBD	10 g groundnuts	• Working gas: air • Power: 60 W • Treatment time: 12 min	HPLC	• 96.8% decrease in AFB1 production from *A. flavus* • Upto 95% decrease in AFB1 from *A. paraticus* • Neither the temperature of the plasma gas or the treated groundnuts was reported • No quality assessment conducted	[23]

Table 3. *Cont.*

Plasma Source	Sample Characteristics	Treatment Conditions	Detection Technique	Significant Findings	References
DBD	20 ng/g of AFB_1 sprayed on 40 g raw hazelnuts	• Working gas: N2 • Flow rate:120 L/min • Power: 1000 W • Treatment time: 12 min	HPLC-MS/MS	• Total aflatoxins and AFB1 reductions over 70% • AFB1 and AFG1 more sensitive to plasma treatments compared to AFB2 and AFG2, respectively • AFB1 more sensitive than AFG1 • Neither the temperature of the plasma gas or the treated hazelnuts was reported • No quality evaluations conducted	[27]
DBD	4 g of unskinned and milled pistachio sample was contaminated with AFB_1 (383.19 μg/kg)	• Working gas: air • Power: 130 W • Voltage: 15 kV • Frequency: 20 kHz • Treatment time: 60, 120, and 180 s	HPLC	• AFB1 reduction: o 60 s: 32.31% o 120 s: 44.77% o 180 s: 52.42% • Plasma treatment led to darker pistachio samples; slightly increased the antioxidant activity of pistachio samples	[21]
DBD	60 μL of 50 μg/mL AFB_1 stock solution spot-inoculated on 15 g raw peanut kernels	• Working gas: 5, 40, and 80% humidity air • Voltage: 90 kV • Treatment time: 2, 5, and 10 min • Post-treatment storage: 0 h and 24 h	ELISA	• 71.3% AFB1 reduction achieved with a treatment of 2 min and 80% RH without post-treatment • Reduction of AFB1 increased with increasing treatment time, air humidity and post-treatment storage • The temperature of gas and treated peanuts was lower than 42 °C	[57]

3.2. Grains

Atmospheric cold plasma treatment has been proven to decontaminate aflatoxigenic fungi (*A. flavus and A. parasiticus*) on corn [50,63,64], rice [65], and wheat [53,63]. Table 4 summarizes the recent research using cold plasma to inactivate *Aspergillus* spp. in grains. A 10 min treatment at 85 kV reduced the number of viable *A. flavus* spores by more than 95% [15]. The optical emission and absorption spectroscopy demonstrate ionization of hydroxyl groups, atomic oxygen and nitrogen, and confirm the production of ROS and RNS, e.g., O_3, NO_2, NO_3, N_2O_4, and N_2O_5. Treated *A. flavus* spores show surface ablation and membrane degradation by scanning electron microscopy. An atmospheric pressure fluidized bed plasma (APFBP) system was used for the decontamination of maize artificially contaminated with *A. flavus* and *A. parasiticus* spores [50]. After treatment of 5 min and using air as the working gas, the maximum reduction was 5.48 and 5.20 log CFU/g for *A. flavus* and *A. parasiticus*, respectively. The temperature of maize after treatment was below 45 °C, which could be considered as a suitable value for "non-thermal" treatment of food materials. But the quality of treated maize samples was not assessed. Another study investigated the use of gliding arc plasma to control the growth of *A. flavus* and *A. parasiticus* on wheat [53]. The plasma treatment was performed using dry air as the working gas, consisting of 21% oxygen and 79% nitrogen, with the injected air containing 1000 ppm of water vapor for humidity inside the reactor. The results showed that plasma treatment led to a significant reduction in fungal growth, with a 78.74% decrease in *A. flavus* and a 68.57% decrease in *A. parasiticus*, respectively. The plasma treatment exhibited its fungal inhibitory effect by extending the lag phase and reducing the growth rate of the fungi. But the temperature and quality of treated wheat have not been evaluated in this study.

A low-pressure cold plasma (LPCP) system was applied to decontaminate *Aspergillus* spp. in wheat, oat, barley, and corn using air and SF_6 as working gases [63]. The running pressure was 66.7 Pa and the applied power was around 300 Watts. A reduction of 3.0 log CFU/g of *Aspergillus* spp. was achieved within a treatment time of 15 min while preserving the germination quality of the seed. The temperature of maize and gas plasma after treatment was not tested. Although the physiological or biochemical basis for inactivation is unknown, air or SF_6 plasma is biocompatible which means that damage to living seeds is avoided and offers a suitable treatment for the surface disinfection of seeds [63]. A 20-min DBD treatment with pure O_2-fed, voltage of 13 kV and frequency of 3 kHz resulted in a reduction of conidial germination and subsequent growth of *A. flavus* on maize by 50–90% [64]. However, the effect of plasma on the temperature and quality of maize has not been evaluated in the study.

Aflatoxin contamination of cereal grains is a global food safety issue. Cold plasma degradation of mycotoxins is a promising approach [66]. The aflatoxin degradation pathways using cold plasma treatment are inevitably related to their molecular structure, the nature of the plasma chemistry, and thus, the species interaction with toxin molecules [31]. Aflatoxin destruction using cold plasma has focused on multiple aspects, including plasma source, working gas, sample type, etc. A summary of research studies demonstrating aflatoxin degradation in grains using cold plasma is provided in Table 5. The reactive gas species such as ozone, hydroxyl, and free radicals (O•, H•, OH•, NO•, etc.) produced by plasma treatment contribute to aflatoxin degradation [28]. AFB_1 is one of the main mycotoxins that are usually found in cereals [67]. Artificially spiked AFB_1 was effectively degraded from corn by using DBD treatment, and it was reported that more than 62% AFB_1 was degraded by 1 min treatment in 40% RH air. In another study, a corona discharge plasma jet was used to degrade AFB_1-spiked food commodities for 30 min and reducing 56.6% and 45.7% from wheat and rice with the initial toxin concentration levels of 7.14 μg/kg and 5.97 μg/kg of AFB_1, respectively [49]. The working gas used was air, with an air blower rotational speed of 3312 rpm, a treatment time of 30 min, and a voltage of 20 kV. The oxidation involving free radicals in plasmas may have played a dominant role in AFB_1 degradation. However, no quality evaluations and temperature measurements were conducted to assess the impact of the treatment on the rice and wheat.

A high-voltage (80 kV) ACP treatment on degradation of aflatoxin in corn (Table 5). Gas type (Air, MA65), relative humidity (5, 40, 80% RH), treatment time (1, 2, 5, 10, 20, and 30 min), mode of reaction, and post-treatment storage were applied and results showed that aflatoxin in corn was degraded by 62% and 82% by 1- and 10-min ACP treatment in 40% RH air, respectively [28]. Higher degradation of aflatoxin was achieved by the treatment with MA65 as a working gas, at higher relative humidities (40% and 80%). The temperature of filled gases changed from 22 °C to around 38 °C. To determine the effect of DBD cold plasma on AFB_1 and the physicochemical characteristics of oats, whole oat grains were subjected to plasma treatment for 1, 3, and 6 min at powers of 32, 37, and 42 W, respectively [68]. Untreated samples were also evaluated as controls. The results showed that DBD treatment had the greatest reduction in 6 min at a power of 37 W which reduced AFB_1 concentration by 82%. Furthermore, there was no significant difference in some qualitative parameters such as protein and pH compared with the control sample, whereas the peroxide value and β-glucan increased and alpha-tocopherol decreased [68]. Another report on the complete degradation of AFB_1 from artificially contaminated corn kernels involved an indirect DBD treatment working in ambient air. Surface analysis of the corn kernels revealed the presence of minor changes in the treated samples, indicating that indirect DBD treatments induced slight oxidation of the corn kernel surface [42].

Table 4. Summary of literature on cold plasma for inactivation of *Aspergillus* spp. in grains.

Plasma Source	Sample Characteristics	Organism(s)	Treatment Conditions	Significant Findings	References
Fluidized bed plasma	10 g maize	*A. flavus* and *A. parasiticus*	• Working gas: air and N2 • Flow rate: 3000 L/h • Voltage: 10 kV • Power: 655 W • Treatment time: 5 min	• Reduction of 5.48 log CFU/g for *A. flavus* and 5.20 log CFU/g for A. parasiticus, respectively • Temperature inside the fluidized bed reactors during the plasma process below 45 °C • No quality evaluations conducted	[50]

Table 4. *Cont.*

Plasma Source	Sample Characteristics	Organism(s)	Treatment Conditions	Significant Findings	References
Plasma jet	Brown rice bar	*A. flavus*	• Working gas: 60% argon -40% air • Voltage: 20 kV • Treatment time: 15 min • Flow rate: 3 L/min gas • Power of 20 W and 40 W • Treatment time 5, 15 and 25 min	• Plasma power of 40 W and exposure time of 20 min required to reduce mold growth of approximately 4.0 log CFU/g • Neither the temperature of the plasma gas nor the treated brown rice bar was reported	[61]
DBD	Maize seeds	*A. flavus*	• Working gas: pure O^2 (air-liquid, 99.999%) and synthetic air (air-liquid, 99.999%) • Voltage: 13 kV • Frequency: 6 kHz • Treatment time: 3 and 20 min	• 50–90% reduction of conidial germination and subsequent growth of *A. flavus* on maize by air-fed plasmas • Neither the temperature of the plasma gas nor the treated groundnuts was reported • No quality assessment	[64]
Gliding arc plasma	Wheat	*A. flavus* and *A. parasiticus*	• Working gas: dry air, consisting of 21% oxygen and 79% nitrogen. • Humidity of the injected air (1000 ppm water vapor concentration) inside the reactor	• 78.74% and 68.57% reduction in the growth of *A. flavus* and A. parasiticus, respectively • Plasma induced a fungal inhibitory effect by prolonging the lag time and decreasing the growth rate • Neither the temperature of the plasma gas or the treated wheat was reported	[53]
Very Low pressure cold plasma (LPCP)	Wheat, oat, barley, corn	*A. flavus*	• Working gas: air, SF6 • Pressure: 66.7 Pa (0.0006 atm) • Frequency: 1 kHz • Voltage: 20 kV • Power: 300 W • Treatment time: 5, 10, and 20 min	• 3.0 log CFU/g reduction achieved within 15 min of treatment • Seed germination retained after treatment • No temperature test after treatment	[63]

Table 5. Application of cold plasma for aflatoxin degradation in grains.

Plasma Source	Sample Characteristics	Treatment Conditions	Detection Technique	Significant Findings	References
Plasma jet	50 g rice (5.97 μg/kg AFB_1)	• Working gas: air • Air blower rotational speed: 3312 rpm • Voltage: 20 kV • Frequency: 58 kHz • Treatment time: 30 min	HPLC	• 45.7% AFB1 reduction in rice • No quality evaluations conducted • Neither the temperature of the plasma jet nor the treated rice was reported.	[49]
Plasma jet	50 g wheat (7.14 μg/kg AFB_1)	• Working gas: air • Air blower rotational speed: 3312 rpm • Voltage: 20 kV • Frequency: 58 kHz • Treatment time: 30 min	HPLC	• 56.6% AFB1 reduction in wheat • No quality evaluations conducted • Neither the temperature of the plasma gas nor the treated wheat was reported	[49]
Gliding arc plasma	Wheat	• Working gas: dry air, consisting of 21% oxygen and 79% nitrogen • Treatment time: 12 min • Power: 10.88 W	HPLC	• 64%, 41%, 59%, 40%, and 61% reduction for AFB1, AFB2, AFG1, AFG2, and total aflatoxin, respectively • Neither the temperature of the plasma gas nor the treated wheat was reported	[53]
DBD	Artificially spiked 100 uL AFB_1 solution in chloroform (50 μg/mL) on 25 g corn kernels	• Working gas: air, and MA65 • Voltage: 90 kV • Relative humidity: 5, 40, 80% RH • Treatment time: 1, 2, 5, 10, 20, and 30 min • Post-treatment storage: 0 and 24 h	Lateral Flow Devices (LFDs)	• AFB1 degraded by 62% and 82% with 1 and 10 min in 40% RH air and 24 h post-treatment, respectively • Higher AFB1 degradation achieved in MA65 at higher relative humidities • The maximum temperature of plasmas was 38 °C • No quality evaluations conducted	[28]
DBD	20 g rice (AFB_1, AFB_2, AFG_1, AFG_2 of 20 μg/kg, 50 μg/kg, 100 μg/kg, 150 μg/kg)	• Working gas: 0%O_2 + 70%N_2 + 30%CO_2, 25%O_2 + 45%N_2 + 30%CO_2, 45%O_2 + 25%N_2 + 30%CO_2, and 65%O_2 + 5%N_2 + 30%CO_2 • Voltage: 70–100 kV • Treatment time: 5–60 min • Input power: 220 V-50 Hz • Relative humidity: 56 ± 1%	UPLC H-Class/ Xevo TQ-S	• Reduced 55.34% of AFB1 and 56.37% of aflatoxins in rice • Degradation rate of aflatoxins increased with the increase in oxygen content • Neither the temperature of the plasma gas nor the treated rice was reported • No significant impact on rice quality except for fat oxidation	[69]

Table 5. *Cont.*

Plasma Source	Sample Characteristics	Treatment Conditions	Detection Technique	Significant Findings	References
DBD	Oat	• Working gas: Argon • Gas flow rate: 1 NL/min • Voltage: 1–10 kV • Treatment time: 1, 3, and 6 min • Power: 32, 37, and 42 W	HPLC	• Treatment for 6 min at a power of 37 W reduced AFB1 concentration by 82% • Increased the amount of peroxide and β-glucan, decreased alpha-tocopherol and viscosity of oat flour • Improved the farinograph properties of oat flour • Neither the temperature of the plasma gas nor the treated oat was reported	[68]
DBD	Corn kernels inoculated with 20 μL AFB_1 (33 μg/mL)	• Working gas: air • Powder: 0.18 and 0.31 W/cm • Treatments time: 30, 60, 120, 240, 480 s	Not available	• 100% AFB1 removed from the contaminated corn kernels • No significant chemical or morphological modifications of the corn kernel surface • Neither the temperature of the plasma gas nor the treated corn was reported.	[42]
DBD	Maize seeds spiked with 0.25 mL of AFB_1 solution (1 μg/mL)	• Working gas: pure O_2 (air-liquid, 99.999%) and synthetic air (air-liquid, 99.999%) • Voltage: 13 kV • Frequency: 6 kHz • Treatment time: 3 and 20 min	HPLC-HRMS	• AFB1 degradation of 23% on kernels achieved after 20 min with pure O_2 process • Neither the temperature of the plasma gas nor the treated maize was reported. • No quality assessment of treated maize	[64]

4. Commercialization of ACP Technology

Cold Plasma technology has a current global market value of 3.2 billion dollars with a Compound Annual Growth Rate (CAGR) of 18% [70]. Much of this market value is in semiconductor manufacturing with growing applications in medicine such as skin rejuvenation and wound healing. The application of cold plasma for the treatment of grains, nuts, fruits, vegetables, and seeds remains a very small global market of approximately 5%, but agricultural applications are growing. Currently, there are several seed treatments utilizing cold plasma including ActivatedAir^{TM} (Zayndu, Ltd., Loughborough, UK) and Clean Crop Technologies (Holyoke, MA, USA).

Commercial cold plasma technology available for mold control and mycotoxins remains limited. For example, NanoGuard Technologies (Louis, MO, USA) claims their cold plasma technology can significantly reduce mold on grains and seeds and reduce aflatoxin by 50% or more [71]. There is no information available on their website regarding whether their technology has been deployed in the grain industry. The Cold Plasma Group (Kingston, ON, Canada) advertises a cold plasma treatment chamber available for use in reducing molds and bacteria on cannabis and cannabis products without reducing the volatile compounds [72]. But their website lacks details on commercialization installation.

In the next five years, exponential growth in cold plasma applications is expected in food safety, food preservation, and food production because of the versatility of the technology to transform air into plant fertilizer, eliminate airborne and surface contaminants, including toxins and molds, while minimally impacting the nutritional value. There are three primary barriers to industry adoption:

(1) The lack of commercial-scale cold plasma equipment available for purchase. Hundreds of research laboratories have developed cold plasma technologies, but food manufacturers want to purchase cold plasma equipment proven to work for their applications. Designing these commercial-scale systems requires manufacturing partners willing to invest in tooling and specialized high-voltage generators with a future return on investment. Existing cold plasma technologies used in medical and semiconductor manufacturing are not suitable for large food processing facilities.

(2) The sequence of steps needed to demonstrate the safety of cold plasma-treated food is highly variable. Currently, regulations under the FDA, USDA, and EPA may apply

to cold plasma technologies. Agricultural and food companies wanting to introduce this new technology into their manufacturing process, have to navigate these uncharted waters. There is significant risk, as with any new technology, in generating adequate data and compelling data to demonstrate a 'de minimis' effect on the food. The regulations are complex beyond the scope of this review, but suffice it to say that they must be understood and followed to ensure a safe food product.

(3) The selectivity of the cold plasma process is very product- and process-dependent. Commercial adoption of cold plasma technology requires a complete understanding of the reactive gas species generated based on the plasma source (gas, voltage, electrodes, treatment time, etc.); the interaction of the reactive gas species with the food; and the reactive gas species reversion process after treatment. For example, what happens with those reactive gas species remaining which do not interact with the food? Are they exhausted on the outside, recaptured, and reused? Are they released inside the facility potentially exposing workers? There are many complex chemistries of gas, solid, and liquids that are needed in the design of a cold plasma treatment for food. Also, the cold plasma process needs to be robust, such that it can guarantee effectiveness regardless of the variability in the incoming grain or seeds.

As more startups and established companies expand into cold plasma equipment design and agricultural applications, it is expected that a cold plasma device manufacturing sector will be established, and as the technology matures, equipment costs will come down and equipment reliability will increase.

5. Conclusions

Aspergillus spp. are commonly found in agricultural products such as grains and nuts, producing aflatoxins, which are harmful mycotoxins. Aflatoxins, particularly B_1, B_2, G_1, and G_2, pose serious health risks to both humans and animals. ACP is gaining attention as a non-thermal, effective method for decontaminating food products. ACP is created by applying an electrical charge to a gas (like air), transforming it into a reactive gaseous state, and generating substances like ozone (O_3), hydroxyl radicals (OH), and nitrogen oxides. These reactive species are highly effective in eliminating *Aspergillus* spp. and breaking down aflatoxins, presenting ACP as a powerful alternative to traditional methods. ACP treatment can significantly reduce aflatoxin contamination, making it a viable alternative to traditional chemical or thermal treatments. ACP offers a safe, environmentally friendly way to meet food safety standards without compromising the quality of the food, making it a promising tool in the fight against mycotoxin contamination. The commercialization of cold plasma treatment is expected to grow in the coming years as the technology matures and more industries begin to recognize its potential benefits. Moreover, further research needs to be conducted on cold plasma to establish process guidance for a number of food products, and understand the chemistry of plasma and its interaction with microbes and food matrices which are essential in optimizing the process for individual applications.

Author Contributions: M.J.: Conceptualization, Literature reviewing, Writing-original draft, Writing-review & editing. W.C.: Conceptualization, Literature reviewing, Writing-original draft, Writing-review & editing. L.T.: Conceptualization, Literature reviewing, Writing-original draft. K.M.K.: Conceptualization, Funding acquisition, Resources, Supervision, Writing-review & editing. All authors have read and agreed to the published version of the manuscript.

Funding: Research funding was provided by the University of Guelph and the Barrett Family Foundation through the Barrett Chair in Sustainable Food Engineering Fund.

Institutional Review Board Statement: Not applicable.

Informed Consent Statement: Not applicable.

Conflicts of Interest: The authors declare no conflict of interest.

Nomenclature

Abbreviation	**Full name**
ACP	Atmospheric cold plasma
AF	Aflatoxin
AFs	Aflatoxins
AFB_1	Aflatoxin B_1
AFB_2	Aflatoxin B_2
AFG_1	Aflatoxin G_1
AFG_2	Aflatoxin G_2
AFM_1	Aflatoxin M_1
AFM_2	Aflatoxin M_2
A. flavus	*Aspergillus flavus*
A. parasiticus	*Aspergillus parasiticus*
CFU	Colony forming unit
CVM	Center for Veterinary Medicine
DBD	Dielectric barrier discharge
DRBC	Dichloran-rose bengal-chloramphenicol
EFSA	European food safety authority
ELISA	Enzyme-linked immunosorbent essay
FAO	Food and agriculture organization
FDA	Food and drug administration
FLD	Fluorescent detector
GC	Gas chromatography
HPLC	High-pressure liquid chromatography
HRP	Horseradish peroxidase
HVACP	High-voltage atmospheric cold plasma
LC-MS	Liquid chromatography-mass spectrometry
LC	Liquid chromatography
LOD	Limit of detection
LFIAs	Lateral flow immunoassays
LOQ	Limit of quantification
MLT	Maximum tolerable level
OD	Optical density
PDA	Photodetector array
RASFF	Rapid alert system for food and feed
RGS	Reactive gas species
RH	Relative humidity
ROS	Reactive oxygen species
RNS	Reactive nitrogen species
TLC	Thin layer chromatography
UV	Ultraviolet
Terminology	**Definition**
Mutagenicity	Mutagenicity is the ability of a chemical or physical agent to cause permanent changes in the structure or amount of an organism's genetic material. These changes can affect a single gene, a block of genes, or chromosomes.
Teratogenicity	Teratogenicity is the ability of a drug to cause fetal abnormalities or deformities.

Acute toxicity	Acute toxicity describes the adverse effects of a substance that result either from a single exposure or from multiple exposures in a short period of time.
Chronic toxicity	Chronic toxicity is the adverse effects that occur after long-term or repeated exposure to a toxic substance.
Epoxidation	Epoxidation is a chemical reaction that converts a carbon-carbon double bond into an epoxide, a cyclic ether with a three-membered ring.
Hepatocellular carcinoma	Hepatocellular carcinoma is the most common type of primary liver cancer in adults and is currently the most common cause of death in people with cirrhosis.
Aflatoxicosis	Aflatoxicosis is a condition caused by consuming food contaminated with aflatoxins, which are toxins produced by fungi.
Genotoxicity	Genotoxicity is *the ability of chemicals to damage the genetic information within a cell* resulting in mutations.

References

1. Rao, W.; Li, Y.; Dhaliwal, H.; Feng, M.; Xiang, Q.; Roopesh, M.S.; Pan, D.; Du, L. The Application of Cold Plasma Technology in Low-Moisture Foods. *Food Eng. Rev.* **2023**, *15*, 86–112. [CrossRef]
2. Kutasi, K.; Recek, N.; Zaplotnik, R.; Mozetič, M.; Krajnc, M.; Gselman, P.; Primc, G. Approaches to Inactivating Aflatoxins—A Review and Challenges. *Int. J. Mol. Sci.* **2021**, *22*, 13322. [CrossRef]
3. Misra, N.N.; Yadav, B.; Roopesh, M.S.; Jo, C. Cold Plasma for Effective Fungal and Mycotoxin Control in Foods: Mechanisms, Inactivation Effects, And Applications. *Compr. Rev. Food Sci. Food Saf.* **2019**, *18*, 106–120. [CrossRef] [PubMed]
4. Jallow, A.; Xie, H.; Tang, X.; Qi, Z.; Li, P. Worldwide Aflatoxin Contamination of Agricultural Products and Foods: From Occurrence to Control. *Compr. Rev. Food Sci. Food Saf.* **2021**, *20*, 2332–2381. [CrossRef] [PubMed]
5. Waliyar, F.; Osiru, M.; Ntare, B.; Kumar, K.V.K.; Sudini, H.; Traore, A.; Diarra, B. Post-harvest Management of Aflatoxin Contamination in Groundnut. *World Mycotoxin J.* **2015**, *8*, 245–252. [CrossRef]
6. Sun, X.D.; Su, P.; Shan, H. Mycotoxin Contamination of Maize in China. *Compr. Rev. Food Sci. Food Saf.* **2017**, *16*, 835–849. [CrossRef]
7. Bailly, S.; Mahgubi, A.E.; Carvajal-Campos, A.; Lorber, S.; Puel, O.; Oswald, I.P.; Bailly, J.D.; Orlando, B. Occurrence and Identification of *Aspergillus* Section *Flavi* in the Context of the Emergence of Aflatoxins in French Maize. *Toxins* **2018**, *10*, 525. [CrossRef]
8. Bhat, R.V.; Vasanthi, S. Mycotoxin Food Safety Risk in Developing Countries: Food Safety in Food Security and Food Trade. 2003. Available online: https://hdl.handle.net/10568/157325 (accessed on 24 October 2024).
9. Callicott, K.A.; Kachapalula, P.; Edmunds, D.; Singh, P.; Jaime, R.; Islam, M.S.; Shenge, K.; Adhikari, B.N.; Arone-Maxwell, L.; Ching'anda, C.; et al. *Brief Protocols for Research on Management of Aflatoxin-Producing Fungi*, 2nd ed.; ARS, USDA: Phoenix, AZ, USA, 2018.
10. Shabeer, S.; Asad, S.; Jamal, A.; Ali, A. Aflatoxin Contamination, its Impact and Management Strategies: An Updated Review. *Toxins* **2022**, *14*, 307. [CrossRef]
11. Yilmaz, S.; Bag, H. Aflatoxin B_1: Mechanism, Oxidative Stress, and Effects on Animal Health. *Insights Vet. Sci.* **2022**, *2*, 017–024. [CrossRef]
12. Liu, Y.; Wu, F. Global Burden of Aflatoxin-Induced Hepatocellular Carcinoma: A Risk Assessment. *Environ. Health Perspect.* **2010**, *118*, 818–824. [CrossRef]
13. Torres, A.M.; Barros, G.G.; Palacios, S.A.; Chulze, S.N.; Battilani, P. Review on Pre- and Post-Harvest Management of Peanuts to Minimize Aflatoxin Contamination. *Food Res. Int.* **2014**, *62*, 11–19. [CrossRef]
14. Laika, J.; Sabatucii, A.; Sacchetti, G.; Michele, D.A.; Hernandez, B.M.J.; Ricci, A.; Rosa, D.M.; Lopez, C.C.; Neri, L. Cold Atmospheric Plasma Inactivation of Polyphenol Oxidase: Focus on the Protective and Boosting Effect of Mono- and Disaccharides. *J. Food Sci.* **2024**, *89*, 9283–9298. [CrossRef]
15. Ott, L.C.; Appleton, H.J.; Shi, H.; Keener, K.; Mellata, M. High Voltage Atmospheric Cold Plasma Treatment Inactivates *Aspergillus flavus* Spores and Deoxynivalenol Toxin. *Food Microbiol.* **2021**, *95*, 103669. [CrossRef]
16. Zheng, H.; Wei, S.; Xu, Y.; Fan, M. Reduction of Aflatoxin B_1 in Peanut Meal by Extrusion Cooking. *LWT* **2015**, *64*, 515–519. [CrossRef]
17. Chen, R.; Ma, F.; Li, P.W.; Zhang, W.; Ding, X.X.; Zhang, Q.; Li, M.; Wang, Y.R.; Xu, B.C. Effect of Ozone on Aflatoxins Detoxification and Nutritional Quality of Peanuts. *Food Chem.* **2014**, *146*, 284–288. [CrossRef] [PubMed]

18. Patil, H.; Shah, N.G.; Hajare, S.N.; Gautam, S.; Kumar, G. Combination of Microwaveand Gamma Irradiation for Reduction of Aflatoxin B_1 and Microbiological Contamination in Peanuts (*Arachis hypogaea* L.). *World Mycotoxin J.* **2019**, *12*, 269–280. [CrossRef]
19. Prakash, S.D.; Siliveru, K.; Zheng, Y. Emerging Applications of Cold Plasma Technology in Cereal Grains and Products. *Trends Food Sci. Technol.* **2023**, *141*, 104177. [CrossRef]
20. Esmaeili, Z.; Hosseinzadeh Samani, B.; Nazari, F.; Rostami, S.; Nemati, A. The Green Technology of Cold Plasma Jet on The Inactivation of *Aspergillus flavus* and the Total Aflatoxin Level in Pistachio and Its Quality Properties. *J. Food Process Eng.* **2023**, *46*, e14189. [CrossRef]
21. Makari, M.; Hojjati, M.; Shahbazi, S.; Askari, H. Elimination of *Aspergillus flavus* from Pistachio Nuts with Dielectric Barrier Discharge (DBD) Cold Plasma and Its Impacts on Biochemical Indices. *J. Food Qual.* **2021**, *2021*, 9968711. [CrossRef]
22. Sohbatzadeh, F.; Mirzanejhad, S.; Shokri, H.; Nikpour, M. Inactivation of *Aspergillus flavus* Spores in A Sealed Package by Cold Plasma Streamers. *J. Theor. Appl. Phys.* **2016**, *10*, 99–106. [CrossRef]
23. Dasan, B.G.; Mutlu, M.; Boyaci, I.H. Decontamination of *Aspergillus flavus* and *Aspergillus parasiticus* Spores on Hazelnuts via Atmospheric Pressure Fluidized Bed Plasma Reactor. *Int. J. Food Microbiol.* **2016**, *216*, 50–59. [CrossRef] [PubMed]
24. Mošovská, S.; Medvecká, V.; Gregová, M.; Tomeková, J.; Valík, L'.; Mikulajová, A.; Zahoranová, A. Plasma Inactivation of *Aspergillus flavus* on Hazelnut Surface in A Diffuse Barrier Discharge Using Different Working Gases. *Food Control* **2019**, *104*, 256–261. [CrossRef]
25. Sen, Y.; Onal-Ulusoy, B.; Mutlu, M. *Aspergillus* Decontamination in Hazelnuts: Evaluation of Atmospheric and Low-Pressure Plasma Technology. *Innov. Food Sci. Emerg. Technol.* **2019**, *54*, 235–242. [CrossRef]
26. Sen, Y.; Onal-Ulusoy, B.; Mutlu, M. Detoxification of Hazelnuts by Different Cold Plasmas and Gamma Irradiation Treatments. *Innov. Food Sci. Emerg. Technol.* **2019**, *54*, 252–259. [CrossRef]
27. Siciliano, I.; Spadaro, D.; Prelle, A.; Vallauri, D.; Cavallero, M.C.; Garibaldi, A.; Gullino, M.L. Use of Cold Atmospheric Plasma to Detoxify Hazelnuts from Aflatoxins. *Toxins* **2016**, *8*, 125. [CrossRef]
28. Shi, H.; Ileleji, K.; Stroshine, R.L.; Keener, K.; Jensen, J.L. Reduction of Aflatoxin in Corn by High Voltage Atmospheric Cold Plasma. *Food Bioprocess Technol.* **2017**, *10*, 1042–1052. [CrossRef]
29. Zhang, Z.H.; Wang, L.H.; Zeng, X.A.; Han, Z.; Brennan, C.S. Non-thermal Technologies and its Current and Future Application in the Food Industry: A Review. *Int. J. Food Sci. Technol.* **2019**, *54*, 1–13. [CrossRef]
30. Pankaj, S.K.; Keener, K.M. Cold Plasma: Background, Applications and Current Trends. *Curr. Opin. Food Sci.* **2017**, *16*, 49–52. [CrossRef]
31. Pankaj, S.K.; Wan, Z.; Keener, K.M. Effects of Cold Plasma on Food Quality: A Review. *Foods* **2018**, *7*, 4. [CrossRef]
32. Fridman, A. *Plasma Chemistry*; Cambridge University Press: Cambridge, UK, 2008.
33. Nikmaram, N.; Brückner, L.; Cramer, B.; Humpf, H.; Keener, K. Degradation Products of Aflatoxin M_1 (AFM_1) Formed by High Voltage Atmospheric Cold Plasma (HVACP) Treatment. *Toxicon* **2023**, *230*, 107160. [CrossRef]
34. Cheng, J.A.; Yamada, T.; Rassier, E.D.; Andersson, C.D.; Westerblad, H.; Lanner, T.J. Reactive Oxygen/Nitrogen Species and Contractile Function in Skeletal Muscle during Fatigue and Recovery. *J. Physiol.* **2016**, *18*, 5149–5160. [CrossRef] [PubMed]
35. Klockow, P.A.; Keener, K.M. Safety and Quality Assessment of Packaged Spinach Treated with A Novel Ozone-Generation System. *LWT-Food Sci. Technol.* **2009**, *42*, 1047–1053. [CrossRef]
36. Tannenbaum, S.R. Nitrate and Nitrite: Origin in Humans. *Science* **1979**, *205*, 1332–1335. [CrossRef] [PubMed]
37. Thomas, M.D.; MacLeod, J.A.; Robbins, R.C.; Goettelman, R.C.; Eldridge, R.W.; Rogers, L.H. Automatic Apparatus for Determination of Nitric Oxide and Nitrogen Dioxide in Atmosphere. *Anal. Chem.* **1956**, *28*, 1810–1816. [CrossRef]
38. Dhawan, V. Reactive Oxygen and Nitrogen Species: General Considerations. In *Studies on Respiratory Disorders*; Springer: Berlin/Heidelberg, Germany, 2014; pp. 27–47.
39. Sewraj, N.; Merbahi, N.; Gardou, J.P.; Akerreta, P.R.; Marchal, F. Electric and Spectroscopic Analysis of a Pure Nitrogen Mono-Filamentary Dielectric Barrier Discharge (MF-DBD) at 760 Torr. *J. Phys. D Appl. Phys.* **2011**, *44*, 145201. [CrossRef]
40. Sies, H.; Stahl, W. Vitamins E and C, Beta-carotene, and Other Carotenoids as Antioxidants. *Am. J. Clin. Nutr.* **1995**, *62*, 1315S–1321S. [CrossRef]
41. Gavahian, M.; Cullen, P. Cold Plasma as An Emerging Technique for Mycotoxin-Free Food: Efficacy, Mechanisms, and Trends. *Food Rev. Int.* **2020**, *36*, 193–214. [CrossRef]
42. Hojnik, N.; Modic, M.; Žigon, D.; Kovač, J.; Jurov, A.; Dickenson, A.; Walsh, J.L.; Cvelbar, U. Cold Atmospheric Pressure Plasma-Assisted Removal of Aflatoxin B_1 from Contaminated Corn Kernels. *Plasma Process. Polym.* **2021**, *18*, 202000163. [CrossRef]
43. Birania, S.; Attkan, K.A.; Kumar, S.; Kumar, N.; Singh, K.V. Cold Plasma in Food Processing and Preservation: A Review. *J. Food Process Eng.* **2022**, *45*, e14110. [CrossRef]
44. Tang, L.; Huang, H.; Hao, H.; Zhao, K. Development of plasma pyrolysis/gasification systems for energy efficient and environmentally sound waste disposal. *J. Electrostat* **2013**, *71*, 839–847. [CrossRef]

45. Winter, J.; Brandenburg, R.; Weltman, K.D. Atmospheric Pressure Plasma Jets: An Overview of Devices and New Directions. *Plasma Sources Sci. Technol.* **2015**, *24*, 064001. [CrossRef]
46. Lu, X.; Laroussi, M.; Puech, V. On Atmospheric-Pressure Non-Equilibrium Plasma Jets and Plasma Bullets. *Plasma Sources Sci. Technol.* **2012**, *21*, 260. [CrossRef]
47. Laroque, A.D.; Seo, T.S.; Valencia, A.G.; Laurindo, B.J.; Carciofi, A.M.B. Cold Plasma in Food Processing: Design, Mechanisms, and Application. *J. Food Eng.* **2022**, *312*, 110748. [CrossRef]
48. Misra, N.N.; Schlüter, O.; Cullen, P.J. *Cold Plasma in Food and Agriculture: Fundamental and Applications*; Elsevier: London, UK, 2016.
49. Puligundla, P.; Lee, T.; Mok, C. Effect of Corona Discharge Plasma Jet Treatment on the Degradation of Aflatoxin B_1 on Glass Slides and in Spiked Food Commodities. *LWT* **2020**, *124*, 108333. [CrossRef]
50. Dasan, B.G.; Boyaci, I.H.; Mutlu, M. Inactivation of Aflatoxigenic fungi (*Aspergillus* spp.) on Granular Food Model, Maize, in An Atmospheric Pressure Fluidized Bed Plasma System. *Food Control* **2016**, *70*, 1–8. [CrossRef]
51. Hadinoto, K.; Niemira, A.B.; Trujillo, J.F. A Review on Plasma-activated Water and Its Application in the Meat Industry. *Compr. Rev. Food Sci. Food Saf.* **2023**, *22*, 4993–5019. [CrossRef] [PubMed]
52. Fridman, A.; Nester, S.; Kennedy, A.L.; Saveliev, A.A.; Mutaf-Yardimci, O. Gliding arc gas discharge. *Prog. Energy Combust. Sci.* **1999**, *25*, 211–231. [CrossRef]
53. Rahnavard, M.A.; Zare, D.; Nassiri, S.M.; Taghvaei, H.; Fazaeli, M. Impact of Gliding Arc Cold Plasma on Deactivating of Aflatoxin and Post-Treatment Fungal Growth on Wheat Grains. *Food Control* **2024**, *164*, 110597. [CrossRef]
54. Du, C.M.; Wang, J.; Zhang, L.; Li, H.X.; Liu, H.; Xiong, Y. The Application of A Non-Thermal Plasma Generated by Gas–Liquid Gliding Arc Discharge in Sterilization. *New J. Phys.* **2012**, *14*, 13010. [CrossRef]
55. Bardos, L.; Barankova, H. *Microwave Plasma Sources and Methods in Processing Technology*; Wiley-IEEE Press: New York, NY, USA, 2022.
56. Sző̋ke, C.; Nagy, Z.; Gierczik, K.; Székely, A.; Spitkól, T.; Zsuboril, Z.T.; Galiba, G.; Marton, C.L.; Kutasi, K. Effect of the afterglows of Low Pressure Ar/N_2-O_2 Surface-wave Microwave Discharges on Barley and Maize Seeds. *Plasma Process. Polym.* **2018**, *15*, 1700138. [CrossRef]
57. Tang, L.; Cao, W.; Keener, M.K. Effect of Dielectric Barrier Discharge High Voltage Atmospheric Cold Plasma on *Aspergillus flavus* Inactivation and Aflatoxin B_1 Degradation on Inoculated Raw Peanuts. *Innov. Food Sci. Emerg. Technol.* **2024**, *97*, 103820. [CrossRef]
58. Lin, C.M.; Patel, A.K.; Chiu, Y.C.; Hou, C.Y.; Kuo, C.H.; Dong, C.D.; Chen, H.L. The Application of Novel Rotary Plasma Jets to Inhibit the Aflatoxin-Producing *Aspergillus flavus* and the Spoilage Fungus, Aspergillus Niger on Peanuts. *Innov. Food Sci. Emerg. Technol.* **2022**, *78*, 102994. [CrossRef]
59. Intanon, W.; Vichiansan, N.; Leksakul, K.; Boonyawan, D.; Kumla, J.; Suwannarach, N.; Lumyong, S. Inhibition of the Aflatoxin-Producing Fungus *Aspergillus flavus* by A Plasma Jet System. *J. Food Process. Preserv.* **2021**, *45*, e15045. [CrossRef]
60. Devi, Y.; Thirumdas, R.; Sarangapani, C.; Deshmukh, R.R.; Annapure, U.S. Influence of Cold Plasma on Fungal Growth and Aflatoxins Production on Groundnuts. *Food Control* **2017**, *77*, 187–191. [CrossRef]
61. Rathod, R.H.; Chaudhari, S.R.; Patil, A.S.; Shirkhedkar, A.A. Ultra-high Performance Liquid Chromatography-MS/MS (UHPLC MS/MS) in Practice: Analysis of Drugs and Pharmaceutical Formulations. *Future J. Pharm. Sci.* **2019**, *5*, 6. [CrossRef]
62. Iqdiam, B.M.; Abuagela, M.O.; Boz, Z.; Marshall, S.M.; Goodrich-Schneider, R.; Sims, C.A.; Marshall, M.R.; MacIntosh, A.J.; Welt, B.A. Effects of Atmospheric Pressure Plasma Jet Treatment on Aflatoxin Level, Physiochemical Quality, and Sensory Attributes of Peanuts. *J. Food Process. Preserv.* **2020**, *44*, e14305. [CrossRef]
63. Selcuk, M.; Oksuz, L.; Basaran, P. Decontamination of Grains and Legumes Infected with *Aspergillus* spp. and *Penicillum* spp. by Cold Plasma Treatment. *Bioresour. Technol.* **2008**, *99*, 5104–5109. [CrossRef]
64. Somma, S.; Masiello, M.; Haidukowski, M.; Ciasca, B.; Sardella, E.; Favia, P.; Palumbo, F.; Roggio, M.; Moretti, A. Decontamination of Maize Kernels and Degradation of Mycotoxins by Means of Cold Plasmas. *LWT* **2025**, *215*, 117205. [CrossRef]
65. Suhem, K.; Matan, N.; Nisoa, M.; Matan, N. Inhibition of *Aspergillus flavus* on Agar Media and Brown Rice Cereal Bars using Cold Atmospheric Plasma Treatment. *Int. J. Food Microbiol.* **2013**, *161*, 107–111. [CrossRef]
66. Panrapee, I.; Phakpoom, K.; Thanapoom, M.; Nampeung, A.; Warapa, M. Exposure to Aflatoxin B_1 in Thailand by Consumption of Brown and Color Rice. *Mycotoxin Res.* **2016**, *32*, 19–25. [CrossRef]
67. Kim, D.; Hong, H.; Kang, S.Y.; Cho, J.W.; Lee, S.M.; Lee, T.K.; Chung, S.H. Simultaneous determination of multi-mycotoxins in cereal grains collected from South Korea by LC/MS/MS. *Toxins* **2017**, *9*, 106. [CrossRef] [PubMed]
68. Dousti, M.; Bashiry, M.; Zohrabi, P.; Siahpoush, V.; Ghaani, A.; Abdolmaleki, K. The Effect of Dielectric Barrier Discharge (DBD) Cold Plasma Treatment on the Reduction of Aflatoxin B_1 and the Physicochemical Properties of Oat. *Appl. Food Res.* **2024**, *4*, 100515. [CrossRef]
69. Zhi, A.; Shi, S.; Li, Q.; Shen, F.; He, X.; Fang, Y.; Hu, Q.; Jiang, X. Aflatoxins Degradation and Quality Evaluation in Naturally Contaminated Rice by Dielectric Barrier Discharge Cold Plasma. *Innov. Food Sci. Emerg. Technol.* **2023**, *88*, 103426. [CrossRef]

70. Cold Plasma Global Market Report 2025. Available online: https://www.thebusinessresearchcompany.com/report/cold-plasma-global-market-report (accessed on 20 January 2025).
71. Nanoguard Technologies Inc. Bringing Clean & Fresh Food to Your Dining Table. Available online: https://nanoguardtechnologies.com/ (accessed on 20 January 2025).
72. Cold Plasma Group Inc. Cold Plasma Sterilization Technologies. Available online: https://www.coldplasmagroup.ca/products (accessed on 16 November 2024).

Disclaimer/Publisher's Note: The statements, opinions and data contained in all publications are solely those of the individual author(s) and contributor(s) and not of MDPI and/or the editor(s). MDPI and/or the editor(s) disclaim responsibility for any injury to people or property resulting from any ideas, methods, instructions or products referred to in the content.

toxins

MDPI

Article

Aflatoxin M1 Determination in Whole Milk with Immersible Silicon Photonic Immunosensor

Dimitra Kourti [1,2], Michailia Angelopoulou [1], Eleni Makarona [3], Anastasios Economou [2], Panagiota Petrou [1], Konstantinos Misiakos [3] and Sotirios Kakabakos [1,*]

1 Immunoassays–Immunosensors Lab, Institute of Nuclear & Radiological Sciences & Technology, Energy & Safety, NCSR "Demokritos", GR-15341 Agia Paraskevi, Greece; d.kourti@rrp.demokritos.gr (D.K.); mikangel@ipta.demokritos.gr (M.A.); ypetrou@rrp.demokritos.gr (P.P.)

2 Analytical Chemistry Lab, Department of Chemistry, National and Kapodistrian University of Athens, GR-15771 Panepistimiopolis Zografou, Greece; aeconomo@chem.uoa.gr

3 Institute of Nanoscience & Nanotechnology, NCSR "Demokritos", GR-15341 Agia Paraskevi, Greece; e.makarona@inn.demokritos.gr (E.M.); k.misiakos@inn.demokritos.gr (K.M.)

* Correspondence: skakab@rrp.demokritos.gr

Abstract: Aflatoxin M1 (AFM1) appears in the milk of animals that have consumed feed contaminated with aflatoxin B1. AFM1 presence in milk is regulated by the European Commission, which has set the maximum allowable limits for adult and infant consumption to 50 and 25 pg/mL, respectively. Here, a rapid and sensitive method for detecting AFM1 in milk based on an immersible silicon photonic chip is presented. The chip features two U-shaped silicon nitride waveguides formed as Mach–Zehnder interferometers. One interferometer is functionalized with AFM1–bovine serum albumin conjugate and the other with BSA to serve as a blank. The chip is connected to a broad-band white LED and a spectrophotometer by a bifurcated optical fiber and an assay is performed by immersing the chip in a mixture of milk with the anti-AFM1 antibody. Then, the chip is sequentially immersed in biotinylated anti-rabbit IgG antibody and streptavidin solutions for signal enhancement. The assay is completed in 20 min and the detection limit for AFM1 in undiluted milk is 20 pg/mL. Given its analytical performance and the absence of pumps and fluidics that lead to a compact instrument design, the proposed immunosensor is ideal for the on-site detection of AFM1 in milk samples.

Keywords: Mach–Zehnder interferometry; aflatoxin M1; immunosensor; whole milk; milk safety

Key Contribution: The aim of this work was to develop a rapid and sensitive method for detecting AFM1 in whole milk using an immersible silicon photonic chip. The chip was designed so as to abolish the need for fluidics and external pumps, thus facilitating the construction of a low-cost portable instrument for the accurate on-site determination of AFM1 in milk samples. The method was evaluated regarding its repeatability, specificity, and accuracy for AFM1 determination in cow, sheep, and goat milk to demonstrate its applicability regardless of the milk origin.

1. Introduction

Mycotoxins are produced by various types of fungi, which grow in soil, hay, decaying vegetation, and grains, especially under conditions of high humidity and temperature, and can contaminate crops, foodstuffs, and animal feed. One mycotoxin that raises high concern due to its mutagenic and carcinogenic function is aflatoxin B1 (AFB1). AFB1 is produced

2. Results and Discussion

2.1. Matrix Effect

Optical transducers that monitor refractive index changes, such as MZIs, can be affected by complex matrices like milk. Additionally, the matrix may interfere with target analyte detection through non-specific interactions with the antibody or the biofunctionalized transducer surface. To assess the interference from the milk matrix on the photonic transducer used in this study, its responses using buffer or milk for the preparation of calibrators were compared. The chips used were spotted with a 100 μg/mL AFM1–BSA solution and the concentration of the antibody was 100 ng/mL. In Figure 3, the net chip response, i.e., the response of the working sensor corrected for the response of the reference sensor, is depicted for the assay buffer (Figure 3a) and milk (Figure 3b). The respective responses of the two sensors per chip, as well as the net chip responses, are provided in Figure S1. As shown in Figure 3a, when buffer was used, the primary immunoreaction, i.e., the reaction of the antibody with the immobilized antigen, generated a signal of approximately 0.26 rad (arrow 1 to 2). This signal increased to 0.72 rad (0.46 rad difference) after reacting with the biotinylated secondary antibody (arrow 3 to 4) and further increased to 2.09 rad (1.37 rad difference) after the streptavidin reaction (arrow 5 to the end).

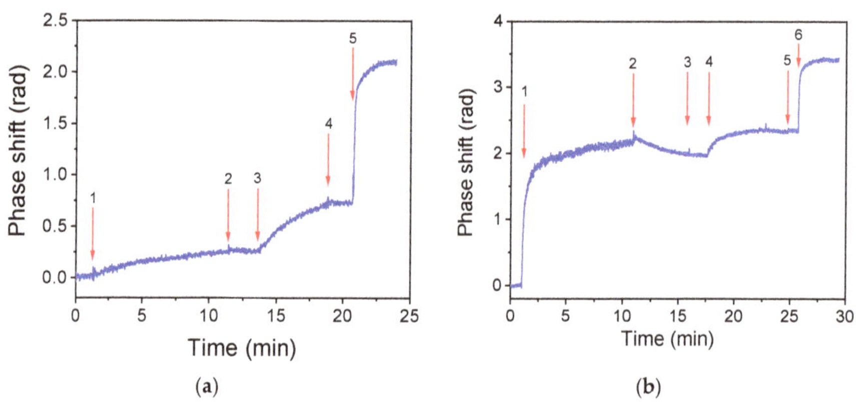

Figure 3. (**a**) Real-time net signal response for the zero AFM1 calibrator in assay buffer. The sequence of solutions is as follows: start to arrow 1: assay buffer; arrow 1 to 2: anti-AFM1 antibody solution/zero calibrator mixture in assay buffer; arrow 2 to 3: assay buffer; arrow 3 to 4: biotinylated secondary antibody in assay buffer; arrow 4 to 5: assay buffer; and arrow 5 to end: streptavidin in assay buffer. (**b**) Real-time net signal response for the zero AFM1 calibrator in undiluted cow milk. The sequence of solutions is as follows: start to arrow 1: assay buffer; arrow 1 to 2: anti-AFM1 antibody solution/zero calibrator mixture; arrow 2 to 3: washing buffer; arrow 3 to 4: assay buffer; arrow 4 to 5: biotinylated secondary antibody in assay buffer; arrow 5 to 6: assay buffer; and arrow 6 to end: streptavidin in assay buffer.

In contrast, when milk was used in the primary immunoreaction, the signal reached 2.16 rad (arrow 1 to 2), indicating a significant contribution from the milk matrix that could not be fully eliminated by subtracting the reference sensor signal. The net signal dropped slightly after washing out the milk (arrow 2 to 3; Figure 3b), and increased by 0.38 rad after the reaction with the biotinylated secondary antibody and by 1.08 rad following the reaction with streptavidin. The differences in the responses observed for the different assay steps between the assay buffer and the milk indicate that components from the milk sample adsorbed onto the chip surface and were removed very slowly during washing. Using diluted samples, the matrix effect could be reduced, however, this would negatively influence the detection sensitivity. Therefore, the subsequent assay optimization study was performed using calibrators prepared in whole cow milk. Additionally, the signal received during the reaction with streptavidin was considered as the analytical signal of the assay to avoid matrix interferences.

2.2. Chemical Activation of the Chip

The chemical activation of the chip aimed to facilitate the attachment of BSA–AFM1 conjugate, preferably on the silicon nitride waveguide area exposed on the working sensor window. Two different chemical activation protocols and BSA–AFM1 immobilization procedures were compared. The first one involved the modification of the chips with 3-aminopropyl-triethoxy silane (APTES) after treatment with Piranha solution (1:1 *v/v* $H_2SO_4/30\%\ H_2O_2$) to clean the surface and introduce silanol groups. The silanol groups reacted with APTES and, during the subsequent thermal curing, stable bonds were formed, as depicted in Figure 4a. Modification with APTES endowed the surface with amine groups which could be used for covalent bonding, but also made the surface suitable for the immobilization of proteins by physical adsorption, which provided higher sensor responses compared to covalent bonding [36]. The second approach consisted of treating the chips with HF solution, during which amine groups were created selectively on the Si_3N_4 [37], followed by the activation of the amine groups with glutaraldehyde to enable the immobilization of BSA–AFM1 conjugate by covalent bonding, as depicted in Figure 4b.

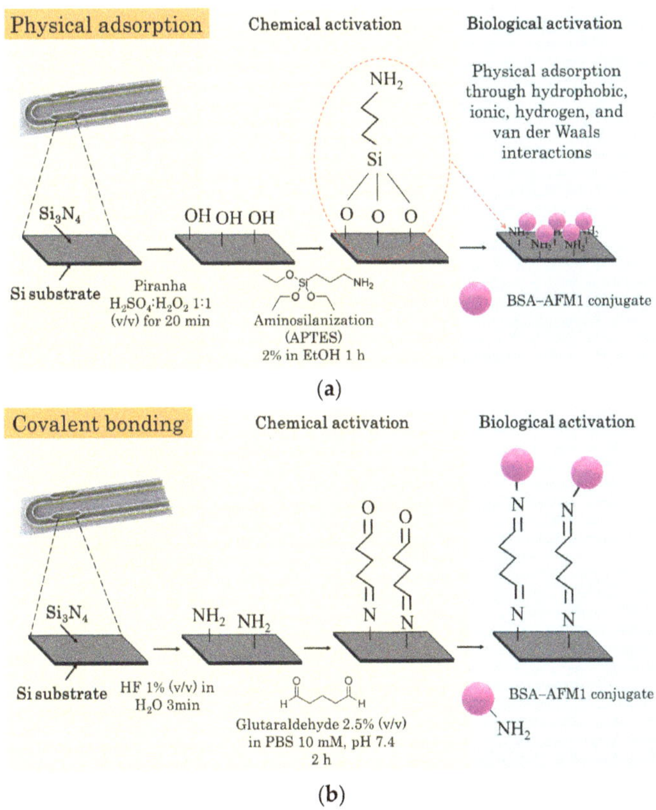

Figure 4. Schematic of the chemical functionalization of the working sensor with (**a**) APTES for BSA–AFM1 conjugate immobilization through physical adsorption and (**b**) HF/glutaraldehyde BSA–AFM1 conjugate immobilization through covalent bonding.

After chemical activation either with APTES or HF/glutaraldehyde, the working sensor was spotted with a 100 μg/mL AFM1–BSA conjugate solution, while the reference sensor was spotted with a 100 μg/mL BSA solution. For the assay, the chip was immersed in a mixture of zero calibrator with 100 ng/mL of anti-AFM1 antibody for 5 min, followed by immersion in a biotinylated anti-rabbit IgG solution for 5 min and a streptavidin solution for 2 min. It was found that the HF/glutaraldehyde activation of the chip provided approximately 15% higher signal values compared to chips modified with APTES. This finding is attributed to the fact that the HF/glutaraldehyde method led to the selective activation of the silicon nitride layer at the window areas of the two MZIs with respect to the surrounding silicon dioxide cladding layer, as opposed to chip modification with APTES. Therefore, the subsequent immunoreactions were also restricted on the sensing window area instead of on a larger area, thus leading to a signal increase. This is depicted in Figure S2a, where fluorescently labeled streptavidin was used to visualize the functionalized area under a fluorescence microscope in the case of activation with HF/glutaraldehyde (left) or APTES (right). Additionally, activation with HF/glutaraldehyde led to homogeneous coverage with the recognition biomolecule, i.e., the AFM1–BSA conjugate, of the silicon nitride area on the two MZI windows, as depicted in the image of Figure S2b. Therefore, treatment with HF/glutaraldehyde was selected for the chemical activation of the chips. The presence of amine groups on Si_3N_4 surfaces treated with HF was quantified following a previously published protocol [38]. In brief, the protocol included the reaction of surface amine groups with sulfo-SHPP [sulfosuccinimidyl-3-(4-hydroxypheynyl) propionate] to convert them into hydroxyphenyl moieties. Then, these moieties were quantified through the BCA protein assay method using an L-tyrosine calibration curve (Figure S3). For comparison reasons, the amine group contents of Si_3N_4 chips modified with APTES were also determined. The amine group content of HF-treated surfaces was determined at $2.23(\pm 0.02) \times 10^{13}$ groups/cm^2, whereas the respective value for APTES treated surfaces was $1.22(\pm 0.03) \times 10^{13}$ groups/cm^2, indicating that the HF treatment led to a higher concentration of amine groups per surface area compared to the APTES modification.

2.3. Assay Parameter Optimization

The antibody dilution buffer composition was one of the critical assay parameters optimized with respect to the zero calibrator signal and the assay sensitivity. Two buffers were tested, 50 mM Tris-HCl, pH 7.8, and 50 mM phosphate-buffered saline (PBS), pH 7.4, both containing 0.9% (*w/v*) NaCl and 0.5% (*w/v*) BSA. The Tris-HCl buffer provided analytical signal values approximately 20% higher than those received for PBS and was, thus, selected for further experiments. Given the low water solubility of AFM1, the impact of adding methanol to the antibody dilution buffer on the zero calibrator signal was investigated. As shown in Figure 5, the addition of 10% (*v/v*) methanol to the antibody dilution buffer had two positive effects. Firstly, the zero calibrator signal increased by approximately 27% compared to that obtained with buffer without methanol. Furthermore, the assay sensitivity was improved, as indicated by the increase in the percentage of signal inhibition obtained for a calibrator containing 0.5 ng/mL of AFM1 from 24.5% in the absence of methanol to 40% in the presence of 10% *v/v* methanol in the antibody dilution buffer. The increased signal can be attributed to the fact that methanol helped to expose the AFM1 moieties immobilized on the sensor surface, as a protein conjugate, making them more accessible for binding to the antibody. Similarly, the enhanced detection sensitivity in the presence of methanol can be ascribed to the in situ extraction of AFM1 molecules from milk proteins, particularly caseins, that increased the concentration of available AFM1 molecules, thus facilitating competition with the AFM1 moieties of the immobilized conjugate for the binding sites of the antibody [39].

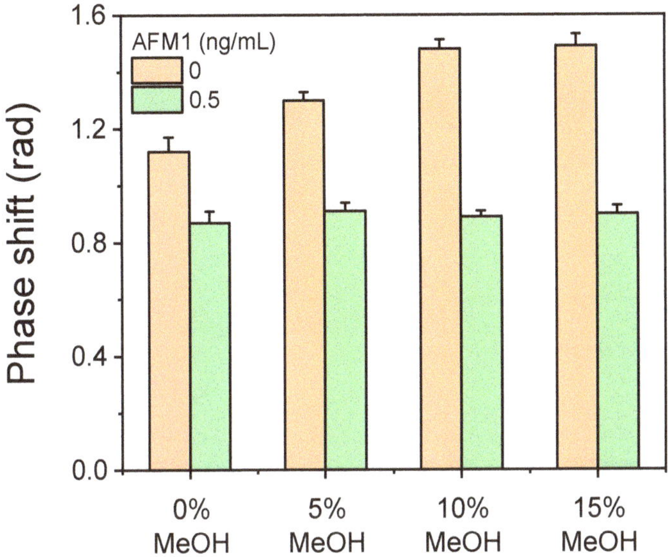

Figure 5. Net signals obtained from chips spotted with a 100 µg/mL BSA–AFM1 conjugate solution for the zero calibrator (orange columns) and a calibrator containing 0.5 ng/mL AFM1 (green columns) with respect to percent volume content of methanol in antibody dilution buffer. The anti-AFM1 antibody concentration used 100 ng/mL and the primary immunoreaction duration was 5 min. Each column corresponds to the mean of 3 chips ± SD.

Based on these findings, we also tested the inclusion of 5% methanol in the buffer used for washing after the primary immunoreaction. It was found that 2 min of washing with this solution reduced the signal received during the primary immunoreaction due to the milk effect by approximately 50%, whereas washing without methanol for 5 min decreased the signal only by 8% (Figure 3b, arrow 2 to 3).

After the selection of the antibody dilution buffer, the concentration of the anti-AFM1 antibody and the primary immunoreaction time were optimized while keeping the concentrations and reaction times constant for the signal amplification steps (i.e., the secondary antibody and streptavidin reactions). As shown in Figure 6a, using either a 50 or 100 ng/mL anti-AFM1 antibody solution, the maximum plateau signal values were obtained for a primary immunoreaction time equal to or higher than 30 min. However, an adequate zero calibrator signal (approximately 1.5 rad) was achieved with either a 6 min reaction using 100 ng/mL or a 9 min reaction using 50 ng/mL of anti-AFM1 antibody. For the AFM1–BSA conjugate used for coating, the maximum zero calibrator signal plateau was reached with a concentration of 50 µg/mL (Figure S4). Thus, using a 50 µg/mL AFM1–BSA conjugate solution for coating, the sensitivity provided by a 6 min reaction using 100 ng/mL of anti-AFM1 antibody or a 9 min reaction using 50 ng/mL anti-AFM1 antibody was tested. As shown in Figure 6b, a 9 min primary immunoreaction duration using a 50 ng/mL anti-AFM1 antibody solution provided 35% inhibition for a calibrator containing 0.2 ng/mL of AFM1, while 22% inhibition was obtained for the same calibrator with a 6 min primary immunoreaction duration using a 100 ng/mL anti-AFM1 antibody solution. Thus, in the final protocol, a 50 ng/mL AFM1 antibody solution and a 9 min primary immunoreaction duration were adopted.

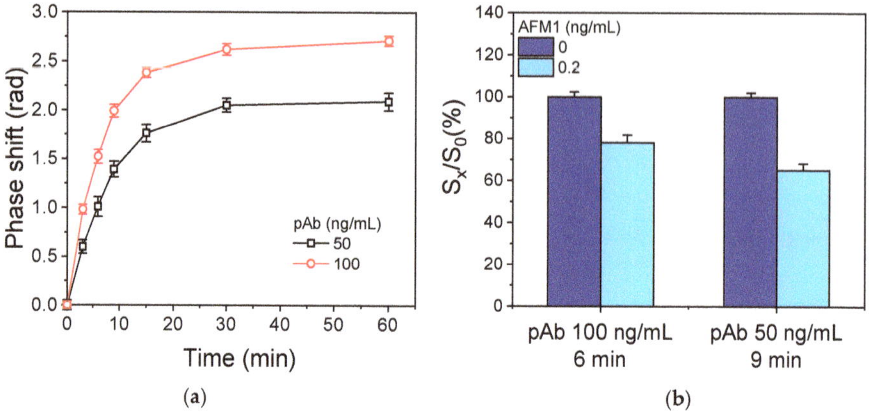

Figure 6. (**a**) Net zero calibrator signals obtained from chips spotted with a 50 μg/mL BSA–AFM1 conjugate and assayed with a 1:1 volume mixture of milk with anti-AFM1 antibody solution with concentration of 50 (black squares) or 100 ng/mL (red circles) for different assay times. (**b**) Percent signal values obtained for a calibrator containing 0.2 ng/mL AFM1 (light blue columns) with respect to zero calibrator (dark blue columns) employing two different combinations of the anti-AFM1 antibody concentration and primary immunoassay duration. Each value corresponds to the mean of 3 chips ± SD.

2.4. Analytical Characteristics

In Figure 7a, the net signals corresponding to different AFM1 calibrators prepared in cow milk are presented. Using the signals corresponding to the reaction with streptavidin, the calibration curve presented in Figure 7b was obtained. Additionally, the calibration curves obtained from calibrators prepared in sheep and goat milk were similar to those obtained from cow milk, demonstrating the potential of the method for AFM1 detection in milk from different animal species (Figure S5).

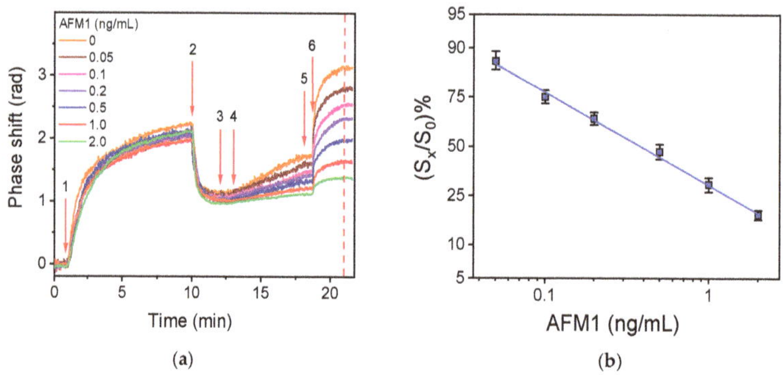

Figure 7. (**a**) Real-time signal responses of calibrators containing 0–2 ng/mL AFM1 in undiluted milk. The sequence of solutions are as follows: start to arrow 1: assay buffer/MeOH; arrow 1 to 2: anti-AFM1 antibody solution/zero calibrator mixture; arrow 2 to 3: washing buffer/MeOH; arrow 3 to 4: assay buffer; arrow 4 to 5: biotinylated secondary antibody in assay buffer; arrow 5 to 6: assay buffer; and arrow 6 to end: streptavidin in assay buffer. The analytical signal is calculated as the difference of response between the vertical red line (indicates the time point of 2 min reaction with streptavidin) and the response at arrow 6 (indicating the start of reaction with streptavidin). (**b**) Typical AFM1 calibration curves obtained with calibrators prepared in undiluted milk. Each point is the mean value of 3 measurements ± SD.

The detection limit (LOD) of the assay was calculated as the concentration corresponding to the percent signal equal to 100-3SD of the mean zero calibrator signal of 10 measurements and was found to be 20 pg/mL AFM1, with a linear dynamic range from 50 to 2000 pg/mL. This LOD is lower than the maximum allowable limits set by the EU for milk intended for both adult and infant consumption. The reproducibility of the method was evaluated through measurements of three fresh cow milk control samples prepared by spiking known amounts of AFM1 that covered the whole range of the calibration curve. The intra-assay coefficient of variation (CV) was determined by triplicate measurements of these samples on the same day, whereas the inter-assay CV was determined by duplicate measurements on 4 different days over a period of 1 month. The assay was reproducible with intra- and inter-assay CVs lower than 7.3 and 9.1%, respectively. Similar experiments were performed for control samples from fresh sheep and goat milk. The intra-assay and the inter-assay CVs were lower than 8.2 and 9.6% for the sheep and 7.7 and 9.3% for the goat milk, respectively. Additionally, the chip-to-chip variation was determined by assaying five chips from the same batch in a single day with respect to the zero calibrator signal. A coefficient of variation of less than 7.5% was determined. Moreover, the batch-to-batch variation was determined by assaying chips (in triplicate) from six different batches prepared over a period of 2 months. The coefficient of variation of the mean zero calibrator value determined for the different batches was up to 10.8%.

The accuracy of the assay was determined by recovery experiments with samples of fresh cow, sheep, and goat milk spiked with AFM1 at three different concentration levels (0.15, 0.4, and 1.5 ng/mL). The %recovery was calculated according to the following equation:

$$\%\text{Recovery} = (\text{AFM1 amount determined})/(\text{AFM1 amount added}) \times 100\%$$

It was found that the assay is accurate, with %recovery values ranging from 86.7 to 112% (Table 1).

Table 1. Recovery of known amounts of AFM1 spiked in full-fat cow, sheep, and goat milk.

Sample	Amount Added (ng/mL)	Amount Determined * (ng/mL)	%Recovery **
Cow milk	0.15	0.16 ± 0.01	107 ± 6.7
	0.40	0.36 ± 0.02	90.0 ± 5.0
	1.5	1.4 ± 0.05	93.3 ± 3.3
Sheep milk	0.15	0.14 ± 0.01	93.3 ± 6.7
	0.40	0.42 ± 0.01	105 ± 2.5
	1.5	1.6 ± 0.07	107 ± 4.7
Goat milk	0.15	0.13 ± 0.01	86.7 ± 6.7
	0.40	0.45 ± 0.03	112 ± 7.5
	1.5	1.4 ± 0.10	93.3 ± 6.7

* Mean value ± S.D. ** Percent mean value ± S.D.

The specificity of the assay was evaluated through cross-reactivity studies. In Figure S6, the AFM1 calibration curve along with inhibition curves obtained with calibrators of substances with a similar chemical structure to AFM1 are presented. The cross-reactivity values determined were 1.4% for aflatoxin B1 (AFB1), 0.30% for aflatoxin B2 (AFB2), and 0.12% for aflatoxin G1 (AFG1), indicating the high specificity of the proposed immunosensor against the targeted analyte.

2.5. Regeneration and Stability of the Sensor

Photonic immunosensor chips can be reused after the completion of an assay cycle by disrupting the immunocomplexes formed and making the immobilized antigen available for a new assay cycle. To achieve this, various solutions, including 50 mM HCl, 40 mM NaOH, and a 0.5% (*w/v*) SDS solution, adjusted at pH 1.3 with 0.1 M HCl (HCl-SDS), as well as combination of HCl with either a NaOH or HCl-SDS solution, were investigated as regeneration buffers. In all cases, the chip was immersed in the regeneration solution for 2 min; when two regeneration solutions were used, a 1 min washing step was introduced between them. To select the optimum regeneration solution, it was necessary to determine the amount of anti-AFM1 antibody that remained on the chip surface after regeneration. For that reason, the residual signal was determined, i.e., the signal received after regeneration, by immersing the chip in a biotinylated anti-rabbit antibody and streptavidin solution. It was found that the use of two-step regeneration with a HCl and SDS-HCl solution almost completely removed the bound anti-AFM1 antibody molecules, as the residual signal was negligible, while, in all other cases, there was a substantial residual signal that would affect the next measurement (Figure 8a). The stability of the signal after multiple measurements/regeneration cycles with HCl/SDS-HCl was also evaluated. As shown in Figure 8b, the chips could be regenerated up to 15 times without any obvious effect on the analytical signal. Further assay/regeneration cycles led to an increased residual signal, which could affect the accuracy of the measurements.

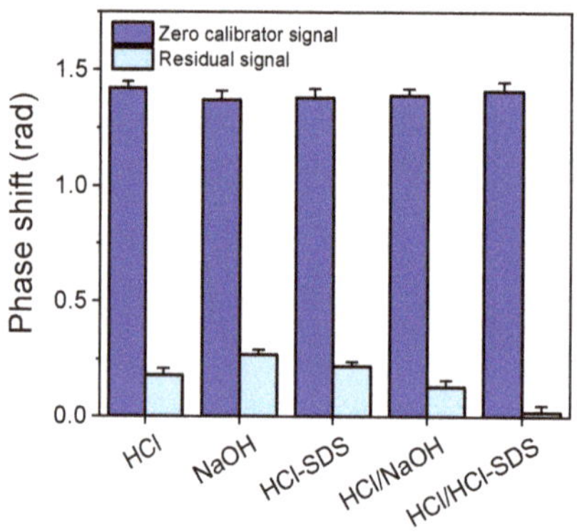

(a)

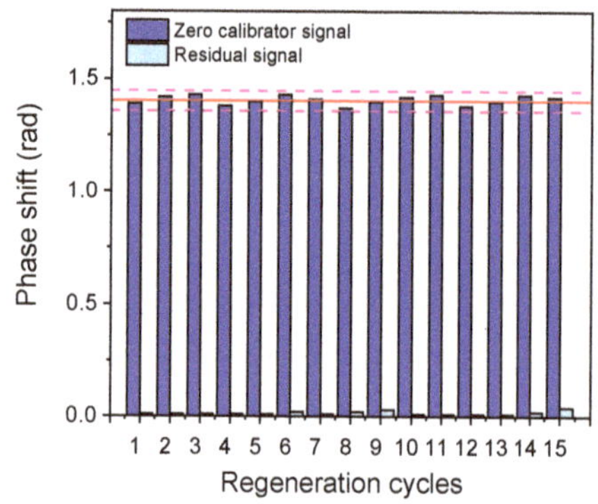

(b)

Figure 8. (**a**) Net zero calibrator signals obtained (blue columns) and the respective residual signals (light blue columns) obtained after treatment with different regeneration solutions. (**b**) Net zero calibrator signals obtained from a single chip after repetitive assay/regeneration cycles. The solid red corresponds to the mean value of the 15 measurements and dashed pink lines to mean value ± SD.

Additionally, the stability of the biofunctionalized chips, which were stored at room temperature (RT), was assessed. The results showed that the biofunctionalized chips could be used for up to 3 months after spotting without any signal loss (Figure S7). In fact, the coefficient of variation of the mean values received from chips assayed in triplicate at different intervals within the 3-month period was 1.51%. Although a more elaborated investigation of sensor performance stability under different storage conditions might be necessary [40], the presented data are very promising, taking into account that no special treatment of the sensor was performed after the biomolecule spotting to help preserve their functionality.

2.6. Comparison with Label-Free Biosensors Reported in the Literature for AFM1 Detection in Dairy Products

A comparison of the developed label-free MZI immunosensor with other label-free electrochemical or optical biosensors reported in the literature for the determination of AFM1 is provided in Table 2. An electrochemical sensor based on interdigitated Pt electrodes (IDEs) modified with an electropolymerized Fe_3O_4/polyaniline layer was functionalized after activation with glutaraldehyde with an aptamer specific for AFM1 [14]. The sensor was applied for direct AFM1 detection in buffer through cyclic and square wave voltammetry, achieving an LOD of 1.98 pg/mL after a 1 h reaction [14]. In another study, two different approaches for the electrochemical detection of AFM1 were investigated [15]. The first method involved gold electrodes modified with a polyamidoamine dendrimer layer to covalently bind an amine-terminated aptamer. The second involved the modification of the electrodes with neutravidin to enable the binding of a biotinylated aptamer [15]. Both approaches yielded similar results in terms of LOD, which was 8.47 pg/L, and AFM1 recovery in spiked milk samples, which was higher than 78%. Another report involved the electropolymerization of the neutral red dye onto glassy carbon electrodes (GCEs) in the presence of a polycarboxylated pillar[5]arene derivative to introduce carboxyl groups onto the electrode for the covalent binding of an aptamer for AFM1 [16]. The sensor was applied for the detection of AFM1 in cow and sheep milk and kefir after their dilution with methanol, achieving an LOD of 40 pg/mL. In another report, screen-printed carbon electrodes were modified with a nanocomposite of molybdenum disulfide (MoS_2) quantum dots (QDs) and a zirconium-based metal–organic framework (UiO-66-NH_2), followed by the covalent immobilization of an AFM1-specific antibody, and used to directly detect AFM1 with an LOD of 60 pg/mL following a 10 min reaction [20]. Furthermore, an impedimetric immunosensor was developed using silver (Ag) wire electrodes modified with 11-mercaptoundecanoic acid and a covalently bound antibody to detect AFM1 in milk samples, with an LOD of 1 pg/mL [21].

Table 2. Comparison of the MZI immunosensor developed with other label-free biosensors for determination of AFM1 in dairy products.

Detection Principle	Biorecognition Element/Assay Type	Sample	LOD (pg/mL)	Assay Duration (min)	Ref.
Electrochemical sensors					
Fe_3O_4/polyaniline film on IDE	Aptamer/direct binding	Buffer	1.98	60	[14]
Dendrimer layer on a gold platform Neutravidin layer on a gold platform	Aptamer/direct binding Biotinylated aptamer/direct binding	Milk (pretreatment required)	8.47 8.52	60	[15]
Poly(neutral red) and carboxylated pillar[5]arene on GCE	Aptamer/direct binding	Cow/Sheep milk Kefir (pretreatment required)	40	60	[16]
MoS_2/UiO-66-modified screen-printed carbon electrodes	Antibody/direct binding	Milk (pretreatment required)	60	10	[20]
Silver wire electrode	Antibody/direct binding	Milk (pretreatment required)	1	20	[21]
Optical sensors					
LSPR	Aptamer/direct binding	Milk (pretreatment required)	40	60	[27]
LSPR based on TIRE	Aptamer/competitive with Au nanoparticles labeled secondary antibody	Buffer	10	15	[30]
Surface Plasmon Resonance (SPR)	BSA–AFM1 conjugate/competitive immunoassay	Milk/ Milk powder (pretreatment required)	100	10	[31]
White Light Reflectance Spectroscopy	BSA–AFM1 conjugate/competitive with biotinylated secondary antibody and streptavidin	Milk	6	25	[32]
Fab′ functionalized asymmetric MZIs	Antibody/direct binding	Milk (pretreatment required)	16.8	1.5	[33]
Broad-band MZIs	BSA–AFM1 conjugate/competitive with biotinylated secondary antibody and streptavidin	Milk Chocolate milk (pretreatment required) Yogurt (pretreatment required)	5 5 10	15	[34]
Immersible photonic chip	BSA–AFM1 conjugate/competitive with biotinylated secondary antibody and streptavidin	Milk	20	20	This work

Regarding the label-free detection of AFM1 with optical sensors, a fiber optic localized SPR aptasensor modified with gold nanoparticle multimers to enhance the signal due to a hot spot effect created on the nanogaps between nanoparticles has been reported [27]. This sensor could detect AFM1 with an LOD of 50 pg/mL. Another sensor that combined

LSPR and total internal reflection ellipsometry (TIRE) enabled the direct detection of AFM1 in buffer, with an LOD of 10 pg/mL [28]. SPR was also employed for AFM1 detection in milk and milk powder with an LOD of 100 pg/mL in 10 min, after the defatting and immunoaffinity isolation of AFM1 [30]. Furthermore, an interferometric immunosensor based on White Light Reflectance Spectroscopy, which monitored the changes in the biomolecular thickness on the transducer surface during reactions, was used to detect AFM1 through a competitive immunoassay, with an LOD of 6 pg/mL [32]. Another interferometric immunosensor, based on integration onto silicon asymmetric Mach–Zehnder interferometers (MZIs) functionalized with the Fab′ fragment of an anti-AFM1 antibody, was used for AFM1 detection in milk within 1.5 min with an LOD of 16.8 pg/mL, after defatting, 20 times pre-concentration, and column purification of the milk samples [33]. Moreover, silicon chips with arrays of ten silicon nitride waveguide MZIs integrated on the same chip along with respective broad-band light sources were used for the detection of AFM1 in plain and chocolate cow milk and yogurt following a competitive immunoassay format [34]. The LODs achieved were 5 pg/mL in plain and chocolate cow milk and 10 pg/mL in yogurt, and the assay was completed, in all cases, in 15 min [34].

Compared to the electrochemical and optical label-free apta- and immunosensors reported in the literature, the proposed immunosensor may not be the most sensitive, but the LOD achieved is lower than the maximum allowable limits of AFM1 in milk for infant and adult consumption. Additionally, one major advantage of the proposed immunosensor is that it does not require any milk sample pretreatment and can be applied to detect AFM1 in milk from different species. Moreover, compared with previous chips developed by our team and used to determine AFM1 in milk and dairy products, the proposed immunosensor has the following advantages. (a) Its immersible form eliminates the need for microfluidics and pumps, allowing for the construction of a small-size portable instrument for on-site determination. The lack of flow, however, might be the reason that the proposed immunosensor has a four times higher LOD than previous ones based either on WLRS [32] or integrated onto silicon MZIs [34]. (b) The instrumentation used is battery-operated, as both the LED and the spectrophotometer are powered by the laptop, ensuring autonomous operation in low-resources environment. This is also a major innovation compared to previous instrumentation set-ups. (c) The chip fabrication cost is reduced by 8–10 times compared to the cost of the chip with the array of 10 integrated MZIs [34]. Finally, since the set-up developed is an analytical platform, it can be employed for the on-site detection of other contaminants such as antibiotics in milk or harmful substances in general in food matrices.

3. Conclusions

The application of an immersible photonic chip sensor for the determination of AFM1 in whole milk from different animal species (cow, sheep, and goat) has been demonstrated. The fact that the sensor developed does not require fluidics and fluid circulation equipment for the performance of the assay simplifies the assay greatly, but mostly the instrumentation required. Thus, the only external components required are optical ones, i.e., a light source′+-, a spectrophotometer, and a bifurcated fiber, leading to a set-up that is really compact and lightweight, making it suitable for point-of-need applications. The developed immersible photonic chip immunosensor is a highly effective tool for detecting aflatoxin M1 (AFM1) in milk, since it achieves a high detection sensitivity involving a compact and robust system. The sensor developed is capable of detecting AFM1 at concentrations as low as 20 pg/mL, with a dynamic detection range from 50 to 2000 pg/mL and a total analysis time of 20 min. In addition, its ability to directly measure in undiluted milk from different species enhances its suitability for the on-site detection of AFM1 at farms and milk collection points. The wide acceptance of the proposed sensor by the dairy industry would be facilitated by the fact that the cost of the chip is comparable to that of the LFIAs currently used for on-site AFM1 determination, with the additional advantage of quantitative results. Thus, although the chip cost is currently about EUR 15 for small-scale production in our research facility, the transfer of production to a silicon foundry could reduce the cost by at least 10-fold, i.e., at EUR 1.5. In addition, for applications such as on-site food analysis, the cost can be further downsized through the regeneration and reuse of a single bio-functionalized chip. Thus, it is expected that the developed immunosensor and measuring system could help to ensure compliance with regulatory limits for AFM1 in dairy products and could evolve into a valuable asset for improving food safety and protecting public health. To accomplish this goal, a thorough investigation of the long-term storage stability and performance of the functionalized chip under real-world testing, as well as all the measuring system tests required for approval by regulatory authorities, should be performed. Furthermore, medium-scale production to obtain a more accurate estimation of the chip's cost is essential in order to commercially exploit the proposed sensor and measuring system.

4. Materials and Methods

4.1. Reagents

Aflatoxin M1 (AFM1), AFM1 conjugate with bovine serum albumin (AFM1–BSA), bovine serum albumin (BSA), goat polyclonal anti-rabbit IgG antibody, 3-aminopropyltriethoxy silane (APTES), glutaraldehyde solution (25% in water), and L-tyrosine were purchased from Sigma-Aldrich (Darmstadt, Germany). Hydrofluoric acid (HF; 50% in water) was obtained from Technic Inc. (Saint-Denis, France). Streptavidin, streptavidin labeled with AlexaFluorTM 546, 3-sulfo-succinimidyl-6-[biotinamido]hexanoate (Sulfo-NHS-LC-biotin), sulfosuccinimidyl-3-(4-hydroxypheynyl) propionate (sulfo-SHPP), sulfuric acid (H_2SO_4), and BCA Protein Assay Kit were from Thermo Fisher Scientific (Waltham, MO, USA). The rabbit polyclonal anti-AFM1 antibody was from AntiProt (Puchheim, Germany). All other chemicals and reagents were obtained from Merck (Darmstadt, Germany). The water used throughout the study was doubly distilled. The pasteurized cow (3.5% fat), sheep (1.7% fat), and goat (3.5% fat) milk were products of Larisa Dairy S.A. "OLYMPUS" (Larisa, Greece) purchased from the local market. Fresh cow, sheep, and goat milk was received from small farms in the Attica area (Greece). Sodium azide was added to all milk samples at a final concentration of 1% *w/v*, and the samples were then aliquoted and kept at −20 °C until use. All milks were tested with AgraQuant® Aflatoxin M1 High Sensitivity ELISA kit (Romer Labs GmbH; Butzbach, Germany) to confirm the absence of detectable amounts of AFM1. The goat anti-rabbit IgG antibody was biotinylated following a previously published protocol [31].

4.2. Instrumentation

The photonic sensor chips were fabricated as previously described [41]. In summary, the fabrication process included the following: (a) the deposition of a silicon nitride (Si_3N_4) layer 150 nm thick on silicon wafers with a 5 μm thick SiO_2 layer (under cladding layer), (b) e-beam lithography and dry etching to define the two U-shaped waveguides and the respective MZIs, (c) the deposition of a 2 μm thick SiO_2 top cladding layer, and (d) the opening of the sensing windows through optical lithography and wet etching. The two sensing windows had a width of 20 μm and lengths of 1 mm and 2 mm, while the final chips had dimensions of 23 mm × 2.1 mm (Figure 9).

Figure 9. Image of the immersible photonic chip on top of a EUR 2 coin.

The optical system comprised a high-brightness, broad-band white light source (Ushio Europe B.V., Oude Meer, The Netherlands) and an external VIS-NIR spectrophotometer (Flame-T-VIS-NIR, Ocean Insight, Duiven, The Netherlands). This set-up connected to the chip via a bifurcated optical fiber with a 200 μm core (Ocean Insight, Duiven, The Netherlands) through the specially designed coupler. A laptop with an in-house developed software was utilized for recording signals and processing data (Figure 3a). During the assay, the dual MZI transmission spectrum was continuously captured and processed using Fast Fourier Transform (FFT) to reveal the wavenumbers of the spectra so as to track the phase shift of each MZI in real time.

4.3. *Chemical Functionalization of the Sensor*

Two different chemical activation protocols and BSA–AFM1 immobilization procedures were examined. The first one involved the modification of chips with 3-aminopropyltriethoxy silane (APTES) and the physical adsorption of the BSA–AFM1 conjugate, and the second treatment involved HF solution and glutaraldehyde followed by the covalent bonding of the BSA–AFM1 conjugate.

4.3.1. Modification with APTES

The chips were treated in an ultrasonic bath for 10 min sequentially with acetone and isopropanol, followed by Piranha treatment (1:1 volume mixture of H_2SO_4 and 30% *v/v* H_2O_2) for 20 min, washing 3 times with distilled H_2O, and drying with a N_2 steam to clean their surfaces. Then, the chips were immersed in a 2% (*v/v*) APTES solution in absolute ethanol solution for 60 min, washed with ethanol, dried with nitrogen, and then heated at 120 °C for 20 min. The chips were kept for at least 48 h in a desiccator prior to biological functionalization.

4.3.2. Treatment with HF and Glutaraldehyde

The chemical functionalization of the chips with HF and glutaraldehyde was performed following a protocol from the literature [37], which led to the selective introduction of amine groups onto the silicon nitride layer at the window areas of the two MZIs with respect to the surrounding silicon dioxide cladding layer. In brief, after cleaning the chips with acetone and isopropanol and conducting the Piranha treatment, as described in Section 4.3.1, the side of the chip with the sensing windows was immersed for 3 min in a 1% *v/v* aqueous HF solution, followed by extensive washing with H_2O. After that, the chips were treated for 2 h with a 2.5% *v/v* glutaraldehyde solution in 10 mM phosphate-buffered saline (PBS), pH 7.4, washed 3 times with PBS solution, and dried with N_2. The modified chips were used immediately for spotting.

4.4. Determination of the Amine Content of the Chemically Functionalized Surfaces

The determination of the amine content of Si_3N_4 surfaces that had been modified either with APTES or with HF was performed following the modification of a previously published protocol [38]. In brief, a 20 mM sulfo-SHPP solution was prepared in 0.1 M sodium bicarbonate buffer, pH 8.5. One milliliter of this solution was added into tubes containing Si_3N_4 chips with dimensions of 5 mm × 15 mm and incubated under gentle shaking for 1 h at RT. Chips with APTES or HF treatment were assayed, whereas chips without any treatment were used as a blank. The chips were washed three times with 2 mL of distilled water and twice with 2 mL of 0.1 M sodium carbonate buffer, pH 11.25. The BCA working solution was then prepared following the instructions of the manufacturer by mixing 50 volumes of Component A with 1 volume of Component B, and 1mL was added to tubes containing the chips. The tubes were then incubated for 60 min at 60 °C under gentle shaking. After that, 100 μL of solution was received from each tube and transferred into a 96-well microtiter plate to determine the optical density at 560 nm using the Victor3 1420 Multilabel Counter (PerkinElmer, Waltham, MO, USA). L-Tyrosine solutions with concentration ranging from 0.1 to 2 mM were also prepared in 0.1 M sodium bicarbonate buffer, pH 8.5, and used to obtain the calibration curve. For this purpose, 50 μL from each L-tyrosine solution was mixed with 1 mL of BCA working solution and incubated for 60 min at 60 °C. Then, 100 μL of each solution was transferred into a 96-well microtiter plate and the optical density at 560 nm was measured. For the calibration curve, the optical density value of the L-tyrosine calibrators was plotted versus the L-tyrosine concentration in the calibrators (4.76–95.2 nmol/tube), and the concentration of the amine groups on the chips' surfaces was determined first as an L-tyrosine concentration in mM and then as the actual number of amine groups per surface area using the following formula:

$$Amine\ groups\ per\ \mathrm{cm}^2 = \frac{L - tyrosine\ concentration\ in\ \mathrm{nmol/tube}}{0.75\ \mathrm{cm}^2} \times 6.023 \times 10^{23}$$

4.5. Biological Functionalization of the Sensor

The 2 mm long sensing arm window of the one MZI of the chemically functionalized chips was modified with a 50 μg/mL AFM1–BSA conjugate solution in 50 mM carbonate buffer, pH 9.2, to be used as a working sensor, while the 1 mm long sensing window at the second MZI was modified with a 50 μg/mL BSA solution in the same buffer to serve as the reference sensor. The two solutions were deposited onto the respective MZIs using a microarray spotter (BioOdyssey Calligrapher Mini Arrayer; Bio-Rad Laboratories Inc.; Hercules, CA, USA) with a solid pin of a 375 μm diameter (Arrayit Corp.; Sunnyvale, CA, USA). After spotting, the chips were incubated overnight under controlled humidity conditions (65–70%), washed, and kept dry until use.

4.6. Assay for Detection of AFM1 with the Sensor

The chip coupled to the bifurcated fiber was immersed in a sequence of solutions placed in microtiter wells (350 μL per well). After equilibration in 50 mM Tris-HCl buffer, pH 7.8, 0.9% *w/v* NaCl, 0.5% *w/v* BSA, 5% *v/v* methanol (assay buffer/MeOH), the chip was immersed for 9 min in a 1:1 volume mixture of calibrators prepared in milk or milk samples with a 50 ng/mL rabbit polyclonal anti-AFM1 antibody solution in 50 mM Tris-HCl buffer, pH 7.8, 0.9% *w/v* NaCl, 0.5% *w/v* BSA, 10% *v/v* methanol (antibody dilution buffer), followed by a 2 min washing step in a 50 mM Tris-HCl buffer, pH 7.8, 0.9% *w/v* NaCl, 5% *v/v* methanol, and 1 min of washing with 50 mM Tris-HCl buffer, pH 7.8, 0.9% *w/v* NaCl, 0.5% *w/v* BSA (assay buffer). Then, the chip was immersed for 5 min in a 10 μg/mL biotinylated secondary antibody solution in assay buffer, washed for 1 min in the same buffer, and then immersed for 2 min in a 10 μg/mL streptavidin solution in assay buffer.

After the end of the assay, the chip was regenerated by immersion for 2 min in a 50 mM HCl solution, and then for 2 min in a 0.5% *w/v* SDS solution in 0.1 M HCl, pH 1.3. Finally, the chip was washed with 50 mM Tris-HCl buffer, pH 7.8, 0.9% *w/v* NaCl, and equilibrated in assay buffer/MeOH, before the next assay cycle. A schematic of the competitive AFM1 immunoassay steps is presented in Figure 3b.

For each calibrator, the net signal was determined by the difference of the signal obtained from the working sensor minus that of the reference sensor multiplied by 2, as the length of the reference sensor window was half (1 mm) compared to the window of the working sensor (2 mm). The signal used to prepare the AFM1 calibration curve was that received during the 2 min reaction with streptavidin (analytical signal). The calibration curve was plotted as the percent ratio of each calibrator signal (S_x) to the zero calibrator signal (S_0) versus the AFM1 concentration in the calibrators.

Supplementary Materials: The following supporting information can be downloaded at: https://www.mdpi.com/article/10.3390/toxins17040165/s1, Figure S1: (a) Real-time signal response for the zero AFM1 calibrator in assay buffer. The sequence of solutions is as follows: start to arrow 1: assay buffer; arrow 1 to 2: anti-AFM1 anti-body solution/zero calibrator mixture in assay buffer; arrow 2 to 3: assay buffer; arrow 3 to 4: biotinylated secondary antibody in assay buffer; arrow 4 to 5: assay buffer; arrow 5 to end: streptavidin in assay buffer. (b) Real-time signal response for the zero AFM1 calibrator in undiluted cow milk. The sequence of solutions is as follows: start to arrow 1: assay buffer; arrow 1 to 2: anti-AFM1 antibody solution/zero calibrator mixture; arrow 2 to 3: washing buffer; arrow 3 to 4: assay buffer; arrow 4 to 5: biotinylated secondary antibody in assay buffer; arrow 5 to 6: assay buffer; arrow 6 to end: streptavidin in assay buffer. In both cases the black line corresponds to the reference MZI response, the red to the working MZI, and the blue line is the difference of the two responses (net chip signal); Figure S2: Fluorescence microscope image depicting the area of the windows over the two MZIs from a chip functionalized with HF/glutaraldehyde after running the AFM1 assay and using, at the final assay step, Alexa FluorTM 546 labelled streptavidin. The image was acquired with an Axioskop 2 Plus epifluorescence microscope (Carl Zeiss; Hamburg, Germany) facilitated with an appropriate filter pair and a MicroPublisher 3.3 RTV CCD camera (QImaging, Surrey, BC, Canada) for image acquisition, and processed with the Image ProPlus software (Media Cybernetics, Inc.; Rockville, MD, USA); Figure S3: L-tyrosine calibration curve obtained with the BCA protein assay method. Each point is the mean value of three replicates ± SD; Figure S4: Net zero calibrator signals obtained from chips spotted with different concentrations of AFM1-BSA conjugate when 50 ng/mL concentration of anti-AFM1 was employed for 9 min. Each column corresponds to the mean of 3 chips ± SD; Figure S5: Typical calibration curve of AFM1 in ewe (blue line) and in goat milk (orange line). Each point is the mean value of 3 measurements ± SD; Figure S6: Calibration curves of AFM1 (black line), AFB1 (purple line), AFB2 (orange line), and AFG1 (blue line) obtained with the immersible MZI immunosensor coated with AFM1-BSA conjugate. The dashed vertical lines correspond to analyte concentration that provides 50% inhibition (horizontal dashed black line); Figure S7: Net zero calibrator signals obtained from chips stored at RT and assayed over a period of 8 weeks. Each column corresponds to the mean of 3 chips ± SD. Horizontal solid lines correspond to mean value ± 3SD.

Author Contributions: Conceptualization, M.A., P.P. and S.K.; methodology, M.A., P.P., A.E., E.M., K.M. and S.K.; formal analysis, D.K., M.A. and K.M.; investigation, D.K. and M.A.; resources, E.M., K.M. and S.K.; data curation, D.K., M.A., A.E, E.M., K.M. and P.P.; writing—original draft preparation, D.K. and M.A.; writing—review and editing, E.M., A.E., P.P., K.M. and S.K.; visualization, D.K. and M.A.; supervision, A.E., K.M. and S.K.; funding acquisition, E.M., P.P. and S.K. All authors have read and agreed to the published version of the manuscript.

Funding: This research has been co-financed by the European Regional Development Fund of the European Union and Greek national funds through the Operational Program Competitiveness, Entrepreneurship and Innovation, under the call RESEARCH–CREATE–INNOVATE (project code:

T2EΔK-01934/FOODSENS), and European Union's Horizon 2020 Research and Innovation program through the Marie Sklodowska-Curie grant agreement No 101007299 (SAFEMILK).

Institutional Review Board Statement: Not applicable.

Informed Consent Statement: Not applicable.

Data Availability Statement: The original contributions presented in this study are included in the article/supplementary materials. Further inquiries can be directed to the corresponding author.

Conflicts of Interest: The authors declare no conflicts of interest.

References

1. Boshra, M.H.; El-Housseiny, G.S.; Farag, M.M.S.; Aboshanab, K.M. Innovative Approaches for Mycotoxin Detection in Various Food Categories. *AMB Express* **2024**, *14*, 7. [CrossRef] [PubMed]
2. Giovati, L.; Magliani, W.; Ciociola, T.; Santinoli, C.; Conti, S.; Polonelli, L. AFM1 in Milk: Physical, Biological, and Prophylactic Methods to Mitigate Contamination. *Toxins* **2015**, *7*, 4330–4349. [CrossRef]
3. Abera, B.D.; Falco, A.; Ibba, P.; Cantarella, G.; Petti, L.; Lugli, P. Development of Flexible Dispense-Printed Electrochemical Immunosensor for Aflatoxin M1 Detection in Milk. *Sensors* **2019**, *19*, 3912. [CrossRef] [PubMed]
4. The Commission of the European Communities. European Commission (EC) Commission Regulation (EU) No 1881/2006 of 19 December 2006 Setting Maximum Levels for Certain Contamination in Foodstuffs as Regards Aflatoxin. *Off. J. EUL* **2006**, *59*, 15–16.
5. Rabie, M.; Movassaghghazani, M.; Afshar Mogaddam, M.R. HPLC-FLD Determination of Aflatoxins M1 and M2 in Raw Cow Milk Samples Using in-Syringe Gas-Controlled Density Tunable Solidification of a Floating Organic Droplet-Based Dispersive Liquid-Liquid Microextraction Method. *RSC Adv.* **2024**, *14*, 5077–5084. [CrossRef]
6. Gallardo-Ramos, J.A.; Marín-Sáez, J.; Sanchis, V.; Gámiz-Gracia, L.; García-Campaña, A.M.; Hernández-Mesa, M.; Cano-Sancho, G. Simultaneous Detection of Mycotoxins and Pesticides in Human Urine Samples: A 24-h Diet Intervention Study Comparing Conventional and Organic Diets in Spain. *Food Chem. Toxicol.* **2024**, *188*, 114650. [CrossRef]
7. Abdul-Hay, M.E.S.I.; Shendy, A.H.; Aboalhassen, A.E.D.F.A.; Gomaa, A.M.; Ismail, E.H. Development and Validation of a Simple Solid-Liquid Extraction Protocol Coupled with LC-ESI-MS/MS for the Determination of Aflatoxin M1 in Products of Colostrum-Based Supplements and Whey Protein-Based Sports Food. *J. Food Compos. Anal.* **2024**, *126*, 105933. [CrossRef]
8. Tarannum, N.; Nipa, M.N.; Das, S.; Parveen, S. Aflatoxin M1 Detection by ELISA in Raw and Processed Milk in Bangladesh. *Toxicol. Rep.* **2020**, *7*, 1339–1343. [CrossRef] [PubMed]
9. Cai, C.; Liu, Y.; Tang, X.; Zhang, W.; Zhang, Q.; Li, P. Development of a Toxin-Free Competitive Immunoassay for Aflatoxin M1 Based on a Nanobody as Surrogate Calibrator. *LWT* **2023**, *182*, 114829. [CrossRef]
10. Shahjahan, T.; Javed, B.; Sharma, V.; Tian, F. Overview of Various Components of Lateral-Flow Immunochromatography Assay for the Monitoring of Aflatoxin and Limit of Detection in Food Products: A Systematic Review. *Chemosensors* **2023**, *11*, 520. [CrossRef]
11. Chen, Q.; Meng, M.; Li, W.; Xiong, Y.; Fang, Y.; Lin, Q. Emerging Biosensors to Detect Aflatoxin M1 in Milk and Dairy Products. *Food Chem.* **2022**, *398*, 133848. [CrossRef] [PubMed]
12. Evtugyn, G.; Hianik, T. Electrochemical Immuno- and Aptasensors for Mycotoxin Determination. *Chemosensors* **2019**, *7*, 10. [CrossRef]
13. Cancelliere, R.; Paialunga, E.; Grattagliano, A.; Micheli, L. Label-free electrochemical immunosensors: A practical guide. *Trend Anal. Chem.* **2024**, *180*, 117949. [CrossRef]
14. Nguyen, B.H.; Tran, L.D.; Do, Q.P.; Nguyen, H.L.; Tran, N.H.; Nguyen, P.X. Label-free detection of aflatoxin M1 with electrochemical Fe_3O_4/polyaniline-based aptasensor. *Mater. Sci. Eng. C Mater. Biol. Appl.* **2013**, *33*, 2229–2234. [CrossRef] [PubMed]
15. Karapetis, S.; Nikolelis, D.; Hianik, T. Label-Free and Redox Markers-Based Electrochemical Aptasensors for Aflatoxin M1 Detection. *Sensors* **2018**, *18*, 4218. [CrossRef]
16. Smolko, V.; Shurpik, D.; Porfireva, A.; Evtugyn, G.; Stoikov, I.; Hianik, T. Electrochemical Aptasensor Based on Poly(Neutral Red) and Carboxylated Pillar[5]arene for Sensitive Determination of Aflatoxin M1. *Electroanalysis* **2018**, *30*, 486–496. [CrossRef]
17. Li, H.; Du, C.; Guo, T.; Zhou, H.; Zhou, Y.; Huang, X.; Zhang, Y.H.; Wang, S.; Liu, X.; Ma, L. Ratiometric Electrochemical Aptasensor Based on Split Aptamer and Au-RGO for Detection of Aflatoxin M1. *J. Dairy Sci.* **2024**, *107*, 2748–2759. [CrossRef] [PubMed]
18. Yang, D.; Hui, Y.; Liu, Y.; Wang, W.; He, C.; Zhao, A.; Wei, L.; Wang, B. Novel Dual-Recognition Electrochemical Biosensor for the Sensitive Detection of AFM1 in Milk. *Food Chem.* **2024**, *433*, 137362. [CrossRef]

19. Jalalian, S.H.; Ramezani, M.; Danesh, N.M.; Alibolandi, M.; Abnous, K.; Taghdisi, S.M. A novel electrochemical aptasensor for detection of aflatoxin M1 based on target-induced immobilization of gold nanoparticles on the surface of electrode. *Biosens. Bioelectron.* **2018**, *117*, 487–492. [CrossRef]
20. Kaur, G.; Sharma, S.; Singh, S.; Bhardwaj, N.; Deep, A. Selective and sensitive electrochemical sensor for aflatoxin M1 with a molybdenum disulfide quantum dot/metal–organic framework nanocomposite. *ACS Omega* **2022**, *7*, 17600–17608. [CrossRef]
21. Bacher, G.; Pal, S.; Kanungo, L.; Bhand, S. A Label-Free Silver Wire Based Impedimetric Immunosensor for Detection of Aflatoxin M1 in Milk. *Sens. Actuator B Chem.* **2012**, *168*, 223–230. [CrossRef]
22. Larou, E.; Yiakoumettis, I.; Kaltsas, G.; Petropoulos, A.; Skandamis, P.; Kintzios, S. High Throughput Cellular Biosensor for the Ultra-Sensitive, Ultra-Rapid Detection of Aflatoxin M1. *Food Control* **2013**, *29*, 208–212. [CrossRef]
23. Angelopoulou, M.; Petrou, P.; Kakabakos, S. Advances in Interferometric Sensors for the Detection of Food Contaminants. *TrAC Trend Anal. Chem.* **2024**, *175*, 117714. [CrossRef]
24. Cai, Y.; Guo, G.; Fu, Y.; Huang, X.; Wang, T.; Li, T. A Fluorescent Aptasensor Based on Functional Graphene Oxide and FRET Strategy Simultaneously Detects Aflatoxins B1 and Aflatoxins M1. *Chin. J. Anal. Chem.* **2024**, *52*, 100408. [CrossRef]
25. Cao, H.; Liang, D.; Tang, K.; Sun, Y.; Xu, Y.; Miao, M.; Zhao, Y. SERS and MRS Signals Engineered Dual-Mode Aptasensor for Simultaneous Distinguishment of Aflatoxin Subtypes. *J. Hazard. Mater.* **2024**, *462*, 132810. [CrossRef] [PubMed]
26. Esmaelpourfarkhani, M.; Ramezani, M.; Alibolandi, M.; Abnous, K.; Taghdisi, S.M. CRISPR-Cas12a-Based Colorimetric Aptasensor for Aflatoxin M1 Detection Based on Oxidase-Mimicking Activity of Flower-like MnO_2 Nanozymes. *Talanta* **2024**, *271*, 125729. [CrossRef]
27. Li, J.; Ni, Y.; Zhang, W.; Nteppe Nteppe, E.L.; Li, Y.; Zhang, Y.; Yan, H. Fiber Optic LSPR Sensing AFM1 in Milk with Enhanced Sensitivity by the Hot Spot Effect Based on Nanogap Construction. *Micromachines* **2024**, *15*, 779. [CrossRef] [PubMed]
28. Ghanim Al-Rubaye, A.; Nabok, A.; Catanante, G.; Marty, J.-L.; Takács, E.; Székács, A. Label-free optical detection of mycotoxins using specific aptamers immobilized on gold nanostructures. *Toxins* **2018**, *10*, 291. [CrossRef]
29. Wang, J.; Chen, Q.; Jin, Y.; Zhang, X.; He, L.; Zhang, W.; Chen, Y. Surface Enhanced Raman Scattering-Based Lateral Flow Immunosensor for Sensitive Detection of Aflatoxin M1 in Urine. *Anal. Chim. Acta* **2020**, *1128*, 184–192. [CrossRef]
30. Karczmarczyk, A.; Dubiak-Szepietowska, M.; Vorobii, M.; Rodriguez-Emmenegger, C.; Dostálek, J.; Feller, K.H. Sensitive and Rapid Detection of Aflatoxin M1 in Milk Utilizing Enhanced SPR and p(HEMA) Brushes. *Biosens. Bioelectron.* **2016**, *81*, 159–165. [CrossRef]
31. Indyk, H.E.; Chetikam, S.; Gill, B.D.; Wood, J.E.; Woollard, D.C. Development and application of an optical biosensor immunoassay for Aflatoxin M1 in bovine milk. *Food Anal. Method* **2019**, *12*, 2630–2637. [CrossRef]
32. Tsounidi, D.; Koukouvinos, G.; Petrou, P.; Misiakos, K.; Zisis, G.; Goustouridis, D.; Raptis, I.; Kakabakos, S.E. Rapid and Sensitive Label-Free Determination of Aflatoxin M1 Levels in Milk through a White Light Reflectance Spectroscopy Immunosensor. *Sens. Actuator B Chem.* **2019**, *282*, 104–111. [CrossRef]
33. Chalyan, T.; Potrich, C.; Schreuder, E.; Falke, F.; Pasquardini, L.; Pederzolli, C.; Heideman, R.; Pavesi, L. AFM1 detection in milk by fab′ functionalized Si_3N_4 asymmetric mach-Zehnder interferometric biosensors. *Toxins* **2019**, *11*, 409. [CrossRef] [PubMed]
34. Angelopoulou, M.; Kourti, D.; Misiakos, K.; Economou, A.; Petrou, P.; Kakabakos, S. Mach-Zehnder Interferometric Immunosensor for Detection of Aflatoxin M1 in Milk, Chocolate Milk, and Yogurt. *Biosensors* **2023**, *13*, 592. [CrossRef]
35. Akgönüllü, S.; Yavuz, H.; Denizli, A.; García, M.; Castro Grijalba, A. Development of Gold Nanoparticles Decorated Molecularly Imprinted–Based Plasmonic Sensor for the Detection of Aflatoxin M1 in Milk Samples. *Chemosensors* **2021**, *9*, 363. [CrossRef]
36. Antoniou, M.; Tsounidi, D.; Petrou, P.S.; Beltsios, K.G.; Kakabakos, S.E. Functionalization of silicon dioxide and silicon nitride surfaces with aminosilanes for optical biosensing applications. *Med. Devices Sens.* **2020**, *3*, e10072. [CrossRef]
37. Bañuls, M.-J.; González-Pedro, V.; Barrios, C.A.; Puchades, R.; Maquieira, A. Selective chemical modification of silicon nitride/silicon oxide nanostructures to develop label-free biosensors. *Biosens. Bioelectron.* **2010**, *25*, 1460–1466. [CrossRef]
38. Kakabakos, S.E.; Tyllianakis, P.E.; Evangelatos, G.P.; Ithakissios, D.S. Colorimetric determination of reactive solid-supported primary and secondary amino groups. *Biomaterials* **1994**, *15*, 289–297. [CrossRef]
39. Chavarría, G.; Molina, A.; Leiva, A.; Méndez, G.; Wong-González, E.; Cortés-Muñoz, M.; Rodríguez, C.; Granados-Chinchilla, F. Distribution, stability, and protein interactions of Aflatoxin M1 in fresh cheese. *Food Control* **2017**, *73*, 581–586. [CrossRef]
40. Cancelliere, R.; Di Tinno, A.; Cataldo, A.; Bellucci, S.; Kumbhat, S.; Micheli, L. Nafion-based label-free immunosensor as a reliable warning system: The case of AFB1 detection in cattle feed. *Microchem. J.* **2023**, *191*, 108868. [CrossRef]
41. Kourti, D.; Angelopoulou, M.; Makarona, E.; Economou, A.; Petrou, P.; Misiakos, K.; Kakabakos, S. Photonic Dipstick Immunosensor to Detect Adulteration of Ewe, Goat, and Donkey Milk with Cow Milk through Bovine κ-Casein Detection. *Sensors* **2024**, *24*, 5688. [CrossRef] [PubMed]

Disclaimer/Publisher's Note: The statements, opinions and data contained in all publications are solely those of the individual author(s) and contributor(s) and not of MDPI and/or the editor(s). MDPI and/or the editor(s) disclaim responsibility for any injury to people or property resulting from any ideas, methods, instructions or products referred to in the content.

toxins

MDPI

Review

Are Aflatoxin Residues in Chicken Products a Real or Perceived Human Dietary Risk?

Madalitso Chelenga [1,*], Limbikani Matumba [2,*], Muloongo C. Sitali [3], Bertha Kachala [2], Verson Nambuzi [1], Merning Mwenifumbo [1], Aggrey Pemba Gama [4], Mulunda Mwanza [5], Maurice Monjerezi [6] and John F. Leslie [7]

1 Department of Veterinary Clinical Studies, Faculty of Veterinary Medicine, Lilongwe University of Agriculture and Natural Resources (LUANAR), Lilongwe P.O. Box 219, Malawi; versonnambuzi@gmail.com (V.N.); mmwenifumbo@luanar.ac.mw (M.M.)
2 FoodPlus Research Group, Faculty of Life Sciences Natural Resources, LUANAR (NRC Campus), Lilongwe P.O. Box 143, Malawi; bkachala@luanar.ac.mw
3 Department of Biomedical Sciences, School of Veterinary Medicine, The University of Zambia, Lusaka P.O. Box 32379, Zambia; muloongosit2013@gmail.com
4 Food Science and Technology Department, Faculty of Food and Human Sciences, LUANAR (Bunda campus), Lilongwe P.O. Box 219, Malawi; agama@luanar.ac.mw
5 Department of Animal Health, Faculty of Natural and Agricultural Sciences, North-West University, Private Bag X2046, Mmabatho 2745, South Africa; mulunda.mwanza@nwu.ac.za
6 Centre for Resilient Agri-Food Systems (CRAFS), University of Malawi, Zomba P.O. Box 280, Malawi; mmonjerezi@unima.ac.mw
7 Department of Plant Pathology, Throckmorton Plant Sciences Center, Kansas State University, 1712 Claflin Rd., Manhattan, KS 66506, USA; jfl@ksu.edu
* Correspondence: mchelenga@luanar.ac.mw (M.C.); lmatumba@luanar.ac.mw (L.M.)

Abstract: Aflatoxin is a health threat to humans and domesticated animals. Chickens are often fed aflatoxin-contaminated grain and may retain some toxins in muscle, eggs, and other tissues. A critical food safety question is whether tissues from contaminated birds pose a threat to the humans that consume them. We evaluated literature published from 1984 to 2023 to determine the level of aflatoxin residues retained in chicken eggs, muscles, livers, gizzards, and hearts. In the studies evaluated (n = 33), ~8100 chickens in 334 trials were fed feed contaminated with 0.1–6400 µg/kg of aflatoxins for 7–180 days. There was a positive correlation between the level of feed contamination and residual aflatoxin concentrations ($r^2 = 0.18$, $p < 0.05$), but <1% of the aflatoxin in the feed carried over to edible broiler tissues. Only 0.6% of the trials reported >20 µg/kg of aflatoxin in the tissues, primarily in the muscle tissue, when the chickens were fed feed contaminated with >300 µg/kg of aflatoxins, which is above the US FDA maximum tolerable limit for components of poultry feeds. These composite results suggest a relatively low risk to public health from consuming chickens fed contaminated feed and a relatively high aflatoxin elimination mechanism in chickens that consume feed containing >300 µg/kg of aflatoxins. The data are consistent with chickens fed feed containing up to 500 µg/kg of aflatoxin being allowed in the human food chain without posing a significant health hazard. In reality, the maximum level of aflatoxin allowed in chicken feed will probably be limited by how much the birds can tolerate and still grow at a suitable rate without deformities rather than the risk that processed birds could present to human health. As chickens effectively act as an absorptive buffer for aflatoxin in contaminated feed, we expect that a contamination level that is acceptable for chicken growth performance is likely to be less than the amount that keeps chicken products safe for human consumption.

Keywords: animal feed; edible tissue; food safety; maximum tolerable level; mycotoxins; poultry

Key Contribution: In 33 studies from 17 countries; total aflatoxin residues in chickens consuming naturally contaminated feed average around 2.0 μg/kg; except for eggs; which average 0.3 μg/kg.

1. Introduction

Health and growth impairment of turkeys consuming contaminated grain [1] led to the original characterization of aflatoxins and their identification as a significant threat to the health of humans and domesticated animals. Extensive efforts have been made to determine safe feed exposure levels and develop regulatory guidelines for these toxic, thermostable secondary metabolites that cause aflatoxicosis, cancer, and liver failure in humans and animals. The primary focus has been on contaminated grains and plant products; however, aflatoxin residues have been identified in animal tissues, and animal products also could be human health threats [2]. Human exposure to aflatoxins can lead to a myriad of health problems including, but not limited to, carcinogenesis, reproductive dysfunction [3], and impairment of the immune system [4]. Humans are primarily exposed to aflatoxins, most commonly aflatoxin B_1 (AFB_1), via the consumption of contaminated cereals (maize, sorghum, rice, wheat) and oilseeds (peanut, sunflower, soybean, cotton) [5]. These same agricultural commodities are also fed to animals directly or as ingredients in formulated feeds. Aflatoxins can carry over to animal products, e.g., milk, eggs, and meat, with contamination of chicken products of particular importance in developing countries. The efficiency of the transfer of aflatoxin from feed to consumable chicken products is a critical factor in determining the suitability of these products for human consumption.

Our objective in this review was to consolidate available data on the efficiency of aflatoxin transfer from feed to chicken products. Our working hypotheses were that aflatoxin contamination in chicken tissues depends (i) on the level of aflatoxin in the feed, (ii) the length of time the contaminated feed was consumed, and (iii) that chickens consuming feed contaminated at levels above that allowed in the feed would result in chicken tissues that pose a hazard to human health. Our results advance the field by identifying that <1% of the aflatoxins in chicken feed are retained in chicken tissues that are consumed by humans.

2. Results

2.1. Characteristics of the Included Studies and the Aflatoxin Exposure Profiles in the Evaluated Chickens

Thirty-three studies conducted in different parts of the world were selected for this study (Figure 1, Table 1). These studies evaluated over 8100 chickens from 342 trials. Four of the 33 studies contained eight trials that did not report the levels of aflatoxins in the feed [6–9] and were subsequently excluded from further analysis. In the remaining 334 trials, an average of 1,700 μg of aflatoxin/kg of contaminated feed (range = 0.1–6400 μg/kg, median = 1680) was fed to chickens for 7–180 days. About two-thirds of the trials in these studies introduced aflatoxin by incorporating fungal-colonized grains on which aflatoxins had been produced into the chicken feed. In the remaining one-third, aflatoxin was introduced through artificial spiking with one or more of aflatoxins B_1, B_2, G_1, or G_2. One study did not report the source of the aflatoxins used in the experiments.

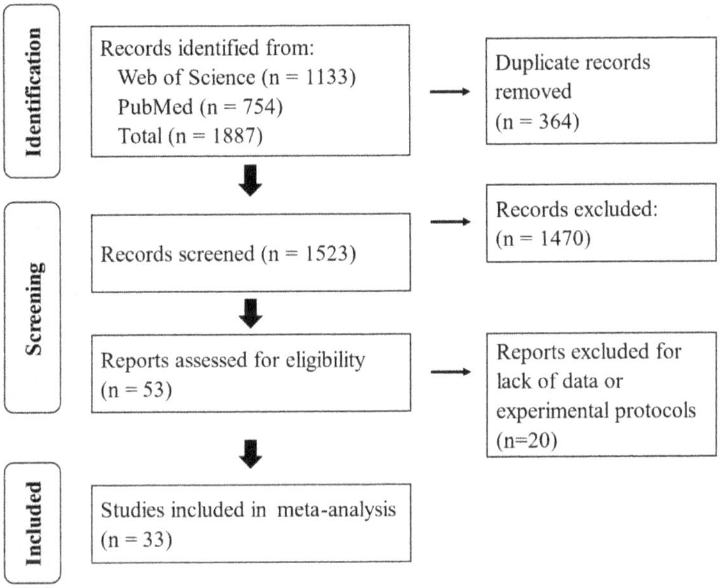

Figure 1. PRISMA flow diagram.

Table 1. Characteristics of the 33 included studies.

COA (μg/kg)	DOF (Days)	Country	Type	N	Tissues Evaluated	Diagnostic Test	Ref.
100, 200, 400	42	Pakistan	Broiler	120	Liver Muscle	HPLC	[10]
250	20	Saudi Arabia	Broiler	360	Liver	HPLC	[11]
300, 600	42	Pakistan	Broiler	180	Kidney Liver	ELISA	[12]
110, 117, 129, 345, 361, 385	21, 28, 35, 42	Saudi Arabia	Broiler	100	Gizzard Kidney Liver Muscle	HPLC	[13]
NR	NR	Iran	NR	50	Eggs Liver	HPLC	[6]
250	35	Egypt	Broiler	150	Liver Muscle	NR	[14]
3000	7	Japan	Broiler Layer	64	Liver Muscle	HPLC	[15]
50, 100	42	Japan	Layer Duck	270	Liver Muscle	HPLC	[16]
1300, 2100	35	USA	Broiler	48	Gizzard Muscle	NR	[17]
1,000	42	Spain	Broiler	120	Liver	HPLC	[18]
2500, 5000	32	Spain	Layer	80	Egg Muscle Liver Gizzard Kidney	TLC	[19]
600, 1200, 1800	7, 14, 21	USA	Broiler	336	Liver Kidney	TLC	[20]
5	21	Spain	Layer	72	Liver Muscle	HPLC	[21]
1600, 3200, 6400	7	Pakistan	Broiler	2,000	Liver Muscle	HPLC	[22]
123	42	China	Layer	336	Eggs	HPLC	[23]
5.8, 8.2	NR	Cameroon	Layer	48	Eggs Liver Muscle	ELISA	[24]

Table 1. *Cont.*

COA (μg/kg)	DOF (Days)	Country	Type	N	Tissues Evaluated	Diagnostic Test	Ref.
NR	NR	Pakistan	Broiler	264	Kidney Liver Muscle	HPLC	[9]
50	36, 64, 169	Italy	Broiler	163	Kidney Liver Muscle Skin	HPLC	[25]
2500	21	Columbia	Broiler	100	Kidney Liver	HPLC	[26]
100, 300, 500	54	Brazil	Layer	93	Eggs Muscle	TLC	[27]
NR	NR	United Arab Emirates	NR	NR	Eggs	LS-MS/MS	[7]
2500, 3900	280	India	Layer	144	Eggs Muscle	NR	[28]
100	22	Argentina	Broiler	25	Liver	HPLC	[29]
45	21	Serbia & Montenegro	Broiler	1,000	Liver	TLC	[30]
2500	28	Italy	Layer	96	Liver	EIA	[31]
NR	NR	Mozambique	Broiler	NR	Gizzard Liver	ELISA	[8]
100	42	Serbia	Broiler	96	Liver Muscle	LC-MS/MS	[32]
5000	28	China	Broiler	NR	Muscle	HPLC	[33]
1700, 3300	28	USA	Layer	16	Eggs	TLC	[34]
1700, 3300	28	USA	Layer	80	Heart Gizzard Kidney Liver Muscle Spleen	TLC	[35]
37, 69, 82, 130	21	China	Broiler	1,200	Eggs Liver	HPLC-MS/MS	[36]
0.1, 25, 52, 77	60	China	Broiler	180	Eggs Heart Liver Muscle	HPLC-MS/MS	[37]
47, 450	30	China	Broiler	75	Liver Muscle	ELISA	[38]

OA—Concentration of aflatoxin in feed; DOF—Duration of Feeding; ELISA—Enzyme-Linked Immunosorbent Assay; HPLC—High-Performance Liquid Chromatography; HPLC-MS/MS—HPLC–tandem mass spectrometry; LC—Liquid Chromatography; LC-MS/MS—LC–tandem mass spectrometry; TLC—Thin-layer chromatography; EIA—Enzyme Immunoassay; NR—Not reported.

2.2. *Aflatoxin Carryover to Various Chicken Tissues*

Generally, the concentration of aflatoxin residues in chicken tissues positively correlated with the concentration in the chicken feeds ($r^2 = 0.15$, $p = 0.007$, Figure 2A). There also was a significant association between aflatoxin accumulation in a tissue and the duration of exposure to aflatoxin ($r^2 = 0.33$, $p < 0.001$, Figure 3). The r^2 values reported suggest that 15% of the variation reported for the aflatoxin retained in the tissue was due to the aflatoxin concentration in the feed and that 33% of the variation observed was due to length of exposure [39]. Nonetheless, the average aflatoxin tissue contamination across all trials and tissues was ~0.5 μg/kg, with the highest average in the gizzards at 1.4 μg/kg of aflatoxins. Thus, we hypothesize that chickens have strong aflatoxin-buffering capability.

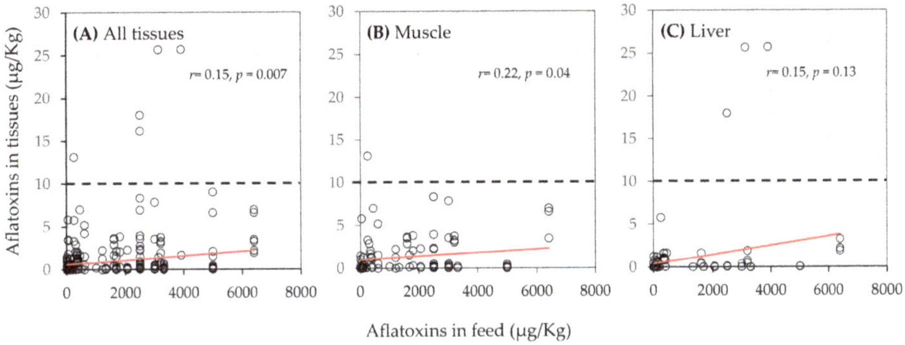

Figure 2. Relationship between the concentration of aflatoxin in chicken feed and aflatoxin residue in (**A**) all chicken tissues, (**B**) muscle, and (**C**) liver. The red lines are the line of best fit. The dotted lines indicate the global median limit of 10 μg/kg for total aflatoxins.

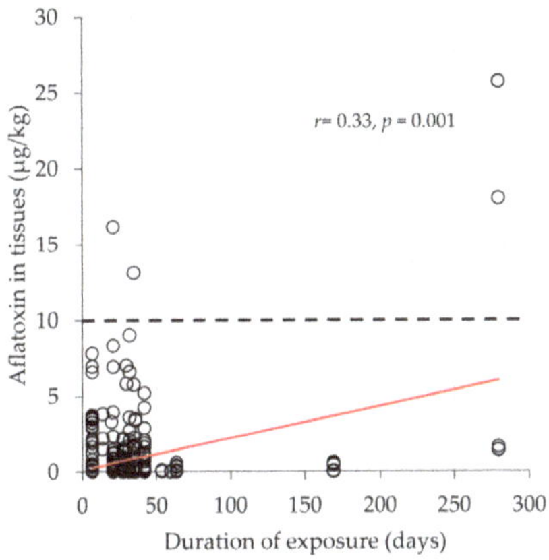

Figure 3. Relationship between the duration of exposure and the accumulation of aflatoxin residues in various tissues derived from chickens fed with aflatoxin-contaminated feed. The red line is the line of best fit. The dotted line indicates the global median limit of 10 μg/kg for total aflatoxins.

When chickens were fed feed contaminated with <100 μg/kg of aflatoxins (n = 80), 91% of the trials found aflatoxin residue levels <1 μg/kg of the aflatoxin in tissues, and none had a residue level higher than 10 μg/kg (Table 2). If the aflatoxin levels in the feed were between 100 and 300 μg/kg, then 77% of the trials reported <1 μg/kg of aflatoxin in the tissues, and none reported >20 μg/kg aflatoxin residue. Most (n = 219) of the studies fed chickens feed contaminated with >300 μg/kg of aflatoxins, which exceeds the FDA's allowable limit for aflatoxin in poultry feed. Seventy-four percent of the trials reported <1 μg/kg of aflatoxin in the tissue, an additional 25% reported <20 μg/kg, and only two trials reported residual levels above the FDA's current regulatory limit of 20 μg/kg [7]. The median amount of aflatoxin residue in all of the tissues was significantly associated (r^2 = 0.15, p = 0.007) with the amount of aflatoxin present in the feed.

2.2.1. Aflatoxin Accumulation in Chicken Muscle

Of the 91 trials reporting aflatoxin levels in feed and residue accumulation in chicken muscles (Table 2), only 38 used feed contaminated with <300 μg/kg of aflatoxin. Amongst these 38 trials, 34 had aflatoxin residues that were <1 μg/kg, and the highest value any of the 4 remaining trials had was <10 μg/kg of aflatoxins. The remaining 53 trials with data for chicken muscle used >300 μg/kg aflatoxin contamination in the feed. Two trials had chickens with residues >20 μg/kg and 40 had chickens with residues that were <1 μg/kg of aflatoxins.

2.2.2. Aflatoxin Accumulation in Chicken Liver

Of the 103 trials reporting aflatoxin levels in feed and subsequent residues in the liver (Table 2), 24 used feed contaminated with <100 μg/kg of aflatoxins with 88% having <1 μg/kg of aflatoxins in the liver and none having >10 μg/kg of aflatoxin residue. In the 12 trials with the feed contaminated with 100–300 μg/kg of aflatoxins, 8 trials reported <1 μg/kg of aflatoxin residue and the other 4 all reported <20 μg/kg residue in the liver. Again, most of the trials (n = 67) used >300 μg/kg of aflatoxin in the contaminated feed with 57% reporting <1 μg/kg of aflatoxin residue and the remaining 43% all below 10 μg/kg of aflatoxin residue in the liver (Table 2). None of these trials reported >10 μg/kg of aflatoxin residues in the liver. The median amount of aflatoxin residue in the liver was not significantly associated ($r^2 = 0.15$, $p = 0.013$) with the amount of aflatoxin present in the feed.

2.2.3. Aflatoxin Accumulation in Chicken Eggs

Fifty-eight trials (Table 2) reported aflatoxin in chicken feed and in eggs laid by chickens that consumed the contaminated feed. If the feed was contaminated with <300 μg/kg of aflatoxin, then the aflatoxin residue was <1 μg/kg in all of the eggs. In the 41 trials that used feed contaminated with >300 μg/kg of aflatoxin, 38 of the trials reported aflatoxin residues <1 μg/kg in the eggs. None of these trials reported aflatoxin residues of >5 μg/kg in the eggs.

2.2.4. Aflatoxin Accumulation in Chicken Kidney

Thirty-five trials reported aflatoxin levels in feed and subsequent residue accumulation in chicken kidneys (Table 2). Twenty-four of the trials used feed contaminated with >300 μg/kg of aflatoxin. Twenty of these trials reported aflatoxin residues <1 μg/kg and none of the trials had kidneys with residues that were >20 μg/kg. If the feed was contaminated with <300 μg/kg of aflatoxin, then 7/11 trials had <1 μg/kg of aflatoxin in the kidney and the remaining 4 trials all had aflatoxin residues that were <5 μg/kg.

Table 2. Total aflatoxin concentration in the feed and subsequent concentration of the residues in various edible tissues.

Tissue (Total Number of Trials)	Aflatoxin in Feeds		Number of Trials (%) with Different Ranges of Aflatoxin Concentration (μg/kg) in Chicken Tissues				
	Concentration (μg/kg)	Number of Trials	<1	1–5	5–10	10–20	>20
	0.1–100	80	73 (91)	6 (8)	1 (1)		
Total (334)	100–300	35	27 (77)	6 (17)	1 (3)	1 (3)	
	>300	219	161 (74)	45 (20)	9 (4)	2 (1)	2 (1)
	0.1–100	52	48 (92)	3 (5)	1 (2)	1 (2)	
AFB_1 only (192)	100–300	25	17 (68)	6 (24)	2 (8)		
	>300	115	65 (56)	38 (33)	8 (7)	2 (2)	2 (2)
	0.1–100	24	21 (88)	2 (8)	1 (4)		
Liver (103)	100–300	12	8 (67)	3 (25)		1 (8)	
	>300	67	38 (57)	23 (34)	6 (9)		
	0.1–100	21	20 (95)	1 (5)			
Muscle (91)	100–300	17	14 (82)	2 (12)	1 (6)		
	>300	53	40 (75)	10 (19)		1(2)	2 (4)
	0.1–100	13	13 (100)				
Eggs (58)	100–300	4	4 (100)				
	>300	41	38 (93)	3 (7)			

Table 2. *Cont.*

Tissue (Total Number of Trials)	Aflatoxin in Feeds		Number of Trials (%) with Different Ranges of Aflatoxin Concentration (μg/kg) in Chicken Tissues				
	Concentration (μg/kg)	Number of Trials	<1	1–5	5–10	10–20	>20
Kidneys (35)	0.1–100	10	7 (70)	3 (30)			
	100–300	1		1 (100)			
	>300	24	20 (84)	2 (8)	1 (4)	1 (4)	
Gizzards (25)	100–300	1	1 (100)				
	>300	24	15 (63)	7 (29)	2 (8)		
Skin (8)	0.1–100	8	8 (100)				
Heart (9)	0.1–100	4	4 (100)				
	>300	5	5 (100)				
Spleen (5)	>300	5	5 (100)				

2.2.5. Aflatoxin Accumulation in Chicken Gizzards

Twenty-five trials reported aflatoxin levels in chicken feed and residues in the chicken gizzards (Table 2). All but one trial used feed contaminated with >300 μg/kg. Sixteen of the trials resulted in aflatoxin residues <1 μg/kg and the remaining nine all had residues <10 μg/kg of aflatoxin in the gizzards.

2.2.6. Aflatoxin Accumulation in Chicken Skin, Heart, and Spleen

A few trials evaluated aflatoxin levels in feed and residues in the chicken skin, heart, and spleen (Table 2). In all of these trials, the aflatoxin residues in the tissues were all < 1 μg/kg regardless of the concentration of aflatoxins in the feed.

2.3. Aflatoxin Buffering Capacity by Tissue

Aflatoxin buffering capacity varied by chicken tissue (Figure 4). Metabolically active organs, such as the liver and kidneys, retained higher levels of aflatoxin and exhibited weaker buffering, while muscle, eggs, and skin exhibited higher buffering capacity. The liver, a primary detoxification organ, had limited buffering capacity as 43% of the trials detected aflatoxin residues >1 μg/kg when feed contamination exceeded 300 μg/kg. Kidneys and gizzards also had moderate buffering capacity, retaining aflatoxins in 17–37% of the trials conducted with feed contaminated with >300 μg/kg of aflatoxins. In contrast, muscle and eggs displayed strong buffering capacity, with no significant aflatoxin residues detected in 76–100% of the relevant trials. Tissues such as skin, heart, and spleen exhibited the highest buffering capacity, with no measurable aflatoxin residue detected in any of the trials.

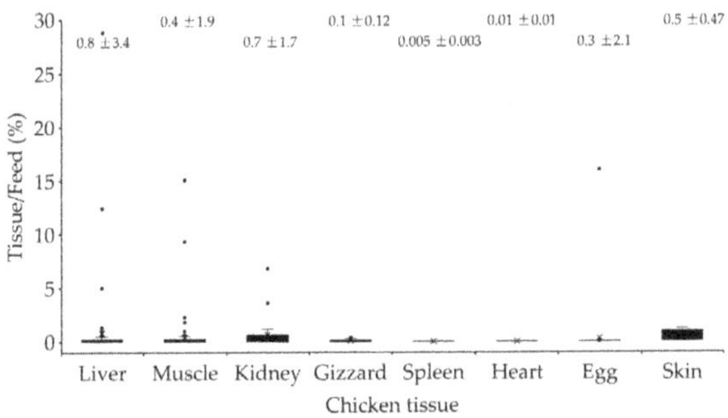

Figure 4. Aflatoxin buffering ability of chickens determined by the percent fold change (tissue: feed ratio) of aflatoxin concentration found in contaminated chicken feeds relative to the concentration of aflatoxin residues accumulated in the various tissues. Numbers above each column are mean ± standard deviation. Solid blocks in each column represent the 25th to 75th percentile range. "X" indicated the median value observed. Bullets are values > 75th percentile.

2.4. *Effect of Aflatoxin Source Used to Contaminate Feed*

When feed was contaminated with naturally occurring aflatoxins, then metabolically active organs generally retained more aflatoxin than when the feed was contaminated with purified aflatoxin (Figure 5A). Aflatoxin residues (Figure 5B) in the liver and kidney were generally the same or higher when the birds were fed spiked feed. For gizzards and muscle, the naturally contaminated feed usually resulted in higher tissue residue values. Eggs generally had low aflatoxin residues regardless of the contamination source. There was insufficient resolvable data for skin, heart, and spleen for a meaningful analysis. The level of contamination in the feed could impact whether natural or spiked contamination was more important in determining the amount of aflatoxin in the tissue residue (Figure 5C). When the feed was contaminated with <100 μg/kg of aflatoxins, then the results were the same for both contamination methods. If the feed contamination level was 100–300 μg/kg, then the tissue residues were higher if the feed was spiked than if it was naturally contaminated. If the feed was contaminated with >300 μg/kg, then the results were reversed, and the aflatoxin residue was higher in birds fed with naturally contaminated feed. Thus, both the method of contaminating the feed and the level of contamination impact the amount of aflatoxin retained in a tissue, and not all tissues behave similarly.

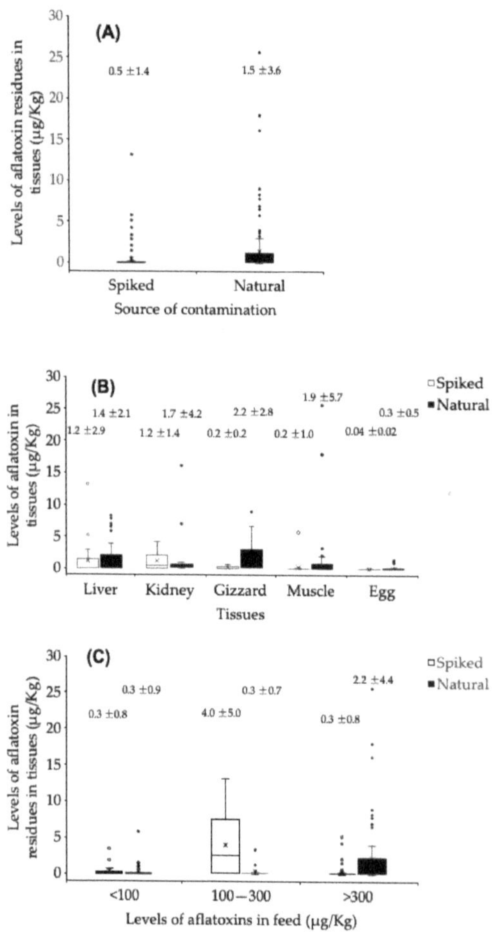

Figure 5. The effects of the source of aflatoxin contamination of feed—spiked versus natural contamination. Figures above each column are mean ± standard deviation. (**A**) Overall comparison of aflatoxin residuals

in tissues from birds fed spiked or naturally contaminated feed. (**B**) Aflatoxin residue concentration in different tissues from birds fed spiked or naturally contaminated feed. (**C**) Overall aflatoxin residue concentration in various tissues of birds fed spiked or naturally contaminated feed stratified by the aflatoxin concentration level in the feed. Solid blocks in each column represent the 25th to 75th percentile range. "X" indicated the median value observed. Bullets are values > 75th percentile.

3. Discussion

Aflatoxin contamination of animal feed raises concerns about animal welfare, productivity, and the impact of residual contamination on secondary consumers, particularly humans. We evaluated residue levels of aflatoxins in chicken products and their association with the feed the animals consume, with a focus on the safety of chicken products for human consumption.

Globally, the regulated limit of total aflatoxin in food ranges from 4 to 30 μg/kg [40], with a median regulatory limit of 10 μg/kg [41]. In the European Union (EU), the limits for aflatoxin B_1 and total aflatoxins are 2 μg/kg and 4 μg/kg, respectively [42], while the US allows up to 20 μg/kg [43–45]. Very few countries have a zero-tolerance policy for aflatoxins because it is virtually impossible to eradicate them. We found that when chickens were fed feed contaminated with aflatoxins ranging from 0.1 to 6400 μg/kg for 7–180 days, <1% of the aflatoxin in the feed is retained as residues in edible tissues despite the significant positive correlation between the concentration of aflatoxin in chicken feeds and the residue levels found in chicken tissues. The FDA set 300 μg/kg of aflatoxin as the maximum level allowed in corn used as an ingredient in chicken feed, although the maximum level allowed in the finished feed is 20 μg/kg [45]. Results from trials in which chickens were fed feed contaminated with >300 μg/kg of aflatoxins resulted in <10 μg/kg of aflatoxin in tissues, which implies that grain contaminated with more than 300 μg/kg can still be used as chicken feed with only minimal risk to humans who consume the birds that ate the contaminated grain.

Reported residue accumulation levels >20 μg/kg FDA regulatory limit were found only in the muscle. Of the trials that reported this level of aflatoxin in the muscle, only 4% of the trials in which chickens were fed feed contaminated with >300 μg/kg of aflatoxin. In the majority of these trials (87/91), <5 μg/kg of aflatoxin was retained in the muscle. In most trials with other edible tissues, e.g., the liver, eggs, kidneys, gizzards, skin, heart, and spleen, all retained <1 μg/kg of aflatoxin. These data suggest that products derived from birds fed feed contaminated with aflatoxin levels exceeding current regulatory levels could still be allowed in the human food chain with only minimal health risks to humans. Note that birds consuming feed that was naturally contaminated, as would occur in a real-world setting, generally had higher aflatoxin residues than those that consumed artificially spiked feed. The average contamination level for all tissues, except eggs, from these birds was around the 2.0 μg/kg level.

In essence, chickens reduce human exposure to aflatoxins when contaminated grain is fed to the chickens rather than being used directly as human food. This relationship is both important and practical since 60–70% of commercial chicken feed usually is grain [46]. If finishing chickens can tolerate up to 300 μg/kg of aflatoxin in chicken feed, then grain containing even higher levels might also be safe for use in chicken feed if the birds can tolerate these levels of contamination without reducing their health and productivity. Using such highly contaminated grain as chicken feed would provide a use for grain that now is often destroyed or allocated to other uses with lower economic values such as ethanol production. This use of contaminated grain as chicken feed could also help increase food security in many developing countries where decreasing aflatoxin in human food and increasing chicken production are both important goals.

Increasing the amount of aflatoxin allowed in chicken feed must consider the impact of these increased levels on animal health and performance. The results from available studies of high levels of aflatoxins on chicken health and performance are inconsistent. One study reported low performance by chickens fed grain containing >500 μg/kg [47] of aflatoxin. A second study found that the productivity of birds fed grain containing 250–1000 μg/kg of aflatoxin B_1 did not significantly differ from that of control birds and that even a dietary aflatoxin B_1 concentration of 24,000 μg/kg in the feed was not lethal [48]. Therefore, further research is necessary to determine the levels of aflatoxin in chicken feed that do not affect animal health and performance and that result in tissue residue levels that are safe for human consumption.

Another consideration is the length of exposure of chickens to aflatoxin in the diet. The duration of exposure to aflatoxin in the feed influences the amount of toxin accumulated in edible tissues (Figure 3). Chen et al. [17] reported that aflatoxin was cleared from chicken tissues within four days when birds were switched to an aflatoxin-free feed. This rapid clearance is consistent with short-term exposure, but longer exposure can lead to increased accumulation. However, diets containing 200 μg/kg and 300 μg/kg of aflatoxin resulted in less than 5 μg/kg and 10 μg/kg of toxin residue, respectively. Also, exposure periods of 200 to 300 days far exceed the typical lifespan of commercial broilers, which usually are slaughtered at 35–50 days of age. This short life span within the commercial production cycle means that even prolonged exposure to relatively highly contaminated feed may result in only minimal toxin retention, and that withdrawal from contaminated feed shortly before slaughter could be an effective strategy for reducing residues. Even in developing countries where chickens may live longer and eat highly contaminated feed, the amount of aflatoxin retained in the tissues will usually be <20 μg/kg, which is below levels considered hazardous for human consumption.

Beyond the inherent aflatoxin-buffering capacity of chickens, dietary supplementation with various additives can further reduce tissue aflatoxin residues, enhancing food safety. In broilers fed 1 mg aflatoxin B_1/kg, curcumin and nano-curcumin reduced liver residues from 10.8 μg/kg to 2.3 μg/kg and 4.65 μg/kg, and muscle residues from 4.7 μg/kg to 1.0 μg/kg and 2.0 μg/kg, respectively [49]. At 250 μg aflatoxin B_1/kg in the feed, Toxfin and Novasil (0.30%) reduced liver residues from 2.9 μg/kg to 0.55 μg/kg and 1.2 μg/kg, and kidney residues from 0.9 μg/kg to 0.3 μg/kg and 0.4 μg/kg [50]. With 1 mg/kg of spiked aflatoxin B_1 in the feed, silymarin (0.6 g/kg), *Spirulina platensis* (1 g/kg), and their combination reduced aflatoxin liver residues to 0.1 μg/kg, 0.1 μg/kg, and non-detectable levels, respectively; all also yielded non-detectable residues in muscle versus 1.1 μg/kg in the controls [51]. In birds fed 485–520 μg/kg of aflatoxin B_1 in the feed, chelated trace minerals lowered liver residues from 1.6 μg/kg to 0.8 μg/kg and kidney residues from 3.2 μg/kg to 1.4 μg/kg [52]. At the EU limit of 20 μg/kg aflatoxin B_1 in the feed [42], turmeric powder (400 mg/kg) reduced liver residues to non-detectable levels compared to 46 μg/kg in the controls [53]. Similarly, other binders and detoxification systems including bentonite [54], multicomponent mycotoxin detoxifying agent [55], and a novel mycotoxin detoxification compound [32] have similar potential to reduce aflatoxin residues in various chicken tissues. These findings collectively affirm that these potential feed additives can synergize with the chicken's endogenous detoxification mechanisms, significantly reducing aflatoxin bioaccumulation even under high-exposure conditions. These additives should also improve the health of the birds that consume these contaminated feeds.

4. Conclusions

There is general consensus across all 33 studies that aflatoxin residues in tissues are much lower, i.e., <1%, than the aflatoxin contamination levels found in the feed in spite of variation in location where studies were conducted, the type of bird used, and the methodology used to contaminate the feed and to detect aflatoxin residues. Aflatoxin residues in chicken products were generally low, suggesting limited risk to human health from eating chicken. The ability of chickens to act as a "biological filter" to remove aflatoxins from the food supply provides a potential alternate use for contaminated grain. In particular, it reduces concerns that chicken consumption in less-developed countries poses a health risk due to aflatoxin contamination, even if the chickens are consuming relatively highly contaminated grain. Including additional additives in the feed has the potential to reduce these residue levels even further. The length of time of consumption of contaminated grain by the chickens is estimated to account for about 33% of the variation observed, and there is a significant chance that birds consuming contaminated feed for a long time, >100 days (Figure 3), will have higher aflatoxin residues in their consumable tissues. These conclusions indicate that further research to establish the maximum tolerable level of aflatoxin in chicken feed is warranted, as is the evaluation of other birds, such as ducks, geese, and turkeys, to determine if they too could serve as "biological filters" for aflatoxins in the food supply.

5. Materials and Methods

5.1. Data Collection and Curation

We systematically searched the Web of Science (Clarivate, London, UK) and PubMed (National Library of Medicine, National Institutes of Health, Bethesda, MD, USA) databases on 5 October 2023 using the following search phrase: ((AF OR AFB1 OR AFB2 OR AFG1 OR AFG2 OR AFL) AND (egg OR meat OR liver OR muscle OR gizzard OR organ OR kidney OR tissue OR residue) AND (poultry OR chicken OR layer OR broiler OR hen) AND feed). This search yielded 1887 articles, with the number reduced to 1523 when duplicates were excluded (Figure 1). Titles and abstracts were screened and 1470 articles were excluded because (i) the study did not involve broiler or layer chickens; (ii) the study did not measure both aflatoxins in the feed and residue in the tissues; (iii) the study was reported in a language other than English and could not be translated by Google translate; (iv) the study was outside the scope of this study, e.g., general and systematic reviews, book chapters, encyclopedias, case reports, and letters to the editor. We sourced, read, and critically analyzed the full texts of the remaining 53 articles, and retained 33 for further evaluation. These 33 articles all: (i) described peer-reviewed experimental research studies in which chickens (any age) were fed aflatoxin-contaminated feed, (ii) included data on aflatoxin residues in various tissues from chickens fed the aflatoxin-contaminated feed, and 3) specified the treatment protocols used.

Two independent reviewers extracted data by using a common data collection template and later resolved disagreements and discrepancies. A third reviewer verified the extracted data and confirmed whether the extraction protocol had been properly implemented. Information was collected, including the types of chickens used in the studies (i.e., broiler and layers), the concentration of aflatoxins in the feeds, the duration of feeding with the contaminated feed, and the concentration of aflatoxin residues or metabolites in multiple tissues after the chickens were sacrificed.

5.2. Data Quality Assessment

Two reviewers independently assessed the quality of the included studies. The potential risk of bias in the studies included in the meta-analysis was evaluated via the SYRCLE risk of bias tool for animal studies [56]. The assessment points were related to sequence generation, baseline characteristics, allocation concealment, random housing, blinding, random outcome assessment, incomplete outcome data, selective outcome reporting, and other potential sources of bias.

5.3. Data Analysis

All data analyses were performed using JMP Pro version 16.0.0 (SAS Institute, Cary, NC, USA). The correlation between the concentration of aflatoxins in chicken feed, the duration of consumption of the contaminated feed, and aflatoxin residues in chickens was determined by using Pearson's correlation test. Some of the studies reported non-detectable levels of aflatoxin residues in chicken tissues. In these cases, we divided the limit of detection for the diagnostic test used by two and used this adjusted result in the analysis. The accepted level of statistical significance was $p < 0.05$. To determine the carryover rate of aflatoxins from feed to chicken tissues, we calculated the percent fold change, i.e., the ratio of aflatoxin in tissue: aflatoxin in feed.

The US Food and Drug Administration (FDA) specifies the action levels for aflatoxin in chicken feeds or ingredients, such as cottonseed meal as 300 μg/kg and as 100 μg/kg for corn and peanut products [45]. We categorized the aflatoxin levels in the chicken feeds into three categories and the subsequent residues in various tissues into five categories (Table 2).

Author Contributions: Conceptualization, L.M.; methodology, M.C., M.C.S., B.K., V.N. and M.M. (Merning Mwenifumbo); validation, L.M. and J.F.L.; writing—original draft preparation, M.C.; writing—review and editing, M.C., A.P.G., M.M. (Maurice Monjerezi), M.M. (Mulunda Mwanza), L.M. and J.F.L. All authors have read and agreed to the published version of the manuscript.

Funding: JFL was funded in part by the USDA National Institute of Food and Agriculture Hatch Multi-State project KS1183A, and the Kansas Agricultural Experiment Station. Manuscript no. 24-220-J from the Kansas Agricultural Experiment Station, Manhattan.

Institutional Review Board Statement: Not applicable.

Informed Consent Statement: Not applicable.

Data Availability Statement: The raw data supporting the conclusions of this article will be made available by the authors on request.

Conflicts of Interest: The authors declare no conflicts of interest.

References

1. Blount, W.P. Turkey X disease. *J. Brit. Turk. Fed.* **1961**, *9*, 52–54.
2. Peles, F.; Sipos, P.; Győri, Z.; Pfliegler, W.P.; Giacometti, F.; Serraino, A.; Pagliuca, G.; Gazzotti, T.; Pócsi, I. Adverse effects, transformation and channeling of aflatoxins into food raw materials in livestock. *Front. Microbiol.* **2019**, *10*, 2861. [CrossRef] [PubMed]
3. Fischer, W.J.; Schilter, B.; Tritscher, A.M.; Stadler, R.H. Contaminants of milk and dairy products: Contamination resulting from farm and dairy practices. In *Encyclopedia of Dairy Sciences*, 2nd ed.; Elsevier: Amsterdam, The Netherlands, 2011; pp. 887–897. [CrossRef]
4. Ráduly, Z.; Szabó, L.; Madar, A.; Pócsi, I.; Csernoch, L. Toxicological and medical aspects of *Aspergillus*-derived mycotoxins entering the feed and food chain. *Front. Microbiol.* **2019**, *10*, 2908. [CrossRef] [PubMed]
5. Nazhand, A.; Durazzo, A.; Lucarini, M.; Souto, E.B.; Santini, A. Characteristics, occurrence, detection and detoxification of aflatoxins in foods and feeds. *Foods* **2020**, *9*, 644. [CrossRef]
6. Amirkhizi, B.; Arefhosseini, S.R.; Ansarin, M.; Nemati, M. Aflatoxin B in eggs and chicken livers by dispersive liquid-liquid microextraction and HPLC. *Food Addit. Contam. Part. B Surveill.* **2015**, *8*, 245–249. [CrossRef]
7. Osaili, T.M.; Al-Abboodi, A.R.; Awawdeh, M.A.; Jbour, S. Assessment of mycotoxins (deoxynivalenol, zearalenone, aflatoxin B_1 and fumonisin B_1) in hen's eggs in Jordan. *Heliyon* **2022**, *8*, e11017. [CrossRef]
8. Sineque, A.R.; Macuamule, C.L.; Dos Anjos, F.R. Aflatoxin B_1 contamination in chicken livers and gizzards from industrial and small abattoirs, measured by ELISA technique in Maputo, Mozambique. *Int. J. Environ. Res. Public. Health* **2017**, *14*, 951. [CrossRef]

9. Khan, M.Z.; Hameed, M.R.; Hussain, T.; Khan, A.; Javed, I.; Ahmad, I.; Hussain, A.; Saleemi, M.K.; Islam, N.-U. Aflatoxin residues in tissues of healthy and sick broiler birds at market age in Pakistan: A one year study. *Pak. Vet. J.* **2013**, *33*, 423–427.
10. Alam, S.; Khan, N.A.; Muhammad, A.; Jan, I.; Hashmi, M.S.; Khan, A.; Khan, M.O. Carryover of aflatoxin B_1 from feed to broilers' tissues and its effect on chicken performance. *Fresenius Environ. Bull.* **2020**, *29*, 214–221.
11. Alharthi, A.S.; Al Sulaiman, A.R.; Aljumaah, R.S.; Alabdullatif, A.A.; Elolimy, A.A.; Alqhtani, A.H.; Al-Garadi, M.A.; Abudabos, A.M. Protective effect of date pits on growth performance, carcass traits, blood indices, intestinal morphology, nutrient digestibility, and hepatic aflatoxin residues of aflatoxin B_1-exposed broilers. *Agriculture* **2022**, *12*, 476. [CrossRef]
12. Ali, A.; Khatoon, A.; Almohaimeed, H.M.; Al-Sarraj, F.; Albiheyri, R.; Alotibi, I.; Abidin, Z.U. Mitigative potential of novel *Lactobacillus plantarum* TISTR 2076 against the aflatoxins-associated oxidative stress and histopathological alterations in liver and kidney of broiler chicks during the entire growth period. *Toxins* **2022**, *14*, 689. [CrossRef] [PubMed]
13. Al-Ruwaili, M.; Alkhalaileh, N.I.; Herzallah, S.M.; Rawashdeh, A.; Fataftah, A.; Holley, R. Reduction of aflatoxin B_1 residues in meat and organs of broiler chickens by lactic acid bacteria. *Pak. Vet. J.* **2018**, *38*, 325–328. [CrossRef]
14. Ashry, A.; Taha, N.M.; Lebda, M.A.; Abdo, W.; El-Diasty, E.M.; Fadl, S.E.; Morsi Elkamshishi, M. Ameliorative effect of nanocurcumin and *Saccharomyces* cell wall alone and in combination against aflatoxicosis in broilers. *BMC Vet. Res.* **2022**, *18*, 178. [CrossRef]
15. Bintvihok, A.; Thiengnin, S.; Doi, K.; Kumagai, S. Residues of aflatoxins in the liver, muscle and eggs of domestic fowls. *J. Vet. Med. Sci.* **2002**, *64*, 1037–1039. [CrossRef]
16. Bintvihok, A.; Kositcharoenkul, S. Effect of dietary calcium propionate on performance, hepatic enzyme activities and aflatoxin residues in broilers fed a diet containing low levels of aflatoxin B_1. *Toxicon* **2006**, *47*, 41–46. [CrossRef]
17. Chen, C.; Pearson, A.M.; Coleman, T.H.; Gray, J.I.; Pestka, J.J.; Aust, S.D. Tissue deposition and clearance of aflatoxins from broiler chickens fed a contaminated diet. *Food Chem. Toxicol.* **1984**, *22*, 447–451. [CrossRef]
18. Denli, M.; Blandon, J.C.; Guynot, M.E.; Salado, S.; Perez, J.F. Effects of dietary afladetox on performance, serum biochemistry, histopathological changes, and aflatoxin residues in broilers exposed to aflatoxin B_1. *Poult. Sci.* **2009**, *88*, 1444–1451. [CrossRef]
19. Fernández, A.; Verde, M.T.; Gascón, M.; Ramos, J.J.; Gómez, J. Aflatoxin and its metabolites in tissues from laying hens and broiler chickens fed a contaminated diet. *J. Sci. Food Agric.* **1994**, *65*, 407–414. [CrossRef]
20. Fowler, J.; Li, W.; Bailey, C. Effects of a calcium bentonite clay in diets containing aflatoxin when measuring liver residues of aflatoxin B_1 in starter broiler chicks. *Toxins* **2015**, *7*, 3455–3464. [CrossRef]
21. Hassan, Z.U.; Khan, M.Z.; Khan, A.; Javed, I.; Hussain, Z. Effects of individual and combined administration of ochratoxin A and aflatoxin B_1 in tissues and eggs of White Leghorn breeder hens. *J. Sci. Food Agric.* **2012**, *92*, 1540–1544. [CrossRef]
22. Hussain, Z.; Khan, M.Z.; Khan, A.; Javed, I.; Saleemi, M.K.; Mahmood, S.; Asi, M.R. Residues of aflatoxin B_1 in broiler meat: Effect of age and dietary aflatoxin B_1 levels. *Food Chem. Toxicol.* **2010**, *48*, 3304–3307. [CrossRef] [PubMed]
23. Jia, R.; Ma, Q.; Fan, Y.; Ji, C.; Zhang, J.; Liu, T.; Zhao, L. The toxic effects of combined aflatoxins and zearalenone in naturally contaminated diets on laying performance, egg quality and mycotoxins residues in eggs of layers and the protective effect of *Bacillus subtilis* biodegradation product. *Food Chem. Toxicol.* **2016**, *90*, 142–150. [CrossRef] [PubMed]
24. Keutchatang, F.D.T.; Tchuenchieu, A.K.; Nguegwouo, E.; Mouafo, H.T.; Ntsama, I.S.B.; Kansci, G.; Medoua, G.N. Occurrence of total aflatoxins, aflatoxin B, and ochratoxin A in chicken and eggs in some Cameroon urban areas and population dietary exposure. *J. Environ. Public. Health* **2022**, *2022*, 5541049. [CrossRef]
25. Micco, C.; Miraglia, M.; Benelli, L.; Onori, R.; Ioppolo, A.; Mantovani, A. Long term administration of low doses of mycotoxins in poultry. 2. Residues of ochratoxin A and aflatoxins in broilers and laying hens after combined administration of ochratoxin A and aflatoxin B_1. *Food Addit. Contam.* **1988**, *5*, 309–314. [CrossRef]
26. Neeff, D.V.; Ledoux, D.R.; Rottinghaus, G.E.; Bermudez, A.J.; Dakovic, A.; Murarolli, R.A.; Oliveira, C.A.F. In vitro and in vivo efficacy of a hydrated sodium calcium aluminosilicate to bind and reduce aflatoxin residues in tissues of broiler chicks fed aflatoxin B_1. *Poult. Sci.* **2013**, *92*, 131–137. [CrossRef]
27. Oliveira, C.A.; Kobashigawa, E.; Reis, T.A.; Mestieri, L.; Albuquerque, R.; Corrêa, B. Aflatoxin B_1 residues in eggs of laying hens fed a diet containing different levels of the mycotoxin. *Food Addit. Contam.* **2000**, *17*, 459–462. [CrossRef]
28. Pandey, I.; Chauhan, S.S. Studies on production performance and toxin residues in tissues and eggs of layer chickens fed on diets with various concentrations of aflatoxin B_1. *Br. Poult. Sci.* **2007**, *48*, 713–723. [CrossRef]
29. Poloni, V.; Magnoli, A.; Fochesato, A.; Cristofolini, A.; Caverzan, M.; Merkis, C.; Montenegro, M.; Cavaglieri, L. A *Saccharomyces cerevisiae* RC016-based feed additive reduces liver toxicity, residual aflatoxin B_1 levels and positively influences intestinal morphology in broiler chickens fed chronic aflatoxin B_1-contaminated diets. *Anim. Nutr.* **2020**, *6*, 31–38. [CrossRef]
30. Resanovic, R.; Sinovec, Z. Effects of limited feeding of aflatoxin B_1 contaminated feed on the performance of broilers. *Mycotoxin Res.* **2006**, *22*, 183–188. [CrossRef]
31. Rizzi, L.; Simioli, M.; Roncada, P.; Zaghini, A. Aflatoxin B_1 and clinoptilolite in feed for laying hens: Effects on egg quality, mycotoxin residues in livers, and hepatic mixed-function oxygenase activities. *J. Food Prot.* **2003**, *66*, 860–865. [CrossRef]

32. Stefanović, D.; Marinković, D.; Trailović, S.; Vasiljević, M.; Farkaš, H.; Raj, J.; Tolimir, N.; Radulović, S.; Nešić, V.; Trailović, J.N.; et al. Evaluation of effectiveness of a novel multicomponent mycotoxins detoxification agent in the presence of AFB1 and T-2 toxin on broiler chicks. *Microorganisms* **2023**, *11*, 574. [CrossRef] [PubMed]
33. Wang, X.H.; Li, W.; Wang, X.H.; Han, M.Y.; Muhammad, I.; Zhang, X.Y.; Sun, X.Q.; Cui, X.X. Water-soluble substances of wheat: A potential preventer of aflatoxin B_1-induced liver damage in broilers. *Poult. Sci.* **2019**, *98*, 136–149. [CrossRef] [PubMed]
34. Wolzak, A.; Pearson, A.M.; Coleman, T.H.; Pestka, J.J.; Gray, J.I. Aflatoxin deposition and clearance in the eggs of laying hens. *Food Chem. Toxicol.* **1985**, *23*, 1057–1061. [CrossRef]
35. Wolzak, A.; Pearson, A.M.; Coleman, T.H.; Pestka, J.J.; Gray, J.I.; Chen, C. Aflatoxin carryover and clearance from tissues of laying hens. *Food Chem. Toxicol.* **1986**, *24*, 37–41. [CrossRef]
36. Yang, J.; Bai, F.; Zhang, K.; Lv, X.; Bai, S.; Zhao, L.; Peng, X.; Ding, X.; Li, Y.; Zhang, J. Effects of feeding corn naturally contaminated with AFB1 and AFB2 on performance and aflatoxin residues in broilers. *Czech J. Anim. Sci.* **2012**, *57*, 506–515. [CrossRef]
37. Zhu, F.; Zhu, L.; Xu, J.; Wang, Y.; Wang, Y. Effects of moldy corn on the performance, antioxidant capacity, immune function, metabolism and residues of mycotoxins in eggs, muscle, and edible viscera of laying hens. *Poult. Sci.* **2023**, *102*, 102502. [CrossRef]
38. Zuo, R.; Chang, J.; Yin, Q.; Wang, P.; Yang, Y.; Wang, X.; Wang, G.; Zheng, Q. Effect of the combined probiotics with aflatoxin B_1-degrading enzyme on aflatoxin detoxification, broiler production performance, and hepatic enzyme gene expression. *Food Chem. Toxicol.* **2013**, *59*, 470–475. [CrossRef]
39. Goodwin, L.D.; Leech, N.L. Understanding correlation: Factors that affect the size of *r*. *J. Exp. Educ.* **2006**, *74*, 249–266. [CrossRef]
40. Mahato, D.K.; Lee, K.E.; Kamle, M.; Devi, S.; Dewangan, K.N.; Kumar, P.; Kang, S.G. Aflatoxins in food and feed: An overview on prevalence detection and control strategies. *Front. Microbiol.* **2019**, *10*, 2266. [CrossRef]
41. FAO. *Worldwide Regulations for Mycotoxins in Food and Feed in 2003*; FAO: Rome, Italy, 2004.
42. EC Commission. Regulation (EU) No 165/2010 of 26 February 2010 amending regulation (EC) No 1881/2006 setting maximum levels for certain contaminants in foodstuffs as regards aflatoxins. *Off. J. Eur. Union* **2010**, *50*, 8–12. Available online: https://eur-lex.europa.eu/eli/reg/2010/165/oj/eng (accessed on 12 February 2025).
43. *FDA Guidance for Industry: Action Levels for Poisonous or Deleterious Substances in Human Food and Animal Feed*; USFDA: Washington, DC, USA, 2000. Available online: https://www.fda.gov/regulatory-information/search-fda-guidance-documents/guidance-industry-action-levels-poisonous-or-deleterious-substances-human-food-and-animal-feed (accessed on 12 February 2025).
44. Dohlman, E. Mycotoxin Hazards and Regulations. In *International Trade and Food Safety: Economic Theory and Case Studies*; Buzby, J.C., Ed.; USDA Agricultural Report no. 828; United States Department of Agriculture: Davis, California, USA, 2003; pp. 97–108. Available online: https://arefiles.ucdavis.edu/uploads/filer_public/2014/05/19/lcfoodsafettrade03.pdf (accessed on 12 February 2025).
45. FDA. *Compliance Policy Guides Sec. 683.100 Action Levels for Aflatoxins in Animal Feeds*; FDA: Washington, DC, USA, 2019. Available online: https://www.fda.gov/regulatory-information/search-fda-guidance-documents/cpg-sec-683100-action-levels-aflatoxins-animal-feeds (accessed on 12 February 2025).
46. National Research Council. *Nutrient Requirements of Poultry*, 9th ed.; The National Academies Press: Washington, DC, USA, 1994. [CrossRef]
47. Yunus, A.W.; Razzazi-Fazeli, E.; Bohm, J. Aflatoxin B_1 in affecting broiler's performance, immunity, and gastrointestinal tract: A review of history and contemporary issues. *Toxins* **2011**, *3*, 566–590. [CrossRef] [PubMed]
48. Fayokun, C.O.; Adegoke, G.O. Studies on residual aflatoxin B_1 in poultry feeds and poultry products. *J. Food Sci. Technol.* **2000**, *37*, 311–314.
49. Hamouda, M.; Fadl, S.E.; Lebda, M.A.; El-Diasty, E.M.; Taha, N.M. Trials for reducing the dangerous effect on poultry fed on aflatoxin contaminated ration using nano-curcumin. *BMC Vet. Res.* **2025**, *21*, 4519. [CrossRef]
50. Alqhtani, A.H.; Alharthi, A.S.; Aljumaah, R.S.; Alabdullatif, A.A.; Elolimy, A.A.; Abudabos, A.M. Evaluating the effectiveness of Toxfin and Novasil as dietary aflatoxin-binding agents in broilers for sustaining hepatic antioxidant capacity and intestinal health status during aflatoxin B_1 exposure. *Mycotox Res.* **2025**, *41*, 145–157. [CrossRef]
51. Hassan, A.M.; Mahmoud, A.S.; Ahmed, N.A.; El-Desoky, N.I. Investigating the potential of silymarin and/or *Spirulina platensis* to attenuate the deleterious consequences of aflatoxin contamination in broilers' feeds. *J. Anim. Plant Sci.* **2024**, *34*, 899–908. [CrossRef]
52. Mohammadi, H.; Khosravi, A.; Seidavi, A.; Ghasemi, H.A.; Chamani, M.; Laudadio, V. Beneficial effects of advanced chelate technology-based 7-minerals in aflatoxin-B_1 challenged broilers: Toxin residue reduction, serum biochemical improvement and modulation of the mRNA expression of NF-κB and Nrf2, and genes within their pathways. *J. Sci. Food Agric.* **2024**, *104*, 1422–1432. [CrossRef]
53. Amminikutty, A.C.; Veldman, A.; Bederska-Lojewska, D.; Szuba-Trznadel, A. Turmeric powder counteracts oxidative stress and reduces aflatoxin B_1 content in the liver of broilers exposed to the EU maximum levels of the mycotoxin. *Environ. Sci. Pollut. Res.* **2023**, *30*, 7761–7772. [CrossRef]

54. Ochieng, P.E.; Croubels, S.; Kemboi, D.; Okoth, S.; De Baere, S.; Cavalier, E.; Kang'ethe, E.; Faas, J.; Doupovec, B.; Gathumbi, J.; et al. Effects of aflatoxins and fumonisins, alone or in combination, on performance, health, and safety of food products of broiler chickens, and mitigation efficacy of bentonite and fumonisin esterase. *J. Agric. Food Chem.* **2023**, *71*, 13462–13473. [CrossRef]
55. Raj, J.; Farkaš, H.; Jakovčević, Z.; Vasiljević, M.; Kumar, R.; Asrani, R.K. Effects of supplemented multicomponent mycotoxin detoxifying agent in laying hens fed aflatoxin B_1 and T2-toxin contaminated feeds. *Poult. Sci.* **2023**, *102*, 102795. [CrossRef]
56. Hooijmans, C.R.; Rovers, M.M.; de Vries, R.B.M.; Leenaars, M.; Ritskes-Hoitinga, M.; Langendam, M.W. SYRCLE's risk of bias tool for animal studies. *BMC Med. Res. Methodol.* **2014**, *14*, 43. [CrossRef]

Disclaimer/Publisher's Note: The statements, opinions and data contained in all publications are solely those of the individual author(s) and contributor(s) and not of MDPI and/or the editor(s). MDPI and/or the editor(s) disclaim responsibility for any injury to people or property resulting from any ideas, methods, instructions or products referred to in the content.

MDPI AG
Grosspeteranlage 5
4052 Basel
Switzerland
Tel.: +41 61 683 77 34

Toxins Editorial Office
E-mail: toxins@mdpi.com
www.mdpi.com/journal/toxins

Disclaimer/Publisher's Note: The title and front matter of this reprint are at the discretion of the Guest Editors. The publisher is not responsible for their content or any associated concerns. The statements, opinions and data contained in all individual articles are solely those of the individual Editors and contributors and not of MDPI. MDPI disclaims responsibility for any injury to people or property resulting from any ideas, methods, instructions or products referred to in the content.

www.ingramcontent.com/pod-product-compliance
Lightning Source LLC
Chambersburg PA
CBHW040839250926
51850CB00007BA/1358
* 9 7 8 3 7 2 5 8 5 0 9 7 6 *